T0380203

DEUS ORDIRI FUNDAMENTUM

God begins, Lays the Order of the Foundation

RICHARD E. FORD

DEUS ORDIRI FUNDAMENTUM
GOD BEGINS, LAYS THE ORDER OF THE FOUNDATION

Copyright © 2024 Richard E. Ford.

All rights reserved. No part of this book may be used or reproduced by any means, graphic, electronic, or mechanical, including photocopying, recording, taping or by any information storage retrieval system without the written permission of the author except in the case of brief quotations embodied in critical articles and reviews.

iUniverse books may be ordered through booksellers or by contacting:

iUniverse
1663 Liberty Drive
Bloomington, IN 47403
www.iuniverse.com
844-349-9409

Because of the dynamic nature of the Internet, any web addresses or links contained in this book may have changed since publication and may no longer be valid. The views expressed in this work are solely those of the author and do not necessarily reflect the views of the publisher, and the publisher hereby disclaims any responsibility for them.

Any people depicted in stock imagery provided by Getty Images are models,
and such images are being used for illustrative purposes only.
Certain stock imagery © Getty Images.

ISBN: 978-1-6632-6650-7 (sc)
ISBN: 978-1-6632-6651-4 (e)

Library of Congress Control Number: 2024918143

Print information available on the last page.

iUniverse rev. date: 10/11/2024

“All men, Socrates, who have any degree of right feeling, at the beginning of every enterprise, whether small or great, always call upon God.”

CONTENTS

Part II: Early Evidence of the Pattern in Civilization

Part III: The Step Pyramid Complex of Saqqara

Part IV: Earth and the Great Pyramid

Part V: The Word of God and Cycles of Time; the Underground Stelae of the Step Pyramid

Appendices

PREFACE

Deus Ordiri Fundamentum started as a quest to find the extent of the astronomical knowledge of the ancient Egyptians, as manifested in the structural components of the Step Pyramid complex of Saqqara. It quickly grew beyond this when it became apparent that there was an underlying Pattern to the various measures of time and distance found in the complex. The Pattern and its origin then became the overarching theme of the book, which still incorporated the earlier subject matter concerning the Step Pyramid. Further expansion on this theme occurred when the Great Pyramid was also examined for its astronomical significance and possible relationship to the Pattern. There has been no effort to attest to the artistic merit of either structure, which is why no photographs of them are included. Both the Step Pyramid complex and the Great Pyramid were rigorously examined regarding their geographic location and orientation, geodesic significance, celestial alignments, measurements of time and distance, cartographic significance, etc. The history of these two renowned monuments is not discussed in any significant detail, except where and when considered necessary in the furtherance of the overall goals of the book.

Many systems of measure seem to derive from this same underlying Pattern, which originates in a simple geometric Pattern. It also seems to permeate all measure, which is broadly traced throughout the first part of the book. The coordination and symmetry in all of this are extraordinary, and cannot have been due to happenstance or a gradual, evolutionary development by mankind. Moreover, there is no record of mankind's having engaged in any such singular, sustained effort, especially one lasting over thousands of years, across many different cultures. What's more, much of it seems to have been with us since the very beginning of civilization.

The book began as, and has remained, a highly technical work out of necessity. Astronomy, geodesy, metrology, cartography, etc. are all hard sciences, which, neither individually nor collectively, can be rendered into easily readable and understandable narrative. There is simply no other way of dealing with issues involving number, except through the medium of number. It is my considered opinion that this is one of the voices of God, and that it is His preferred voice when dealing with such matters. Unlike the spoken word, or language, where meaning is often subject to argument and interpretation, number is invariably clear and precise, and its meaning seldom in doubt, even across vast distances of time. And, unlike the spoken word, the language of mathematics has existed, unaltered, since the very beginning of time.

Once these various scientific issues were examined in detail, there arose a glaring need to address their cosmogonic, cosmographic, and cosmologic significance, i.e. their relationship to the origin and nature of the universe, behind all of which is the unmistakable presence of God. *Deus Ordiri Fundamentum* did not start off with a religious tone, but such a tone seemed to arise naturally and unavoidably as the book developed. There was no single moment of revelation for the writer, but a gradual awakening to a great truth. This awakening also coincided with the selection of a title for the book, which up until that moment had only various unsatisfactory working titles, none of which seemed to adequately capture the subject matter of the book or its significance.

Essentially, *Deus Ordiri Fundamentum* recovers a lost science, though it was never really lost so much as forgotten. While modern science is rightly focused on extreme accuracy in its measures, this forgotten science focused on the unifying principals embodied in the patterns and geometric shapes underlying cosmic order. Knowledge of these underlying and unifying principals were not slowly developed over time by a patient and persistent humanity. Instead, all of it was given to mankind at the very beginning. This gift of knowledge was encoded into the monumental structures of the ancient Egyptians, and undoubtedly those of other early civilizations as well, to memorialize it and to teach it to mankind.

Deus Ordiri Fundamentum is most assuredly not a cover-to-cover read, one that can be enjoyed at leisure; there is simply too much number in it for that. Instead, the best approach is to sample and browse it.

Acknowledgements and Credits

I am forever grateful to the following sources and contributors: Wikipedia, through whose many, on-line portals vast quantities of diverse scientific knowledge could be easily and readily accessed; to the U.S.'s National Aeronautics and Space Administration (NASA)'s web site, through whose portals much of the essential astronomical detail used in this book originated; Martino Publications, publishers of Firth and Quibell's, *Excavations at Saqqara, the Step Pyramid,* for their kind permission to copy Jean-Phillipe Lauer's drawings, which appear in Appendix 3 of this book, and his photographs of the Step Pyramid stelae, which were rendered into the drawings of the stelae made by Betsy Miller, that appear in Part V ; to Dr. Zahi Hawass, Secretary General, Supreme Council of Antiquities [of Egypt] for his kind permission to copy and use J.P Lauer's paper on the measurements of the Step Pyramid complex, which is found in Appendix 2 of this book, and to Jo Ann Fisher for her translation of the text from French to English; to Livio Catullo Stecchini for his great knowledge in all matters pertaining to measure, especially regarding ancient structures, as found in his Monograph that appears as an Appendix to Peter Tompkins' seminal work, *Secrets of the Great Pyramid*; to Betsy Miller and McKenna Giamfortone for the art work appearing on the cover; to Theresa Byers for her assistance with several of the graphics; and finally to Plato for preserving and publishing much of the ancient knowledge knowledge in his *Timaeus*, which was frequently consulted as a resource, a guide, and an inspiration throughout the many, many long hours involved in the writing of this book.

Author's biography

Richard E. Ford is a retired Coast Guard officer, well-versed in the art and practice of celestial navigation—a timeless, practical form of astronomy. He is also a life-long student of ancient Egyptian and Middle Eastern studies.

INTRODUCTION

DEUS ORDIRI FUNDAMENTUM

God Begins, Lays the Order of the Foundation.

Everything in nature appears to be in perfect harmony with everything else around it, but closer observation reveals minor inconsistencies, imperfections, and contradictions. The closer the examination, the more glaring these flaws appear. Have we not been lead to believe that the Creator of all is perfect and if so, why does His handiwork not reflect His perfection? Questions from time immemorial. But, maybe we have misunderstood, erred in our thinking, mistakenly accepted what was before us as evidence of an imperfect author.

Maybe it is His plan or design that is perfect, but the execution of it by the lesser gods, or nature, or heaven's emissaries—decidedly less so. This would explain the imperfections in nature. But if there is a plan, where is it, and why are we not aware of it? Why can't we see it? Then again, maybe it has been right in front of our eyes all along and we are fully aware of it, but have not understood its meaning. There is, indeed, such a plan. A plan that is just as valid today as it was on the day of its inception. A plan universal in its application and with the exact same meaning in all times and all places. A plan that cannot be corrupted, lost or destroyed. A plan written in a language that is timeless and pure, whose meaning is universally understood—mathematics.

The plan is a geometric progression or Pattern, and everything proceeds from and follows this Pattern. The Pattern is perfect; nature's fidelity to it decidedly less so. This is the singular argument of this book and everything that follows will demonstrate and support the argument's validity. The language of the argument is mathematics. A substantial number of the principal measures of the universe that surrounds us, crossing all fields of scientific discipline, will be examined and rigorously compared with the numbers from the Pattern. The measures examined are from recognized scientific bodies, including, but not limited to, the International Astronomical Union (IAU), the U.S. National Aeronautics and Space Administration (NASA), and the Système International (SI). This is not a history of measure; though measure will be ever present throughout the argument.

The argument does not question nor does it doubt any measurement examined, and accepts all of them as entirely accurate, without reservation. However, the argument's process whereby actual measurements are compared to Pattern numbers, and the differences are attributed to imperfections of the underlying phenomena that is measured, may prove disconcerting at first, but in no case does this call into question the measure itself. Modern science can justly be proud of the accuracy of their measures and their ability to use them to humanity's advantage, but the underlying phenomena that are measured, are flawed in their execution of the Pattern, which is perfect. The Pattern is more than a mere plan, though. It is an irresistible impulse, an imperative that affects everything, and which has existed since the beginning, continues in undiminished force today, and will continue in the future.

The Pattern also synchronizes and harmonizes all measures, regardless of their dimension. This has been a goal of metrology, the science of measurement, for centuries, and it was a principle motivating factor in establishing the modern metric system, with its easy to use decimal format, which has been almost universally adopted. There have been discussions to decimalize the arc measures of the circle and time itself, but so far there have been no serious efforts to do this. In eighteenth-century, revolutionary France there was actually talk of implementing a 10-day week, with a day of 10-metric hours, a metric hour of 100 metric minutes, and a metric minute of 100 metric seconds.[1] In any case, the old measurement regimes for time and arc still prevail, and these do synchronize and harmonize with one another through the Pattern, as do the International foot and mile, the old English measures of distance. Of central import to this book's thesis, though, is that it is the Pattern that underlies many of the disparate dimensions and measures found in nature, and it is the Pattern that defines them and binds them together.

Modern measurements all begin with the Système International (SI) seven base units, established by the International Bureau of Weights and Measures. As a general rule, all measures discussed in this book are reduced to the SI Base Units, their derivatives, or their equivalent measures in the U.S. Customary System. The SI Base Units are;

The second of 1/86,400 of the length of the mean solar day, for measuring time[2]
The meter, for measuring distance
The kilogram for measuring mass
The amp for measuring electric current
The kelvin for measuring temperature
The mole for measuring the amount of a substance
The candela for measuring luminous intensity

---

[1] Whitelaw, Ian, *A Measure of All Things, The Story of Measurement Through the Ages*; David & Charles, an F + W Publications, Inc. Company, 4700 East Galbraith Road, Cincinnati, Ohio, USA, 2007; page 90

[2] Ibid. p 90. This is no longer the accepted definition, but it still remains a useful measure for day-to-day purposes.

These measures and their derivatives will be used for all analysis purposes. However, the International foot and mile are used for all measures of distance, in place of their equivalent meter measure, and the pound avoirdupois is used for measures of mass. The English measures of distance and mass will be shown to be primordial measures, and the fact that they are identified with England has nothing to do with their origins and original meaning.

The pound avoirdupois, the pound of mass, is the primordial measure of mass. However, in the various analyses of measures of mass, there is another measure of mass provided for that is based on volume. This issue is further detailed in the chapter on SI Measures. However, all recognized measurements of mass in terms of kilograms are converted into lbs. avdp. throughout this work, for review and analysis purposes, without exception.

"Wikipedia, the Free Encyclopedia" is frequently cited as a reference. There are two reasons for this. The background material on specific measurements being reviewed and analyzed may prove daunting to those not familiar with it, but it is fairly basic and straightforward to those who are; therefore, a general, universally available source was considered satisfactory. For those who are not familiar with the subject matter and who require further explanation or just have a need to further explore the subject matter, Wikipedia provides ready access to such material for them; and that it is available online is an added plus. However, where reference to specialized sources is clearly called for, such material is used and appropriately cited.

The book is laid out in six parts. Part one describes the Pattern and details its manifestations in nature, which, for the purposes of this book, includes the solar system. Part two describes the appearance of the Pattern in early human civilization. Part three describes the numerous measures and astronomical cycles, which are related to the Pattern, and that are incorporated in the Step Pyramid in Saqqara, Egypt. Part four describes the numerous measures of the Earth and time incorporated in the Great Pyramid at Giza, Egypt. Part five interprets the six relief panels found beneath the Step Pyramid complex and their relationship to the Pattern. And part six briefly summarizes the origin of the Pattern, and proposes a rationale for how it is transmitted throughout nature.

This book is not a cover-to-cover read. Part one, in particular, is meant to be sampled so as to familiarize the reader with all of the areas of measurement through which the Pattern manifests itself in nature. Reading the introduction and chapter one are essential to understanding the Pattern. Thereafter, the reader can browse the following chapters on measurements, paying particular attention, though, to the tables that can be found throughout the chapters, which compare the actual measurements to the Pattern numbers.

PART I

THE UNDERLYING PATTERN OF DIMENSION AND MEASURE

CHAPTER 1

THE PATTERN DESCRIBED

The Pattern is a geometrical progression of the numbers 2 and 3, and their respective multiples. The multiples of 2 are: 4, 8, 16, 32, etc.; the multiples of 3 are: 9, 27, 81, etc. The hi-lighted numbers are exact progressions of the $2 \cdot 3 = 6$ Pattern, e.g. $36 = 6^2$. All other numbers are even integer multiples of the Pattern, e.g. 72 is $2 \cdot 6^2$, 846 is $4 \cdot 6^3$. Each number of the Pattern is a 'root number.' All orders of magnitude (± 10) of the individual root numbers in the Pattern are included, e.g. 36 includes: .036, .00036, 360, 3,600,360,000 etc.

X 2↓ X 3→

1	3	9	27	81	243	729	2,187	6,561
2	6	18	54	162	486	1,458	4,374	13,122
4	12	36	108	324	972	2,916	8,748	26,244
8	24	72	216	648	1,944	5,832	17,496	52,488
16	48	144	432	1,296	3,888	11,664	34,992	104,976
32	96	288	864	2,592	7,776	23,328	69,984	209,952
64	192	576	1,728	5,184	15,552	46,656	139,968	419,904
128	384	1,152	3,456	10,368	31,104	93,312	279,936	839,808
256	768	2,304	6,912	20,736	62,208	186,624	559,872	1,679,616
512	1,536	4,608	13,824	41,472	124,416	373,248	1,119,744	3,359,232

In reviewing actual measures from the natural order around us, the Pattern number is always considered the standard for comparison purposes. The actual value of a particular measure is always accepted as accurate without question or reservation. Measures from the metric system are frequently used for analysis purposes and they are also accepted as accurate. This book does not challenge the metric system and accepts its universal application without question.

A particular measure from nature is considered to be a manifestation or expression of the Pattern if it satisfies a specific criterion. This criterion is arbitrary, but considered desirable to control for error. The criterion that a measure must satisfy require it be proximate to a Pattern number within a deviation factor of ± .009. If it is outside of this range—as indeed a number of measures are, though most of these have deviation factors of less than ±.02—the measure still may be an expression of the Pattern, but the issue may require further analysis. Only measures that satisfy this criterion are considered to be manifestations of the Pattern in nature. However, ALL measures reviewed are considered to be wholly accurate and valid, regardless of whether or not they are considered to be expressions of the Pattern in nature, without exception.

A number of measures differ in their expression of the Pattern, in that they are numbers proximate to Pattern numbers formed wholly from either 2 or 3, or their respective multiples, but not from both, as is the case for most other numbers in the Pattern. Such numbers include: 27, 32, 64, 81, and 243, etc. These numbers appear to be associated with certain contexts, and a possible explanation for this will be discussed later. However, these numbers are considered to be numbers from the Pattern and measures approximating them are considered to be manifestations of the Pattern in nature.

Measures from a number of fields of scientific discipline will be reviewed in the following chapters and compared to the numbers from the Pattern. In many cases, measures from these disciplines will be found to be close or exact matches to Pattern numbers and a manifestation of the Pattern in nature, while other measures exceed the range of deviation established and are not considered to be manifestations of the Pattern, although in many cases they are close. It is concluded, therefore, that the Pattern does not manifest itself in all measures within a particular field of scientific discipline. Furthermore, if a measure that is found to be a manifestation of the Pattern in nature is acted upon mathematically, i.e. it is multiplied, divided, added, or subtracted, then the Pattern will likely not be manifested in the result.

The numbers from the Pattern that are used for comparison purposes are limited to those shown in the foregoing matrix. This is an arbitrary decision, but it is based on the fact that the Pattern perpetrates itself ad infinitum and a number in the desired form could likely be found at some point if this were done. Therefore, the range of numbers are limited to those shown.

The heading above the Table is styled, "The Pattern of **Dimension** and Measure." This was by design, as the Pattern appears to govern not only the measure of a particular phenomenon, but in a number of important cases the underlying physical reality or dimension as well. This will become apparent as specific phenomenon are reviewed and analyzed. The question of how this may occur will be discussed later.

An example of the analysis of several specific phenomena:

Astronomical Unit (AU): 92,955,807 miles.

There is a Pattern number that very closely approximates this number: 93,312,000 ($2 \cdot 6^6 \cdot 10^3$). Its Pattern pair is: (128/729). (The Pattern pair is: the multiple of 2 and multiple of 3, respectively, that locates the number in the Pattern matrix.)

Speed of Light: 186,234 miles/second.

There is a Pattern number that very closely approximates this number: 186,624 ($4 \cdot 6^6$). Its Pattern pair is (256/729). (The speed of light travels 93,117,000 miles in 500 seconds, a distance within approximately .002% of the value of the Astronomical Unit, 92,955,807 miles.)

Defined Value	Pattern Number	Difference Δ	Dev. ± from Pattern
92,955,807 miles *	93,312,000 ($2 \cdot 6^6 \cdot 10^3$)	356,193	-.0038
186,234 miles/second *	186,624 ($4 \cdot 6^6 \cdot 10^3$)	389	-.0021

*Both of these numbers are considered to be manifestations of the Pattern in nature. Both are constants recognized by the International Astronomical Union (IAU). See chapter on "Astronomical Constants."

a. Identical Root Numbers of Some Measurements

There are often identical root numbers in a number of disparate measures, which the reader should be alert to. Using the length of the day, as an example, 1 day is 86,400 seconds, ½ day is 43,200 seconds and 1/4th day is 21,600 seconds. These same three measures of time have identical root numbers to the measures of distance as follows:

- The diameter of the Sun is almost exactly 864,000 international miles.
- The diameter of Earth's electromagnetic core is almost exactly 4,320 international miles.
- The diameter of the Moon is almost exactly 2,160 international miles.

For some numbers, 2,160 for example, identical root numbers are quite often present across many disparate measures:

- There are 21,600 seconds in 1/4th of a day.
- The diameter of the Moon is almost exactly 2,160 international miles.
- There are 21,600 arcminutes in the measure of the circle
- There are 2,160 tropical years in the measure of one great age of a Platonic Year.

All of this was by specific design.

CHAPTER 2

THE CIRCLE AND ITS MEASURE

a. Measures of the Circle

The circle is considered to be a perfect figure; it has no angles and has no beginning or end. This is also true of the sphere, its solid manifestation. In nature, the Sun and Moon present universally available images of the circle. Less obvious, the daily apparent journeys of the Sun and Moon across the heavens, evokes images of a circular pathway. The measure of the circle begins with the Pattern, and the circle's measures are related to many other measures that also derive from the Pattern. The circle's measures are called angular or arc measures. They are:

Measures of the Circle:

1 circle = 360 degrees

360 degrees = 21,600 arcmins. = 1,296,000 arcsecs.

1 arcdegree = 60 arcmins. = 3,600 arcsecs.

1 arcmin. = 60 arcsecs.

All of the numbers found in the system of angular measures for the Earth and celestial sphere are exact matches to numbers that are manifestations of the Pattern in nature. Their pattern pairs are:

6 (2/3)
36 (4/9)
216 (8/27)
1,296 (16/81)

Defined Value*	Pattern Number	Difference Δ	Dev. ± from Pattern
360	360 ($6^2 \cdot 10$)	0	0
21,600	21,600 ($6^3 \cdot 10^2$)	0	0

1,296,000	1,296,000 ($6^4 \cdot 10^3$)	0	0
60	60 ($6 \cdot 10$)	0	0

*These numbers are all manifestations of the Pattern in nature.

b. Angular Measures of the Earth

The Earth is a sphere, whose broad outline has been traditionally defined in terms of angular measures. These measures derive from the angular measure system for the circle of 360 degrees of 60 arc minutes each, of 60 arc seconds each. Traditionally, its origins are traced to ancient Babylonia. On Earth, the system is used to define angular measures, commensurate with the four cardinal directions. The system is generally referred to as the **equatorial coordinate system**, and is composed of lines of latitude and longitude, and is appropriately compared to a net, gathered together at the poles, through which the axis about which the Earth rotates passes.

Proceeding north or south from the equator, the system consists of a series of evenly-spaced, parallel lines, called lines or **parallels of latitude**, denoted as north or south, starting with the equator or 0, and ending at 90 degrees north at the North Pole or 90 degrees south at the South Pole.

Proceeding east or west, the system consists of a series of lines that run in a north-south direction, from Pole to Pole, starting from a reference point referred to as the prime meridian or 0 meridian or longitude (currently the prime meridian passes through Greenwich, England), and continuing with lines drawn at even intervals along the equator, measured east and west of the prime meridian, and ending in a meridian referred to as the International Dateline, or 180 degrees east/west of the prime meridian. **Lines of longitude or meridians** are not parallel to one another, as their separation narrows proceeding from the equator until they intersect together at the poles.

This would be a perfect system if the Earth were a perfect sphere, but the Earth actually bulges at the Equator and is somewhat flattened at the Poles. This has little effect on the measure of longitude, but it does cause distortions in the measure of latitude, particularly near the Poles.

c. Angular Measures of the Celestial Sphere.[3]

Looking heaven-ward, the immediate impression is that the Earth is surrounded by a dome or sphere that lies at an evenly spaced distance above the Earth. Of course, this is an illusion, but the imagery that it evokes is extremely useful in measuring the heavens. The system that is used to measure the heavens is an exact replica of the Earth's system of lines of latitude

[3] Wikipedia contributors, "Celestial Sphere" Wikipedia, the Free Encyclopedia; https://en.wikipedia.org/wiki/Celestial_sphere (accessed May 16, 2016)

and longitude. The equator of the celestial sphere is an extension of the Earth's Equator and the celestial poles are extensions of the Earth's axis, running through the North and South Poles. to the North celestial and South celestial poles, respectively. However, instead of lines of latitude and longitude the lines of angular measure defining the celestial sphere are referred to as declination and right-ascension, respectively. The lines of declination run parallel, north and south of the celestial equator, ending at the celestial poles. Lines or right-ascension run east of the Earth's prime meridian. (There is no equivalent to the International Dateline on the celestial sphere, and the hour angles are measured simply as degrees of right-ascension, from 1 to 360 degrees. The Sun, Moon and all of the planets and stars can be located on the celestial sphere. However, this is a simplification as the apparent movement of the celestial sphere is not perfect. The reasons for this will be discussed in the Chapter titled, "Other Long Duration Time Cycles and their Measure."

d. Angular Measures of the Celestial Sphere of the Plane of the Ecliptic.[4]

There is yet another celestial coordinate system that is based on the plane of the ecliptic. The plane of the ecliptic is the plane followed by the apparent path of the Sun across the Earth. It is inclined some 23.4 degrees from the plane of the celestial equator. This inclination is commonly referred to as the **obliquity of the ecliptic**.

The plane of the ecliptic cuts the equator at the equinoxes, as the sun tracks the ecliptic north to the Tropic of Cancer (23.4 degrees North) or south to the Tropic of Cancer (23.4 degrees South), respectively, where it turns and proceeds in the opposite direction. The ecliptic poles track the Earth's polar circles in apparent motion. The polar circles are located at 66.6 degrees North and 66.6 degrees South, respectively.

The ecliptic is a circle of 360 degrees across the celestial sphere of the ecliptic. It is divided into 15-degree increments that correspond to the hours of the day, or into 30-degree increments that correspond to the twelve signs of the zodiac. The ecliptic is surrounded by a band that is defined by the motions of the planet Mercury, which appears to move some 7 degrees north and south of the ecliptic, making the band of the ecliptic some 14 degrees in width. Within the limits of this band, the Moon and all of the planets of the solar system appear to move around the Earth.

The coordinate system of the celestial sphere uses the same naming convention as the equatorial coordinate system, with lines of latitude and longitude. Lines of latitude run parallel to the ecliptic, terminating at the ecliptic poles at 90 degrees North and 90 degrees South, respectively. Lines of longitude run perpendicular to the ecliptic, and rise to the ecliptic poles. The longitudes of the ecliptic sphere start with 0-degrees longitude at the vernal equinox, or "First Point of Aries" and rising in evenly spaced degree increments, proceeding eastward,

---

4 Wikipedia contributors, "Ecliptic", Wikipedia the Free Encyclopedia; https://en.wikipedia.org/wiki/Ecliptic accessed May 17, 2016

until the circle is complete at 360 degrees. The band of the ecliptic is not stationary, but appears to drift westward with time. This movement is associated with a much longer time cycle called, the precession of the equinoxes, which will be reviewed in a later chapter, "Other Long Duration Time Cycles and their Measure."

Since almost all of the measures of the celestial sphere of the ecliptic are identical to the celestial sphere of the equator, these measures are numbers from the Pattern and are manifestations of the Pattern in nature. The only exceptions to this are: the 14-degree width of the band of the ecliptic, the 23.4-degree measure of axial tilt, and the 15-degree division of the ecliptic. However, 15 is half of 30 (3 · 10), which is a number from the Pattern.

Defined Value	Pattern Number	Difference Δ	Dev. ± from Pattern
14	15 (1/2 of 30)	1	-.066
23.4	24	.6	-.025
15*	15 (1/2 of 30)	0	0

*This is a manifestation of the Pattern in nature.

CHAPTER 3

EARTH'S ROTATION AND THE MEASURE OF TIME

Today, almost all daily references to time involve clock-time, which is a mechanical or electronic means for tracking the progression of the day. It is measured, predictable and reliable. However, at its foundation clock-time is a means to track the Earth's rotational movements about the Sun, which are not always even or smooth. For example, high noon does not always occur at precisely 12:00 AM, even though our clock says that it is high noon, an event that varies throughout the year depending on the location of the Earth along its elliptical orbit about the Sun. This difference is reflected in the terms 'mean time', which is clock-time, as opposed to 'apparent solar time', or solar time.

The precise origins of the measure of time, using a sexagesimal base of 6 or 60 for its measure, is lost to history, but has been traditionally credited to the ancient Babylonians. The measure of time and of the circle are rooted in the same sexagesimal base and in all likelihood came into being at the same time. The two systems hold many terms in common, including the minute and the second, as they are both based on the circle, the one on the geometric figure of the circle and the other on the apparent daily revolution of the Sun about the Earth. Both systems of measure are largely abstractions, in that there are no real-world manifestations of their existence in nature. There is one exception, which is a possible link between the measure of the minute of time and the angular diameter of the Sun, which has a mean angular measure of 32 arc minutes.

The apparent movement of the solar orb across the heavens is a readily available method for the measure of time. The apparent movement of the solar orb across a distance equal to its angular radius of 16 arc minutes very closely approximates one minute of time, but it is some 4 seconds longer than the present measure of the minute, which is 1/1,440 of a day, or 60 seconds—with the second defined, until recently, as 1/86,400 of a day. However, if we are constructing a measurement system for the measure of the day from scratch, this is not an insurmountable impediment. This assumes, of course, that the measure of the circle is available. Also, all measures would have to be made on the equator at the equinox, with the Sun at its mean angular measure, 32 arcminutes in diameter or 16 arcminutes in radius.

In constructing such a system, if one minute is equal to the apparent movement of the solar orb across a distance equal to the measure of its radius, then the minute must equal 64 seconds of time. To convert the measure of the day into 64-second minutes, multiply 1,440 minutes by 60/64, which then defines length of the day in 64-second minutes as 1,350 minutes long. This figure approximates 1,352, which is the multiple of 26 · 52. If the length of the second is adjusted slightly, so that the 64-second minute becomes 63.905 seconds long (using our present definition of the second), then we have a day of 26 hours, of 52 minutes each, of 63.905 seconds each, which equals 86,399.6 seconds for the length of the day, a figure that compares favorably to the present measure of the length of the day in seconds, 86,400 seconds. The 64-second duration of the minute can be preserved if the present value of the second is adjusted slightly so as to make it to equal .9985 of a present second. This system for the measure of time is a very viable option for defining the day, and it is based on a readily observable natural phenomenon for the measure of the length of the minute.

So, what is the point of all this? Most assuredly not to advocate for the adoption of such a system, but only to point out that a readily available alternative to the present system was available in nature, yet it was not adopted. Instead, a system based on an abstraction was selected. This was not an arbitrary or unfortunate decision. The numbers for our hypothetical alternative system: 1,352, 26, and 52 are <u>not</u> numbers from the Pattern, while their counterparts from the present system: 1,440, 24, and 60 are. Also, the minute is exactly equal to the <u>apparent</u> movement of the leading edge of the solar orb across 15 arcminutes on the celestial sphere.

The upper or lower edge or limb of the Sun could serve the same purpose, and in fact may be a far superior means for making the observation that the Sun moves at a rate of 15 arcminutes in one minute of time, by using an opaque square that almost touches the upper and lower limbs of the Sun to track its movement across the top/bottom of the square. This procedure would measure the passage of one Sun diameter, or two minutes of time. The time that the sun's upper and limb appears at top of the near edge of the square to the time it reaches the center of the square is one minute of time, and the time from the center to the far edge is also one minute.

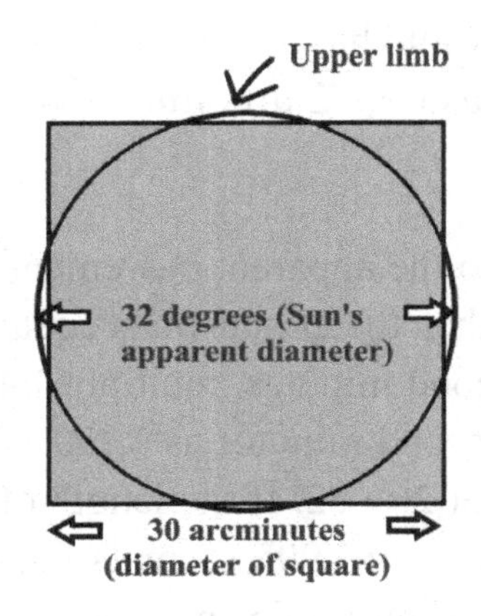

1 Minute of Time = 15 arcminutes of the Sun's apparent movement across the celestial sphere

In 1,440 minutes—length of the day in minutes—the edge of the solar orb crosses 21,600 arcminutes—the circumference of the circle in arcminutes (360° · 60 arcminutes/degree). All these numbers are also numbers from the Pattern. This is by design. The measure of time has always been connected to the measure of the circle. Furthermore, the measure of time is also connected to the measure of distance on the Earth, through many of these same Pattern numbers.

The Sun's apparent motion across the celestial sphere translates into corresponding measures of arc (longitude) and distance on the Earth, which are directly related to the measures of time. The corresponding distance measures on Earth are always denoted by the term 'geographic.'[5] With the Sun traveling directly above the equator, the leading edge of the Sun will move across 15 geographic miles (90,000 geographic feet) on the Earth in one minute of time. In 1 second, it will travel across 1,500 geographic feet, and in 4 seconds it will travel 6,000 geographic feet or 1 geographic mile. Accordingly, the following time measures and their longitudinal and geographic measures of distance equivalents can be derived:

1) 24 hours = 360 arcdegrees of longitude, or 21,600 geographic miles on the equator which is equal to the equatorial circumference of the Earth
2) 60 minutes = 15 arcdegrees of longitude or 900 geographic miles on the equator
3) 15 minutes = 1 arcdegree of longitude or 60 geographic miles on the equator
4) 1 minute = 15 arcminutes of longitude or 15 geographic miles on the equator
5) 4 seconds = 1 arcminute of longitude or 6,000 geographic feet or 1 geographic mile on the equator
6) **1 second = 15 arcseconds of longitude, or 1,500 geographic feet on the equator**

[5] The geographic foot and mile are discussed in further detail in the chapter, "Earth and its Dimensions."

All of the numbers mentioned in the foregoing paragraphs are Pattern numbers, except for 15, although ½ of 30, a Pattern number, is 15.

The beauty of this system is that the measures of time in longitude are valid at all parallels of latitude on the Earth, since these measures automatically scale down in direct proportion to the Earth's slower rotation rates at higher latitudes. And once their corresponding distance measures are defined, these, too, scale down in the same manner. All distance measures in this system including the Earth's circumference of 21,600 miles and the mile of 6,000 feet, are held constant, while the measure of the geographic foot varies as the cosine of the degree of latitude. Thus, all parallels of latitude are defined as exactly 21,600 geographic miles of 6,000 geographic feet. The key measure in this system is the geographic foot.

Today, geographic measures are seldom, if ever, used, but the geographic mile is a defined unit under the Système International (SI) system of measures, and its equivalent value in U.S. Customary Units is 1.1528 International mile or 6,087.0 feet (assuming a mile of 5,280 feet and a foot of 304.8 mm). The measure of the geographic foot is derived by dividing the measure of the geographic mile 6,087.0 feet by 6,000 feet and multiplying the result by the measure of the foot (304.8 mm).

$$6{,}087 \div 6{,}000 = 1.0145 \cdot 304.8\text{mm} = \underline{\text{one geographic foot of } 309.22 \text{ mm}}$$

This measure of the geographic foot as 309.22 mm is valid only on the equator. As noted above, the measure of this foot varies with higher parallels of latitude, as 309.22 · cosine θ (degree of latitude). Thus, the geographic foot, as adjusted for latitude, will automatically adjust all of the other geographic measures for latitude that are based on it, while their time equivalents will remain constant, e.g. 1 hour will always equal 15 degrees of longitude or 900 geographic miles, regardless of the latitude of any given location.

If there were a physical standard made of the geographic foot, as measured at the equator (309.22 mm), it could be readily marked with the necessary adjustments for higher latitudes. To do this, the standard is used as the hypotenuse of a right triangle, while the angle subtended with the side adjacent of the triangle is made equal to the degree of latitude. The measurement of the side adjacent is then determined as described earlier (309.22 · cosine θ), which defines the adjusted geographic foot for latitude. Once the adjusted geographic foot has been defined, its measure can be marked on the standard and denoted with the degree of latitude to which it applies. This can be readily done, without the need for any astronomical observations. The standard of the geographic foot, once marked for various degrees of latitude, can then be used anywhere.

1 second of time = 15 arcseconds of longitude, or 1,500 geographic feet

(This is a primordial measure of time)

a. The Daily Measures of Time.

All of the subdivisions of daily time are numbers that are exact matches to numbers which are manifestations of the Pattern in nature.

1 day = 24 hrs. = 1440 mins. = 86,400 secs.
½ day = 12 hrs. = 720 mins. = 43,200 secs.
¼ day = 6 hrs. = 3,600 mins. = 21,600 secs.
1 hr. = 60 mins. = 3,600 secs.
1 min. = 60 secs.

The Pattern pair for each number is:

24 (8/3)
1,440 (16/9)
86,400 (32/27)
12 (4/3)
720 (8/9)
43,200 (16/27)
6 (2/3)
36 (4/9)
21,600 (8/27)

Defined Value*	Pattern Number	Difference Δ	Dev. ± from Pattern
24 hrs. / day	24 ($4 \cdot 6$)	0	0
1,440 mins. / day	1,440 ($2 \cdot 6^2 \cdot 10$)	0	0
86,400 secs. / day	86,400 ($4 \cdot 6^3 \cdot 10^2$)	0	0
12 hrs. / ½ day	12 ($2 \cdot 6$)	0	0
720 mins. / ½ day	720 ($2 \cdot 6^2 \cdot 10$)	0	0
43,200 secs. / ½ day	43,200 ($2 \cdot 6^3 \cdot 10^2$)	0	0
6 hrs. / ¼ day	6	0	0
3,600 secs. / ¼ day	3,600 ($6^2 \cdot 10^2$)	0	0
21,600 secs. / ¼ day	21,600 ($6^3 \cdot 10^2$)	0	0

*These numbers are all manifestations of the Pattern in nature

Yearly, monthly and weekly measures of time. There are several measures of time that do not satisfy the criteria set forth for considering their measure a manifestation of the Pattern in nature.

These are the measures for the tropical year of 365.24 days, the month of 30.5 days, and the week of 7 days.

b. Weekly, Monthly and Yearly Measures of Time

Tropical year = 365.24 days
Solar Month = 30.5 days
Week = 7 days

Defined Value*	Pattern Number	Difference Δ	Dev. ± from Pattern
365.24 days	360 ($6^2 \cdot 10$)	5.24	+ .014
30.5 days	30 ($5 \cdot 6$)	.5	+.016
7 days	None	N/A	N/A

*These numbers are not manifestations of the Pattern in nature

CHAPTER 4

MEASURES OF THE FOOT, THE MILE, AND THE POUND

a. The International Foot and International Mile

The origins of the English statute mile and foot (now referred to as the International mile and International foot)[6], are vague, but the measure of the foot was formalized sometime around 1300 and the measure of the mile was formalized by statute in 1593, as 5,280 feet. The modern measure of the foot is given as 304.8 mm for convenience sake, but its original definition was 12 inches, with each inch being of 3 barleycorns. The reason that the origins of the measure of the foot and mile are considered vague is that they appear to pre-date the arrival of the Romans, in terms of the furlong measure—a very stable measure—which was used for land surveys and tax purposes, and survived in use afterwards. The furlong is now measured as 660 feet and there are 8 furlongs in a mile. Without belaboring the issue, the foot and mile are believed to have retained their measure from time immemorial, regardless of how they were translated into other measures throughout the centuries.

Regardless of the stability of their measure and long-term use in England, they are likely not of English origin. They are, instead, the universal measures of distance, and the proof lies with the measures of daily time, and several measures of distance of key dimensions in the solar system with remarkably similar numbering form, all of which are rooted in numbers from the Pattern.

The measures of time and distance at issue are (distance figures are from the Chapters cited in the parentheses; all were determined to be manifestations of the Pattern in nature):

1) 86,400 secs. in 1 day and 864,937 mi. in Sun's dia. (Chapter: "Astronomical Constants")

---

[6] Whitelaw, Ian; pp 18 and 19. Also, Wikipedia contributors, "United States Customary Units" Wikipedia, the Free Encyclopedia https://en.wikipedia.org/wiki/United_States_customary_units (accessed May 21, 2016). The International mile and International foot are recognized units of the U.S. Customary Units maintained by the U.S. Institute of Standards and Technology. They are, respectively, a mile of 5,280 feet, and a foot of 304.8 mm.

2) 43,200 secs. in ½ day and 4,328 mi. in the dia. of Earth's core ("Earth and its Dimensions")
3) 21,600 secs. in 1/4th day and 2,159.4 mi. in the dia. of the Moon ("Astronomical Constants")
4) The speed of light travels 186, 234 miles in 1 second ("Astronomical Constants")
5) The speed of light travels roughly 92,955,807 miles in 500 seconds, a distance that is equal to the astronomical unit

The odds of this randomly occurring are staggering. However, as Albert Einstein once said, "God does not play dice with the universe." But if we're considering the odds, we'll make it simple and use only numbers with the root 432 and reformat the others to this same root, using their Pattern measures. Accordingly, there are: 1) 4,320 miles in the diameter of the Earth's core, 2) 2 · 432,000 miles in the Sun's diameter, 3) ½ · 4,320 miles in the Moon's diameter, 4) 432 increments of 432 miles each in the speed of light in 1 second, and 5) 432 increments of 216 miles each in the astronomical unit. Keep in mind that all measures are, in theory, totally random events, which would make the odds of their occurring at best 432^5, as there are five random events to consider. That's less than 1 chance in 15 trillion! The odds are actually far worse, when one considers that several figures contain orders of magnitude. It is simply not possible that these are the result of purely random events.

It is possible that someone could construct a mile of 5,280 feet of 304.8 mm each, so as to achieve a perfect measure of the diameter of the Moon at 2,160 miles, and harmonize the measure with the measure of 1/4th day, 21,600 seconds. However, it staggers the imagination to think that this same individual did so in the full knowledge that this measure would also harmonize with the diameter of the Sun, the diameter of the Earth's core, the speed of light, and the astronomical unit, all of which use the same measure. Again, this is simply not possible.

The only reasonable conclusion is that the mile of 5,280 feet, of 304.8 mm each, are primordial measures, that were determined and set at the very beginning. To ask how or why, is to drift off into endless speculation. The same questions could be asked regarding the second of time, even though the second can be precisely defined in its application to measure. How and why it was set at 1/86,400 of a solar day are unknown and likely unknowable.

b. The Pound of Volume and the Cubic Foot.

A cubic foot of water has a mass of approximately 62.42 lbs. avoirdupois (28.316 kg/cubic foot of water, .4536 kg/lb. = 62.427 lbs.) as defined in the U.S. Customary Units of mass. The cubic foot contains 7.48 gallons or 60 pints, in the U.S. Customary Units system. Thus, a pint in a cubic foot of water contains 1.0405 lbs. (62.427 lbs./60 pints) or slightly more than a pound.

The measure of the mass of a cubic foot of water as 62.427 pounds or approximately 62.5 pounds facilitates a "double-my-predecessor" system for measuring 'goods of weight', which is the literal meaning of the term 'avoirdupois.' (62.5 doubles to 125, then to 250, then to 500, then to 1,000, then to 2,000 etc.) Accordingly, 8 cubic feet would contain 500 lbs. avoirdupois of water, 16 cubic feet would contain 1,000 lbs., etc.) If "double-my-predecessor" is important to facilitating the measure of large units of mass, then a similar system, "halving-my-predecessor", is equally important for measuring smaller units. Cubing is important in both systems, as demonstrated by the value of 8 cubic feet, which itself is a cube measuring 2 $ft.^3$ containing 500 lbs. avoirdupois.

If we symmetrically sub-divide the volume of the cubic foot into a whole number of smaller cubes of equal dimension, such that the total number of these smaller cubes is equal to or closely approximates the number 62.427—a number which does not lend itself to such subdivision—then there should be 64 (4^3) cubes (pints?), not 62.427 or 60. Using such symmetrical sub-division, there would be exactly 8 'gallons' of 8 'pints' each, for a total of 64, in a cubic foot of water, and each 'pint' would have a symmetrical volume of exactly 3" X 3" X 3" or 27 in^3 and a mass of **.975 lbs.** (62.427/64), less than the mass of the lb.(avdp.). Also, a kilogram of such pounds would equal: **2.204 lbs. (avdp.)/.975 lbs. = 2.26 lbs. (vol.).** In any case, the defined mass of the cubic foot of water remains intact, 62.427 lbs. avdp. or approximately 62.5 lbs. (avdp.)

1 $ft.^3$ water = 62.427 lbs. (avdp.) = 64 lbs. (vol.)
1 lb. (avdp.)/.975 =1 lb. (vol.)

This symmetrical subdivision and the 'pound (vol.)' will be used in several ensuing calculations. I have chosen to call such a pound the 'pound of volume', recognizing that it may not have been used anywhere in history.[7] The cubic foot and the volume of water that it contains are considered fundamental units of both mass and volume. The lb. (avdp.) is the unit of mass used for all purposes in this work, except where the lb. (vol.) is specifically introduced.

The symmetrical division of the cubic foot of 12" x 12" x 12" yields several measures that are all exact Pattern numbers.

1 "pint" of 3" x 3" x 3" is equal to 27 in^3
1 "gallon" of 6" x 6" x 6" is equal to 216 in^3 (8 "pints")
1 cubic foot contains 12" x 12" x 12" or 1,728 in^3 (8 "gallons")

The numbers 27, 216, and 1,728 are all exact Pattern numbers. Their Pattern pairs are: 27 (27), 216 (8/27), and 1,728 (64/27).

[7] Klein, Herbert Arthur; *The Science of Measurement, A Historical Survey*; Dover Publications, Inc. New York, 1988; Part I, pp 28-52. Also, see Wikipedia contributors, "Pound (Mass)" Wikipedia, the Free Encyclopedia: https://en.wikipedia.org/wiki/Pound_(mass)

Defined Value	Pattern Number	Difference Δ	Dev. ± from Pattern
27 in^3 in a pint *	27 (3^3)	0	0
216 in^3 in a gallon *	216 (6^3)	0	0
1,728 in^3 in a cu. foot *	1,728 (6^4)	0	0

*All of these measures are considered to be manifestations of the Pattern in nature.

It should be noted that there has been constant tension throughout history regarding the measure of mass. Should it be based on volume, as I've described in the foregoing, or an artifact with no or only nominal connection to volume? The measure of the kilogram has undergone such turmoil. The original definition of the kilogram defined it as 10 cm^3 of water or exactly one liter, a perfect volumetric measure. However, this is no longer the definition of a kilogram, and while it does closely approximate the mass of water contained in one liter or 10 cm^3, the kilogram is now defined by the mass of an international prototype or artifact maintained in Paris.

c. The Pound Avoirdupois

There may be an impression that the pound avoirdupois is not a manifestation of the Pattern in nature and is, therefore, not a natural measure. Neither statement is true. The lb. (avdp.) is based on one of the oldest measures of mass—the grain. In a lb. (avdp.) there are 7,000 grains, which is remarkably close to a Pattern number, 6,998.4. The pattern pair of this number is (32/2,187).

Defined Value	Pattern Number	Difference Δ	Dev. ± from Pattern
7,000 grains	6,998.4	1.6	-.00023

The Pound Avoirdupois of 7,000 grains is a Manifestation of the Pattern in Nature

The divisions of the lb. (avdp.) reflect the unending tension between measures of mass and measures of volume. In a lb. (avdp.), there are 16 ounces, of 16 drams each, of 27 11/32 grams each, for a total of 7,000 grains. If we use the definition for a lb. (vol.) as .975 lb. (avdp.), then a lb. (vol.) is 6,825 grains, which is not a number from the Pattern. Or, if we use the subdivisions of the lb. (vol.), then, in each lb. (vol.), there are 27 $inches^3$ of 27 $barleycorns^3$ or 729 $barleycorns^3$ in a lb. (vol.), none of which are manifestations of the Pattern in nature.

CHAPTER 5

EARTH AND ITS DIMENSIONS

a. The International Foot

The International foot[8]of .3048m is a U.S Customary unit and is recognized by the Système International (SI); its basis is the English statute foot, a measure of vague origins and a confusing history. The International mile[9] of 1.609344km, or 5,280 feet, is also U.S. Customary Unit and is recognized by the SI; it is based on the English statute mile. These two measures, regardless of their history, are fundamental to all measures of distance in the cosmic order. They are referred to hereafter as the **International foot** and the **International mile**, though their origins and applications are decidedly universal in nature.

The primordial measures of distance:
International foot = 304.8 mm
International mile = 5,280 English feet

Both are primordial measures of distance, and they directly relate to other units of distance measure that define the dimensions of the Earth. These measures and their modern definitions are: the nautical mile of 1.852km or 6,076 feet (see footnote 1 below), and the geographic mile[10] of 1.8553km or 6,087 feet. The nautical mile is used in conjunction with measures in degrees of latitude and the geographic mile is used in conjunction with measures in degrees of longitude. For purposes of what follows, both the nautical mile and geographic mile are redefined as 6,000 nautical feet and 6,000 geographic feet, respectively, which requires that the dimension of the foot be redefined specifically for the nautical mile and geographic mile. The measures of each foot are, of necessity, different.

---

[8] Wikipedia contributors, “Foot”, Wikipedia the Free Encyclopedia: https://en.wikipedia.org/wiki/Foot_(unit) (accessed April 28, 2016)

[9] Wikipedia contributors, “Mile”, Wikipedia the Free Encyclopedia: https://en.wikipedia.org/wiki/Mile (accessed April 28, 2016)

[10] Wikipedia contributors, Wikipedia, the Free Encyclopedia, “Geographic Mile: https://en.wikipedia.org/wiki/Geographical_mile (accessed May 3, 2016)

b. The geographical foot.

The geographical foot is defined as:

6,087 Int. ft. / 6,000 geo. ft. · 304.8 mm (Int. ft.) =
1 Geographic Foot of 309.22 mm

c. The nautical foot.

The nautical foot is defined as:

6,076 Int. ft. / 6,000 naut. ft. · 304.8 mm (Int. ft.) =
1 Nautical Foot of 308.66 mm

The nautical foot and geographic foot can also be derived with specific conversion factors that are rooted in arithmetic progressions, based on the numbers 8 and 9, both of which are Pattern numbers. These progressions are, for the number 8: 64, 72, 80; and, for the number 9: 63, 72, 81. The specific numbers are selected members of these two progressions; their derivation is based on several triangles that will be discussed in depth in a later chapter. However, the numbers 80 and 81 derive from the right triangle with sides 2 and 1 and hypotenuse √5 discussed earlier. The numbers 72 and 73 are based on a second right triangle with sides 1 and √3 and hypotenuse of 2. **The numbers 64, 72, 80, and 81 are numbers from the Pattern**. Using these numbers, the following conversion factors are defined:

1) Geographic foot = 73/72, or 1 + 1/72, of the international foot = 309.03 mm
2) Nautical foot = 81/80, or 1 + 1/80, of the international foot = 308.61 mm
3) (There is a third measure of the foot that derives from these two arithmetic series, which will be introduced in a later chapter.)

Geo. ft. 309.03mm (Pattern); 309.22 mm actual
Naut. ft. 308.61 mm (Pattern); 308.66 mm actual

Defined Value	Pattern Number	Difference Δ	Dev. ± from Pattern
309.22 mm (geo. foot)	309.03 mm*	.19	+.0006
308.66 mm (naut. foot)	308.61 mm*	.05	+.0002

*These numbers are derivatives of a number from the Pattern; 1+ 1/**72** for the nautical foot, and 1 + 1/**80** for the geographic foot. Both of these measures are considered to be manifestations of the Pattern in nature. These measures were defined in terms of millimeters to make them accessible and understandable. The actual measures begin with the international foot, or 1 foot, to which is added the fraction 1/72 for the geographic foot and 1/80 for the nautical foot.

If the modern definitions for the lengths of the geographic mile and nautical mile are compared to their calculated estimates, then the geographic mile is: 309.03/309.22 · 6,087 Int. feet = 6,083 Int. feet, or 4 Int. feet short of the modern figure. The nautical mile is 308.61/308.66 · 6.077 = 6,075 Int. feet, or 2 Int. feet short of the modern figure. The modern figure for the nautical mile is based on a mean measure, as the actual measure, which is based on one arcminute of latitude, varies considerably with increasing latitude toward the poles.

The nautical mile of 6,000 nautical feet of 308.61mm and the geographic mile of 6,000 geographic feet of 309.03 mm will be used in the following calculations of the Earth's dimensions; they are based on the circumference measure of 21,600 miles. The nautical mile is still a viable measure of distance and is used in the great circle navigation of ships and aircraft. The geographic mile is seldom, if ever, used anymore.

d. The Meridional circumference of Earth

The meridional circumference (measured through the poles) of the Earth is: 21,600 nautical miles · 6,000 nautical feet · 308.61mm/nautical foot = 39,995,856,000 mm or 39,995.856 km or 24,852.2 International miles. This compares to the modern estimate of approximately 24,859.7 International miles.[11] However, only the modern estimate will be analyzed as a possible manifestation of the Pattern in nature, as further analysis of the conversion factors and their relationship to the aforementioned triangles is required.

Earth's Meridional Circumference: 24,859.7 International Miles

e. The Equatorial Circumference of Earth

The equatorial circumference of the Earth is: 21,600 geographic miles · 6,000 geographic feet · 309.03mm/geographic foot = 40,050,288 km or 24,886.05 International miles. This compares to the modern measure of approximately 24,901.5 International miles.[12] However, only the modern estimate will be analyzed as a possible manifestation of the Pattern in nature, as further analysis of the conversion factors and their relationship to the aforementioned triangles is required.

---

[11] See Wikipedia entry for Meridian Arc: https://en.wikipedia.org/wiki/Meridian_arc#Numerical_expressions, and in particular, 'Numerical Expressions' using WGS-84 ellipse model.

[12] Wikipedia contributors "Equatorial Circumference", Wikipedia, the Free Encyclopedia: https://en.wikipedia.org/wiki/Equator#Exact_length, (Accessed May 7, 2016), see in particular, section headed, 'Exact Length.' This length is based on the International Astronomical Union (IAU) figure. This figure is also incorporated in the WGS-84 ellipsoid.

Earth's Equatorial Circumference: 24,901.5 International Miles

There is a Pattern number that very closely approximates the measures for both the meridional and equatorial circumferences. The Pattern number is 24,8832, or 24,883.2 $\cdot 10^{-1}$; its Pattern pair is (1,024/243).

Defined Value	Pattern Number	Difference Δ	Dev. ± from Pattern
Meridional: 24,859.7 *	24,883.2 ($6^5 \cdot 32 \cdot 10^{-1}$)	23.5	-.0009
Equatorial: 24,901.5 *	24,883.2 ($6^5 \cdot 32 \cdot 10^{-1}$)	18.3	+.0007

*These are not Pattern numbers, as they do not satisfy the parameters set for accepting them as manifestations of the Pattern in nature. However, both are based on a combination of Pattern numbers, including 32 and 6. The significance of numbers such as these will be discussed in a later chapter.

Arc measures of longitude and their relationship to geographic measures of distance:

360° of longitude = 21,600 geographic miles on the equator (equatorial circumference)
15° of longitude = 900 geographic miles on the equator
1° of longitude = 60 geographic miles on the equator
15' (arcminutes) of longitude = 15 geographic miles on the equator
1' of longitude = 6,000 geographic feet or 1 geographic mile on the equator
15" (arcseconds) of longitude = 1,500 geographic feet on the equator

These measures relate directly to the measure of time. In fact, the primary usefulness of geographic measures of distance is their relationship to time.

Arc measures of latitude and their relationship to nautical measures of distance:

360° of latitude = 21,600 nautical miles on an arc of meridian (meridional circumference)
15° of latitude = 900 nautical miles on an arc of meridian
1° of latitude = 60 nautical miles on an arc meridian
15' of latitude = 15 nautical miles on an arc of meridian
1' of latitude = 6,000 nautical feet or 1 nautical mile on an arc of meridian
15" of latitude = 1,500 nautical feet on an arc of meridian

An arc of meridian is a single line of longitude. The foregoing measures relate directly to the measure of distance, particularly along a great circle, of which an arc of meridian is one. However, it should be recalled that the nautical mile is a mean measure of distance along an arc of meridian. At higher lines of latitude, large errors rates are possible, using this mean measure.

The measure of the International mile can be related to the measure of the nautical and geographical miles. However, only the nautical mile will be discussed for the sake of brevity, but all of the following principals can be applied to the geographic mile as well.

The nautical and International mile relate to one another by means of the right triangle with sides 2 and 1 and hypotenuse of $\sqrt{5}$, and its further refinements: 1) a right triangle with sides 8 and 4 and hypotenuse of approximately 9, and 2) a right triangle with sides of 64 and 16 and hypotenuse of approximately 80-81. Both of these refinements are necessary factors in the conversion of the nautical mile to the International mile, and the International mile to the nautical mile.

Using the 8-4-9 refinement, the International mile relates to the nautical mile as, respectively, side of 8 to hypotenuse of 9 or 8 to 9, or, inversely, the nautical mile relates to the International mile as 9 to 8. However, in both instances a correction factor must be applied, as the square of number $\sqrt{5}$ on the hypotenuse of the base triangle, when multiplied by 4, ($\sqrt{5} \cdot 4 \approx 8.9$) does not directly relate to the number 9 on the hypotenuse of the refinement triangle. If we take the difference between the relationship of sides 9/8 (≈ 1.12) of the first refinement triangle and the relationship of sides $\sqrt{5}/2$ (≈ 1.11) of the base triangle, we arrive at a difference of $1.12 - 1.11 = .01$ or 1/100. This is the conversion factor.

To convert the nautical mile measure of 6,000 feet to the International mile measure of 5,280 feet, first apply the conversion factor to the measure of the nautical mile ($.01 \cdot 6{,}000) = 60$, then subtract the result from the nautical mile $(6{,}000 - 60) = 5940$, which when multiplied by $8/9 = (5{,}940 \div 9) \cdot 8 = 5{,}280$. To convert from the measure of the International mile of 5,280 to the nautical mile of 6000, proceed in reverse order. First, multiply the International mile 5,280 by 9/8, or $(5{,}280 \div 8) \cdot 9 = 5{,}940$, and then apply the conversion factor, $.01 \cdot 5{,}940 = 59.4$, which when added together yields $5{,}940 + 59.4 = 5{,}999.4$. Of course, these procedures only relate the broad measures of the two miles to one another. The specific measure of the nautical foot and International foot also need to be applied, which, as earlier discussed, relate to one another as 1 to (81/80), which is where the second refinement to the hypotenuse of the base triangle comes into play. As noted in a foregoing paragraph, the nautical foot derives from the International foot of 304.8 mm by multiplying the International foot by 81/80, which yields 308.61 mm. While millimeters are used to define these measures, the proper manner for describing the difference is that the nautical foot is 81/80 of the International foot, which is a primordial measure.

The International mile of 5,280 English Feet is a primordial measure and factors in many manifestations of Patter numbers in nature. Its precise measure is of critical importance in all distance measures, particularly those beyond the Earth. As will be demonstrated in coming chapters, the diameters of the Sun, Earth's magnetic core (see following section), and Moon, which are, respectively, manifestations of the Pattern numbers 864, 432, and 216, ALL are measured in International miles. Thus the importance of a precise definition of this measure, and for having a means to relate it to the Earth measures of distance, including the nautical and geographic miles.

f. Earth's Electro-magnetic Core and its Measure

The Earth's interior is made up of a number of distinct layers, much like the layers of an onion. However, the only structure that will be examined here is the Earth's core, which lies beneath the Earth's mantle at the core-mantle boundary (CMB)[13], located some 1,793 miles beneath the surface of the Earth.

The core consists of an outer layer of electro-magnetically active, molten matter and an inner core of solid nickel-iron. The radius of the core structure as a whole, including the inner and outer structures, is determined as follows. Starting with the Earth's average radius[14], 3,959 International miles, subtract the average depth of the CMB, 1,795 (range is from 1,790 – 1,800) International miles = 2,166 International miles, or a Core diameter of 4,328 International miles.

Diameter of the Earth's Core = 4,328 Int. miles

There is a Pattern number that very closely approximates this figure. The number is 4,320 and its Pattern pair is (16/27).

Defined Value	Pattern Number	Difference Δ	Dev. ± from Pattern
4,328 miles*	4,320 ($2 \cdot 6^3 \cdot 10$)	8	+.0018

*This is a manifestation of the Pattern in nature.

g. Volume of the Earth.

The volume of the Earth[15] is estimated at: $108.321 \cdot 10^{10}$ km^3. Converting this figure to miles^3, yields: $25.9874 \cdot 10^{10}$ miles^3.

Volume of the Earth = $25.9874 \cdot 10^{10}$ Int. miles^3

There is a Pattern number that very closely approximates this figure. The number is 25.92 and its Pattern pair is (32/81).

Defined Value	Pattern Number	Difference Δ	Dev. ± from Pattern
25.9874 *	25.92 ($2 \cdot 6^4 \cdot 10^{-2}$)	.0674	+.0026

*This is a manifestation of the Pattern in nature.

---

13 Wikipedia contributors, "Earth, Physical Characteristics", Mean Radius; Wikipedia, The Free Encyclopedia: https://en.wikipedia.org/wiki/Earth (Accessed May 18, 2016)

14 Wikipedia contributors, "Earth, Physical Characteristics", Mean Radius; Wikipedia, The Free Encyclopedia: https://en.wikipedia.org/wiki/Earth (Accessed May 18, 2016)

15 NASA Earth Fact Sheet, web site: http://nssdc.gsfc.nasa.gov/planetary/factsheet/earthfact.html

h. Mass of the Earth.

The estimate for the mass of the Earth is $5.9722 \cdot 10^{24}$ kg x 2.2046 = $13.166 \cdot 10^{24}$ lbs. (avdp.).

$$\text{Mass of the Earth} = 13.166 \cdot 10^{24} \text{ lbs. (avdp.)}$$

There is a Pattern number that very closely approximates this figure, 13.122; its Pattern pair is (2/6,561).

Defined Value	Pattern Number	Difference Δ	Dev. ± from Pattern
$13.166 \cdot 10^{24}$ *	13.122 ($6 \cdot 2.187 \cdot 10^{24}$	.044	+.003

*This measure of the mass of the Earth is a manifestation of the Pattern in nature.

CHAPTER 6

EARTH-MOON DISTANCE AND THE BARYCENTER

The distance between the Earth and the Moon is always measured from the center of each body. It is usually cited as the mean measure of the distance between the Moon's mean perigee (closest approach to Earth and its apogee (furthest distance from the Earth. This distance is 238,855 International miles. However, if the difference between the extreme apogee and the extreme perigee is taken as the simple average distance, the measure is slightly different. At extreme apogee, the distance is: 406,700km (252,711 International mile) and at extreme perigee, the distance is: 356,400km (221,456 International miles), which means the **average distance is: 381,550km (237,084 International miles).**

There is a Pattern number that very closely approximates this number; the Pattern number is 236,196 and its Pattern pair is: (4/59,049).

Defined Value	Pattern Number	Difference Δ	Dev. ± from Pattern
237,084 Int. miles *	236,196 ($81^2 \cdot 6^2$)	888	+.0037

*This is not a manifestation of the Pattern in nature, as the Pattern number associated with it does not satisfy the parameters set for accepting it as a manifestation of the Pattern in nature. However, the Parameter number is based on a combination of Pattern numbers, including 81 and 6. The significance of numbers such as this will be discussed in a later chapter.

The closest approach of the surface of the Moon at extreme perigee (221,456 International mile) to the surface of the Earth is: 221,456 – 3,693 (Earth radius) = 217,493 International mile – 1,080 (Moon radius) = 216,413 International mile.

Closest Approach of Moon = 216,413 Int. miles

There is a Pattern number that very closely approximates this number. The Pattern number is 216,000 and its Pattern pair is: (8/27)

Defined Value	Pattern Number	Difference Δ	Dev. ± from Pattern
216,413 Int. miles *	216,000 ($6^3 \cdot 10^3$)	413	+.002

*This measure of the closest approach of the Moon to the Earth is a manifestation of the Pattern in nature.

The barycenter is the center of mass of two or more bodies about which their centers orbit in a circle.[16] For the Earth-Moon pair, the barycenter is located within the Earth, some 1,100 International miles below its surface. To find the distance from the Earth's center to the Earth-Moon barycenter, use the following formula.

$$r_1 = a \cdot m_2 / (m_1 + m_2)$$

Where: r_1 is the distance to the barycenter from the Earth's center, a is the average distance between the Earth and Moon—measured center-to-center—m_2 is the mass of the Moon proportional to the Earth (1/81, or .012345679), and m_1 is the Mass of the Earth (1).

$$r_1 = 237{,}084 \text{ Int. miles} \cdot (.012345679)/1.012345679)$$
$$= 2891 \text{ Int. miles})$$

Subtracting the barycenter's distance from the center, 2,891 miles, from the earth's radius, 3,963 miles, gives the barycenter's distance from the surface of the Earth as, 1,072 miles.

Barycenter's distance beneath Earth surface is: 1,072 International miles.

There is a Pattern number that very closely approximates this number. The Pattern number is 1,080 and its Pattern pair is: (4/27)

Defined Value	Pattern Number	Difference Δ	Dev. ± from Pattern
1,072 Int. mi. *	1080 ($3 \cdot 6^2 \cdot 10$)	8	-.0074

*This measure of the Earth-Moon barycenter at the closest approach of the Moon to the Earth is a manifestation of the Pattern in nature.

[16] See Wikipedia entry for Barycenter: https://en.wikipedia.org/wiki/Barycenter . Also, see Wikipedia entry for Moon: https://en.wikipedia.org/wiki/Moon .

CHAPTER 7

OTHER MEASURES UNIQUE TO THE EARTH

There are a number of measures that are unique to the Earth. Recognizing that there are likely a large number of such measures, the following were reviewed and are described below.

a. Standard Atmospheric Pressure

The Standard Pressure of the atmosphere[17] is measured at sea-level, at 0°C. The Pascal is the SI derived unit measure for standard pressure and temperature (STP). STP is 101.325 kPa and the equivalent in pounds-force units is 14.696 pounds (avdp.)-force per in.^{2} (abbreviated as psi). 14.696 lbs. (avdp.)-force per in.^{2} converts to 15.0728 lbs. (vol.)-force per in.^{2}. (The lb. (vol.) is described in the chapter on "The Mile, the Foot, and the Pound.") Alternatively, another equivalent in pounds-force, is: 2,170.483 lbs. (vol.)-force per ft.^{2}.

Standard Atmospheric Pressure is:
2,170.48 lbs. (vol.)-force per ft.^{2}

There is a Pattern number that very closely approximates 2,170.486 lbs. (vol.) per ft.^{2}. The Pattern number is, 2,160 and its Pattern pair is (8/27).

Defined Value*	Pattern Number	Difference Δ	Dev. ± from Pattern
2,170.48 lbs. (vol.)/ ft^2	2,160 ($6^3 \cdot 10$)	10.486	+.0048

*The measure of Earth's standard atmospheric pressure is considered to be a manifestation of the Pattern in nature.

b. The Speed of Sound.[18]

This is a very tenuous measure as it depends on many factors, including temperature, the density of the medium through which it travels, atmospheric pressure, etc. Therefore, it is not

[17] Whitelaw, Ian; pp 120-121
[18] Klein, Herbert Arthur; pp 579-580

a constant. However, the speed of sound is generally recognized as: 1,087.32 ft. per second, in dry air, at sea level, at 0° C.

Speed of Sound is 1,087.32 feet per second

There is a Pattern number that very closely approximates 1,087.32 ft. per sec. The Pattern number is 1080 and its Pattern pair is (4/27).

Defined Value*	Pattern Number	Difference Δ	Dev. ± from Pattern
1,087.32 ft. per sec.	1,080 ($3 \cdot 6^2 \cdot 10$)	7.32	+.0067

*The measure of the speed of sound is considered to be a manifestation of the Pattern in nature.

c. Genetic Code.[19]

There are 64 groups in the genetic code for the amino acids that are similar among all living organisms. The code consists of 3 nucleotides for each group, which are called nucleotide triplets or codons. The codons determine the sequence followed by living cells for adding amino acids during protein synthesis.

There is much more to this topic, but it is of interest that the numbers 3 and 64 are organizing factors in the genetic code. However, they are not measures. These numbers are numbers from the Pattern, although they do not satisfy the criteria for considering them a manifestation of the Pattern in nature. Their Pattern pairs are: 64 (64) and 3 (3).

There are 64 codons of 3-nucleides each in the genetic code

Defined Value*	Pattern Number	Difference Δ	Dev. ± from Pattern
64 codons	64	0	0
3 nucleotides/codon	3	0	0

*These figures are considered to be manifestations of the Pattern in nature.

d. Music Theory.

There is an extraordinary confluence of the mathematics of music theory with the numbers of the Pattern. There are two tuning systems used in western music.

[19] Wikipedia contributors, "Genetic Code", Wikipedia, the Free Encyclopedia, https://en.wikipedia.org/wiki/Genetic_code (Accessed May 24, 2016)

The first is equal temperament tuning, which is based on the octave, with equal-tempered intervals between the octaves, generally either 6-tone intervals or 12[20] semi-tone intervals. At each octave the frequency or pitch doubles, in an ascending scale, and halves in a descending scale. The octaves above a given pitch are called super-octaves, and the first super-octave is double the frequency, with each succeeding super-octave doubling the frequency of its predecessor. In a descending scale, the octaves below the pitch are called sub-octaves, and the first octave below the pitch halves the frequency of the pitch, with each succeeding sub-octave halving the frequency of the sub-octave preceding it. However, between each octave, whether super-octave or sub-octave, are equal intervals of 6 or 12 steps. These intervals are typified by the familiar scale; doh, re, me, fa .so, la, tee, doh. This doubling or halving the pitch frequency is similar to the doubling in the Pattern matrix, moving down the 2 side. Moving across the 3 side, produces different doubling numbers.

The second system is just tuning, or Pythagorean tuning.[21] This is a system based on the ratio 3:2. This system is a pure manifestation of the Pattern. The following is quoted directly from cited the Wikipedia source. The purpose is to demonstrate the flow of the numbers and their generation using the ratio 3:2, and to invite a comparison between these numbers and those in the Pattern matrix. All are exact matches.

"Given D tuned to 288 Hz, the following semi-tone notes are derived:

Eb—Bb—F—C—G—**D**—A—E—B—F#—C#—G#

"...the A is tuned such that the frequency ratio of A and D is 3:2—if D is tuned to 288 Hz, then A is tuned to 432 Hz. The E above A is also tuned in the ratio 3:2—with the A at 432 Hz, this puts the E at 648 Hz, 9:4 above the original D. When describing tunings, it is usual to speak of all notes as being within the octave of each other, and as this E is over an octave above the original D, it is usual to halve its frequency to move it down an octave. Therefore, the E is tuned to 324 Hz, a 9:8 above the D. The B at 3:2 above that E is tuned to the ration 27:16 and so on. Starting from the same point working the other way, also from D to G is tuned as 3:2. With D at 288 Hz, this arrives at G as 192 Hz, or, brought into the same octave, to 384 Hz."

Note	Interval from D	Formula	Frequency Ratio
Ab	Diminished Fifth	$(2/3)^6 \times 2^4$	1024/729
Eb	Minor second	$(2/3)^5 \times 2^3$	256/243
Bb	Minor sixth	$(2/3)^4 \times 2^3$	128/81
F	Minor third	$(2/3)^3 \times 2^2$	32/27
C	Minor seventh	$(2/3)^2 \times 2^2$	16/9

[20] Wikipedia contributors, "Music and Mathematics", Wikipedia, the Free Encyclopedia, https://en.wikipedia.org/wiki/Music_and_mathematics (Accessed may 23, 2016)

[21] Wikipedia contributors, "Pythagorean Tuning", Wikipedia, the Free Encyclopedia, https://en.wikipedia.org/wiki/Pythagorean_tuning (Accessed May 24, 2016)

G	Perfect fourth	2/3 x 2	4/3
D	Unison	1/1	1/1
A	Perfect fifth	3/2	3/2
E	Major second	$(3/2)^2$ x ½	9/8
B	Major sixth	$(3/2)^3$ x ½	27/16
F#	Major third	$(3/2)^4$ x $(1/2)^2$	81/64
C#	Major seventh	$(3/2)^5$ x $(1/2)^2$	243/128
G#	Augmented fourth	$(3/2)^6$ x $(1/2)^3$	729/512

Just-tuned Music Perfectly Reflects the Pattern

It is tempting to think that because Pythagorean, or just tuning is perfectly matched to the Pattern that it would be ideal for music. In fact, it is not. It works well for music that does not change key very often or is not harmonically adventurous. This restricts the range of such music and it is the reason that equal-tempered music is much preferred over just-tuned music. Maybe this is a reflection of the universal fact that the 'perfect' is seldom if ever present in the world of nature. Then again, maybe Man, being an imperfect being, can perceive the Pattern through just-tempered music, but cannot understand it perfectly; therefore, he prefers what is most pleasing to him—equal-tempered music—which is not perfect, but works well in the world of nature.

CHAPTER 8

SYSTÈME INTERNATIONAL D'UNITÉS (SI) UNITS

The purpose of this chapter is to demonstrate that the Pattern also exists in modern measurements. The measurements that are dealt with are the 'base units' of the Système International d'Unités, or SI. There are seven base units, from which all other units derive, and include:

Second (s), measure of time.
Meter (m), measure of distance.
Kilogram (kg), measure of mass.
Kelvin (K), measure of thermodynamic temperature.
Mole (mol), measure of amount of substance.
Ampere (A), measure of electric current.
Candela (cd), measure of luminous intensity.

There are some 22 derived units recognized by SI that are given unique names. These include, among others, the newton (N), the SI measure of force, and the Joule, the SI measure of energy or work. The newton and Joule are emphasized here because they are frequently used in the derivation of other SI units.

All units, regardless of their specific applications, can be traced to the two classic units of measure, time and distance. Even mass derives from distance in that it is a factor of volume, a derivation of distance. I have elsewhere dealt with what I will refer to here as the 'primordial measures of time and distance', which include:

Minute of 60 seconds, measure of time.
Foot of .304.8 m <u>and</u> mile of 5,280 ft., measures of distance.

To these, there is added the pound avdp., a measure of mass, which derives from the cubic foot. Also, since mass and distance are determining factors in the derivation of the newton, it, too, is the subject of detailed analysis in a dedicated section of this chapter.

All units will be defined in terms of the foot and the pound avdp. The value of the pound is adjusted for volume so that it is geometrically symmetrical with the cubic foot, i.e., the cubic foot is divided into 64, 3" x 3" x 3" cubes, each of which is assigned the weight of one 'pound (vol.)', which will be discussed in greater detail in a following section.

Four base units of the SI system are each the subject of dedicated sections in this chapter, including: the Kelvin, the ampere, the mole and the candela, in order to ascertain if their respective measures are influenced by the Pattern.

a. The Newton and the Measure of Force.

The newton[22] is the SI unit of force. It is a derived unit, not a base unit, even though it factors in a number of other derived units. It is defined as the amount of force necessary to increase the velocity of 1 kilogram of mass, by one meter a second, in one second. It is based on the constant acceleration of gravity (g_n), 32.174 ft./sec.^{2} or 9.81 meters/sec.^{2}

Factors:
Force (f) = mass · acceleration(a)
1 newton (N) = 1 kg. · m/sec.^{2}
At the surface of the earth, 1 kg. exerts 9.8066 N of force due to gravity (g_n).

Calculations using lb. (avdp.):
9.8066 newton (N) = 1 kg. · m/sec.^{2} = 2.2046 lbs. (avdp.) · m/sec.^{2}
1 newton (N) = 2.2046 lbs. (avdp.) · m/sec. ÷ 9.8066 = 0.2248 lbs. (avdp.)-force
= 7.23 poundal (The poundal is the unit of measure for force where acceleration is involved, and is equal to .2248 lbs. (avdp.)-force ÷ 1/32.2 ft. sec.$^{-1}$. The poundal is defined as the amount of force necessary to accelerate a mass of one pound by one foot per second per second or lb.· ft. · sec.^{2}.)

= 1 slug (32.17 lbs. (avdp.) · .2248 lb.-f. The slug (32.17 lbs. avdp.) is that mass which when a force of one pound-force is applied to it will accelerate it by one pound per second, per second. It is similar to the poundal in that it accounts for acceleration in the equation f = m · a, except that it multiplies by 32.17 instead of dividing.

1 Newton = .2248 lbs.-force
= 7.23 poundal (.2248 lb.-f ÷ 1/32.17)
= 1 slug (32.17 lbs.) ·.2248 lb.-f

[22] Whitelaw, Ian; *A Measure of All Things; The Story of Measurement Throughout the Ages*; David & Charles, Cincinnati, Ohio, USA; 2007; pages 112-113. See also, Wikipedia contributors, "Newton (unit)" Wikipedia, the Free Encyclopedia: https://en.wikipedia.org/wiki/Newton_(unit) (Accessed: May 30, 2016)

There are numbers from the Pattern that closely approximate all three of these measures. For the lb. force, the number is .2239488, which has a pattern pair of (1,024/2,187). For the poundal, the number is 7.2, which has a pattern pair of (8/9). For the slug, the number is 32, which has a pattern pair of (32/1).

Defined Value	Pattern Number	Difference Δ	Dev. ± from Pattern
1 N = .2248 lb. (avdp)-f*	.2239 ($8 \cdot 6^7 \cdot 10^{-7}$)	.0009	+.004
1 N = 7.23 poundal*	7.2 ($2 \cdot 6^2 \cdot 10^{-1}$)	0.03	+.004
1 N = 32.17 lbs. · .2248*	32	0.17	+.005

All three of these measures of the Newton are considered to be manifestations of the Pattern in nature.

(*These are the values of the newton (N), as measured in lb. (vol.)-force, poundal and slug, that will be used for all further purposes in this work, where force is a factor.)

b. Gravity and Standard Acceleration.

The figure for standard acceleration due to Earth's gravity is an SI derived unit, but it does not have a special name. As noted in the previous section, it is 9.8066 m · $sec.^{-2}$ or 32.174 ft. · $sec.^{-2}$.

The latter figure, 32.174, is approximately equal to 32, which is a number from the Pattern and has a pattern pair (32/1). However, it is a defining number and not a number that fits the defining parameters for parameter numbers.

Earth's gravity (g_n) = 32.17 ft. $sec.^{-2}$

Defined Value	Pattern Number	Difference Δ	Dev. ± from Pattern
32.174 ft. · $sec.^{-2}$	32	0.174	+.005

The measure of g_n is considered to be a manifestation of the Pattern in nature.

c. The Joule

The Joule (J) is the SI unit for force, and is equal to 1 N · m or the amount of energy necessary to move one kg, one meter, with a force of one newton. It is an SI derived measure. Converting the joule's SI measures to lbs. (avdp.) and international feet, yields: 7.233 poundal · 3.2808 ft./meter = 23.73 poundal feet. There is no number from the Pattern that approximates this number. The closest one is 24, but the difference between the two numbers is .27 and the deviation from 24 is .0112—well beyond the range of deviation of ±.009 established in this work for such numbers.

Defined Value	Pattern Number	Difference Δ	Dev. ± from Pattern
23.73 ft. · $sec.^{-2}$	24	0.27	+.011

The measure for the joule is not a manifestation of the Pattern in nature.

The reason that the measure of the joule does not appear to be a manifestation of the Pattern in nature is not known. However, there may be a physical point at which the Pattern no longer clearly manifests itself or disappears all together. This may explain why the dimensions of the other planets of the Solar System did not appear to be manifestations of the Pattern in nature. Several, including Mars and Jupiter were close to the Pattern, but their measures, like the joule, did not fall within the established range of acceptable deviation.

Interestingly, if James Prescott Joule's original definition of the unit that bears his name is used, then the Pattern does manifest itself. Joule's definition was based on the observed fact that a mass of 772 lbs. (avdp.), falling one (international) foot, while turning a paddlewheel immersed in one pound (avdp.) of water, would raise the temperature of the water by one-degree F.[23] There are 23.997 slugs (32.17 lbs. (avdp.) per slug) in the number 772. The number 23.997 very closely approximates the number 24, a number from the Pattern with a pattern pair of (8/3). This means that the mass of 772 lbs. (avdp.) will generate a force equal to 173.5456 N, which compares to a Pattern number of 172.8, that has a pattern pair of (64/27). The difference between these two numbers is .7456, which means that 173.5456 N deviates from the Pattern number 172.8 by +.004. However, all of this is little more than a historical note. The modern measure of the joule is unquestionably the amount of energy necessary to move 1 kg of mass, one meter, with a force of one newton.

d. The Kelvin and Temperature.

The Kelvin scale is an absolute, thermodynamic scale used to measure temperature from the triple point of water, 273.16 K, to absolute 0 K, a point at which all thermal motion at the atomic level ceases. The Kelvin is one of the seven currently recognized SI base units.

The Kelvin scale is related to all other temperature scales, including: Celsius, Fahrenheit, and Rankin. One Kelvin is equal to one degree Celsius.

The current definition of the kelvin is the fraction 1/273.16 K. However, there is an effort underway at SI to redefine the Kelvin in terms of force or energy, using Boltzmann's Constant (1.3806505 X 10^{-23}) in terms of the joule. The proposed redefinition is: 1.3806505 · 10^{-23} ·joule/kelvin. Boltzmann's constant (k_B) can be derived from the gas constant (R) (8.3144598 J/K) divided by Avogadro's Constant (N_a) (6.022140 · 10^{24}) = R/N_a = 1.3806505 · 10^{-23} J/K).

[23] Whitelaw, page 128.

$$1.3806505 \cdot 10^{-23} \cdot \text{J/K}^{24} =$$
$$1.3806505 \cdot .2248 \text{ lb. (avdp.)-f.} \cdot 3.2808 \text{ ft./m} =$$
$$1.01826 \text{ lbs. (avdp.)-force ft./Kelvin}$$

There is a number, 1.0368, from the Pattern that approximates this number; its pattern pair is (128/81). However, the difference between these numbers is .01854, and the deviation from 1.0368 is -.018, a number that is well outside the range of acceptable deviation. The likely reason for this was set forth in the foregoing section on the joule.

Defined Value	Pattern Number	Difference Δ	Dev. ± from Pattern
1.01826 lbs.(avdp.)-f/K	**1.0368 ($8 \cdot 6^4 \cdot 10^{-4}$)**	**.01854**	**- .018**

The measure of the Kelvin does not appear to be a manifestation of the Pattern in nature.

e. The Ampere and Electric Current.

One Ampere or Amp. is defined as: "That constant electrical current which, if maintained in two straight parallel conductors (wires) of infinite length, of negligible cross-section, and placed one meter apart in vacuum, would produce between these conductors a force equal to 2 x 10^{-7} newton per meter of length.' The Amp is one of the seven SI basic units.[25]

$$1 \text{ Amp} = 2 \cdot \text{N} \cdot 10^{-7} \text{ per meter}$$
$$= 2 \cdot 0.2248 \text{ lb. (avdp.)-f.}^{*} \cdot 10^{-7} \text{per m}$$
$$= 0.4496 \text{ lb. (avdp.)-force} \cdot 10^{-7} \text{ per m.}$$

(* See the section on the newton, where 1 newton (N) is defined as .2248 lb. (avdp.)-force.) This is a number from the Pattern; its pattern pair is (512/9).

Defined Value	Pattern Number	Difference Δ	Dev. ± from Pattern
.2248 lb. (vol.)-f./ft.	.2239 ($8 \cdot 6^7$)	.0009	+.004

While the measure of the Amp appears to be a manifestation of the Pattern in nature, this cannot be definitively stated, as the distance between the two wires is one meter. Converting this distance to one foot will not work, because it would change the definition altogether.

[24] Wikipedia contributors, "Kelvin", Wikipedia, the Free Encyclopedia, https://en.wikipedia.org/wiki/Kelvin (Accessed May 31, 2016)

[25] Whitelaw, Ian; *A Measure of All Things; The Story of Measurement Through the Ages*; David & Charles, Cincinnati, Ohio; 2007; page 132. See also Wikipedia contributors, "Ampere", Wikipedia, the Free Encyclopedia,) https://en.wikipedia.org/wiki/Ampere (Accessed May 31, 2016

f. Mole and Atomic Mass.

The mole is the atomic mass of a substance measured in grams, and the unified atomic mass unit is defined as 1/12 of the mass of a carbon-12 atom or 1 gram of carbon-12. Thus 1 mole of carbon-12 has a mass of 12 grams. The mole is one of the seven SI base units. The mole is based on the fact that equal volumes of gases at the same temperature and pressure contain equal numbers of molecules. Thus 1 mole of any given element contains exactly 6.02214199 $\cdot 10^{23}$ molecules, a measure named, "Avogadro's constant."[26]

On the face of it, Avogadro's constant would easily satisfy the deviation standard from a pattern number, which in this case would be the number, 6. The deviation from the pattern number, 6, is less than .0034. There is a potential problem in that the reference measure of the mole is the gram, a measure not found in the primordial system of measures. There are, however, several primordial measures of mass that the gram can be expressed in that are within the allowable deviation for associated pattern numbers.

1 g. = .035 oz. (avdp.)/.975 = .03589 oz. (vol.) *
1 g. = 15.432 grains

(*See the derivation of the lb. (vol.) in the section headed, "The Cubic Foot and the Pound of Volume." As was explained in this section, the weight of 1 cubic foot of water (62.43 lbs. (avdp.) is divided by 64 to arrive at the lb. (vol.), which is 62.43/64 = .975 of a lb. (avdp.). This same factor, .975, is also used to derive the ounce (vol.) from the ounce (avdp.).)

There are pattern numbers that very closely approximate these two measures. The number .03589 closely approximates the pattern number .036, which has a pattern pair of (4/9). The number 15.432 closely approximates the pattern number 15.552, which has a pattern pair of (64/243).

Defined Value	Pattern Number	Difference Δ	Dev. ± from Pattern
0.03589 oz. (vol.)/g.	0.036 ($6^2 \cdot 10^{-3}$)	0.00011	- .003
15.432 grains/gram	15.552 ($2 \cdot 6^5 \cdot 10^{-3}$)	0.120	-.008

There is, however, some reason to question the oz. (vol.)/g. measure. The gram is clearly a measure of mass and, as such, it should be converted into the oz. (avdp.) measure and not the oz. (vol.) measure. The only possible justification for using the oz. (vol.) measure, instead, is that the gram very closely approximates the volume measure 1 mm^3, and, therefore, the corresponding oz. (vol.) measure should be used for conversion purposes. This is, admittedly,

[26] Whitelaw, Ian; A Measure of All Things; The Story of Measurement Through the Ages; David and Charles, Cincinnati, Oho, USA; 2007; pages 74-75.
See also, Wikipedia contributors, "Mole (Unit)", Wikipedia the Free Encyclopedia, https://en.wikipedia.org/wiki/Mole_(unit) (Accessed May 30, 2016)

a very weak argument for using the oz. (vol.) measure, and were it the only justification for considering Avogadro's Constant as a possible manifestation of the Pattern in nature, then it would fail the test. The further support from the gram's grain measure equivalent justifies the oz. (vol.) measure's inclusion, though, in any case, the gram's grain measure equivalent alone provides justification for considering Avogadro's Constant.

Avogadro's Constant very closely approximates a number from the Pattern, 6, which has a pattern pair of (2/3).

Defined Value	Pattern Number	Difference Δ	Dev. ± from Pattern
6.02214 molecule/mol.	6	0.022149	+.004

Avogadro's Constant is a manifestation of the Pattern in nature.

g. Intensity of Light and the Candela.

The candela is defined as: "The luminous intensity, in a given direction, of a source that emits monochromatic radiation of frequency $540 \cdot 10^{12}$ hertz and that has a radiant intensity in that direction of 1/683 watt per steradian." [27]The candela is one of the seven SI basic units.

Defined Value	Pattern Number	Difference Δ	Dev. ± from Pattern
540 hertz	540 (9 · 6 ·10)	0.00	0.00

[27] Whitelaw, Ian; A Measure of All Things; The Story of Measurement Through the Ages; David & Charles, Cincinnati, Ohio, 2007; page 135. See also, Wikipedia contributors, "Candela": https://en.wikipedia.org/wiki/Candela (Accessed May 30, 2016)

CHAPTER 9

ASTRONOMICAL CONSTANTS

The International Astronomical Union (IAU)[28] is the international governing body for the science of astronomy. Among its other tasks, the IAU defines various fundamental and physical constants for the science of astronomy. A list of selected constants set by the IAU its 2009/2012 System of Astronomical Constants[29] is maintained by the U.S. Naval Observatory, and this list is widely referenced throughout this chapter.

Not all constants are dealt with. Also, a number of constants pertinent to the Earth and Moon are dealt with in chapters that specifically deal with these bodies, and are not repeated here. However, the constants dealing with equatorial radii are repeated for comparison purposes. The following constants are discussed in specific sections.

a. **Speed of Light**, A Natural Defining Constant
b. **Astronomical Unit**, An Auxiliary Defining Constant
c. **Constant of Gravitation**, a Natural Measurable Constant
d. **Mass Ratios**, Other Constants:

    i. Earth to Moon
    ii. Sun to Earth
    iii. Earth and Moon

1) **Equatorial Radii** constants from the IAU WG on Cartographic Coordinates and Rotational Elements 2009:
    Radius of Earth
    Radius of the Sun
    Radius of the Moon

[28] See IAU web site at: http://www.iau.org/

[29] Online reference: "K6 Selected Astronomical Constants" maintained by the U.S. Naval Observatory. The IAU 2009 System of Astronomical Constants (1) as published in the Report of the IAU Working Group on Numerical Standards for Fundamental Astronomy (NSFA, 2011) and updated by resolution B2 of the IAU XXVIII General Assembly (2012), (2) planetary equatorial radii, taken from the report of the IAU WG on Cartographic Coordinates and Rotational Elements 2009 (2011), and lastly 3) other useful constants.

Radius of Venus
Radius of Mars

a. The Speed of Light

The speed of light[30], *c,* is a Natural Defining Constant in the 2009/2012 IAU System of Astronomical Constants. It is 299,792,458 m sec^{-1} or 186,234.7 miles $sec.^{-1}$, measured in a vacuum. The speed of light is the maximum speed at which all physical matter travels, and is also the speed at which electromagnetic energy travels. The speed of light relates space and time in the mass-energy equation $E=mc^2$.

There is a number from the Pattern, 186,624, that approximates this number; its pattern pair is (729/256).

Defined Value	Pattern Number	Difference Δ	Dev. ± from Pattern
186,234.7 miles $sec.^{-1}$ *	186,624 ($6^6 \cdot 4$)	389.3	-.0021

*This is a manifestation of the Pattern in nature.

b. The Astronomical Unit

The astronomical unit[31], *au,* is an Auxiliary Defining Constant of the IAU 2009/2012 System of Astronomical Constants. It the approximate distance between the Earth and the Sun. Because the Earth travels in an elliptical pattern around the Sun, it was originally defined as the average of the Earth's aphelion and perihelion distances, but is now defined as exactly 149,597,870,700 meters or 92,932,010.6 miles. It is used primarily to define distances within the Solar System.

Defined Value	Pattern Number	Difference Δ	Dev. ± from Pattern
92,932,010.6 miles	93,312 ($6^6 \cdot 2 \cdot 10^3$)	379,989.4	-.004

c. The Constant of Gravitation.

The constant of gravitation[32], *G,* is a Natural Measurable Constant of the IAU 2009/2012 System of Astronomical Constants. It is the constant of proportionality between two bodies. The attractive force between them is directly proportional to the product of their masses and inversely proportional to the square of the distance between them.

[30] Wikipedia contributors: "Speed of Light", Wikipedia the Free Encyclopedia, https://en.wikipedia.org/wiki/Speed_of_light (Accessed May 31, 2016)
[31] Wikipedia contributors, "Astronomical Unit", Wikipedia, the Free Encyclopedia, https://en.wikipedia.org/wiki/Astronomical_unit (Accessed May 31, 2016)
[32] Wikipedia contributors, "Constant of Gravitation", Wikipedia, the Free Encyclopedia, https://en.wikipedia.org/wiki/Gravitational_constant (Accessed May 31, 2016)

$$F = G \cdot (m_1 \times m_2) \div r^2$$

G is 6.67428 x 10^{-11} m^3 kg^{-1} s^{-2}, with a range of uncertainty of: ± 6.7 x 10^{-15}.

G in lbs. (avdp.) and ft. = 519.6 x 10^{-11} ft^3 lb. $(avdp.)^{-1}$ s^{-2}.

There is a number from the Pattern that very closely approximates the number 519.6; this number is 518.4 and its pattern pair is (64/81).

Defined Value	Pattern Number	Difference Δ	Dev. ± from Pattern
519.6 ft^3 lbs. $(avdp.)^{-1}$	518.4 ($4 \cdot 6^4 \cdot 10^{-1}$)	1.2	+.002

The constant of gravitation (G) is a manifestation of the Pattern in nature.

d. Equatorial Dimensions of Selected Bodies.

The equatorial radii of the following selected bodies are given along with their equivalents in International mile. All figures shown for the radii in kilometers are from the "Constants from the IAU WG on Cartographic Coordinates and Rotational Elements 2009." Figures for the equatorial diameters in International mile are also shown.

Celestial Body	Radius (km)	Radius (International miles)	Diameter (International miles)
Sun	696,000	432,473.52	864,937.04
Mercury	2,439.7	1,516	3,032
Venus	6,051.8	3,760.4	7,520.8
Earth	6,378	3,963.2	7,926.4
Moon	1,734.4	1,079.6	2,159.14
Mars	3,396.2	2,110.2	4,220.4
Jupiter	71,492	44,423	88,846

Only the measurements for the Sun and Moon align with the Pattern; the measurements for the other bodies do not, nor do the figures for the remaining planets. For example, the figure for the diameter of Mars is 4,220 International miles, but it deviates from the Pattern number, 4,320, by -.023, far outside the range of range of deviation established earlier. A theory as to why the Pattern does not seem to manifest itself beyond the radii of the Sun and Moon is given in a later chapter. The Earth is a special exception as the Pattern is present in its electro-magnetic core (see chapter on Earth dimensions).

Defined Value	Pattern Number	Difference Δ	Dev. ± from Pattern
Solar Dia. 864,937.04	864,000 ($6^3 \cdot 4 \cdot 10^3$)	937.04	+.001
Lunar Dia. 2,159.14	2,160 ($6^3 \cdot 10$)	0.86	-.0004

Mass Ratios of Selected Bodies. The ratios of the mass values of the Earth to Moon (E to M), and Sun to Earth & Moon (S to E & M) manifest the Pattern. These ratios are constants recognized by the IAU.

2.25 pt	Pattern Number	Difference Δ	Dev. ± from Pattern
E to M: 81.3	81	0.3	+.004
S to E & M: 328,900.6	326,592 ($7 \cdot 6^7$)	2,308.6	+.007

Both of these measures are considered to be manifestations of the Pattern in nature.

CHAPTER 10

ASTRONOMICAL CYCLES WITH LONG TIME PERIODS

There are a number of measurable astronomical cycles that involve long time periods of decades or more. Among these are: 1) the Saros eclipse cycle, which is used to record and forecast solar and lunar eclipses, 2) the Venus transit cycle, which is used to record and forecast transits of the Sun by the planet Venus, 3) the precession of the Earth's equinoxes, which is used to track the Platonic ages that are based on the astronomical signs of the zodiac, and 4) the phoenix cycle, a legendary cycle, which will be shown to be connected to the Saros eclipse and Venus transit cycles. There are a number of other astronomical cycles, but only the aforementioned ones will be described and analyzed here for their relationship to the Pattern.

a. The Saros cycle.

This is a very reliable and widely used cycle for tracking, recording and forecasting both solar and lunar eclipses. It has been used to track not only currently active Saros series, but also as a reliable tool for hindcasting earlier eclipse events for historical and chronological purposes.[33]

An eclipse, broadly speaking, is an astronomical event involving two large, heavenly bodies of similar angular dimension that align in conjunction with one another, whereby one obscures or eclipses the other, as viewed from a third heavenly body that is also in alignment. The eclipse events with which we are most familiar involve the Sun, Earth, and Earth's Moon. When the Sun, Moon and Earth are in alignment, in that order with the Moon between the Earth and the Sun, the Moon eclipses the Sun and there is a solar eclipse. When the Sun, Earth, and Moon are in alignment, in that order with the Earth between the Sun and the Moon, the Earth eclipses the Moon and there is a lunar eclipse. A solar or lunar eclipse can occur whenever the Moon crosses the ecliptic, or apparent path of the Sun about the Earth. The two points where the Moon's orbit intersects the ecliptic are called the nodes—the ascending node and descending node. While eclipses routinely occur at these two nodes, only certain ones with

[33] NASA web site, "Periodicity of Lunar Eclipses", http://eclipse.gsfc.nasa.gov/LEsaros/LEperiodicity.html . The NASA web site is a great resource for eclipses and there is a wealth of information on this site.

shared characteristics comprise a specific Saros series. These shared characteristics, and a harmonic involving several measures of the lunar month, including the synodic, anomalistic, and draconic months, endure for centuries. Several of these measures of the lunar month are manifestations of the Pattern; others are not. All are measured in days.

Synodic: 29.53 (full moon-to-full moon)
Draconic: 27.21 (ascend. node-to-ascend. node)
Anomalistic: 27.55 (perigee-to-perigee)

All three of these measures are close to a pattern number. The synodic month is close to 30, while the draconic and anomalistic months are close to 27. The pattern pair for 30 is 3, and the ssspattern pair for 27 is 27.

Defined Value	Pattern Number	Difference Δ	Dev. ± from Pattern
29.53 days (synodic)	30 (3 · 10)	.47	-.0156
27.21 days (draconic)	27 (3^3)	.21	+.0081
27.55 day (anomalistic)	27 (3^3)	.55	+.02

Only the measure for the draconic month, 27.21 days, is considered to be a manifestation of the Pattern in nature. The measures for the synodic and anomalistic months are not considered to be manifestations of the Pattern in nature, because their deviation is outside of the established range.

Saros series members recur with great accuracy, with an almost exact 18.03 tropical-year period between succeeding members. However, succeeding members are longitudinally displaced westward some 120°, because of the precession of the lunar nodes, but the series returns roughly to the original location almost exactly 54.1 years later. This longer, but related, 54.1-year period is called the exeligmos cycle, which is Greek, meaning. 'turn-of-the-wheel.'

Saros cycle is approximately 6,585.32 days,
or 18.03 tropical years.
Exeligmos cycle is 54.1 tropical years.

These numbers are very close to numbers from the Pattern. The measure of the Saros cycle in days is very close to the number, 6,561, which has a pattern pair of (6,561). The measure of the Saros cycle in years is very close to the number, 18, which has a pattern pair of (2/9). The measure of the exeligmos cycle in years is very close to the number 54, which has a pattern pair of (2/27).

Defined Value	Pattern Number	Difference Δ	Dev. ± from Pattern
6,585.32 days (Saros)	6,561 (3^8)	24.32	+.0037
18.03 yrs. (Saros)	18 ($3 \cdot 6$)	.03	+.0016
54.1 yrs. (exeligmos)	54 ($9 \cdot 6$)	.1	+.0018

All three of these measures are manifestations of the Pattern in nature.

The average Saros series endures for approximately 1,298 years. Some are shorter and some are considerably longer. The differences are largely determined by the elliptical orbits of the Earth and Moon. There are approximately 73 eclipses in the average Saros series.

Duration of average Saros series is 1,298 years
Number of eclipses in average Saros series is 73

These numbers are very close to numbers from the Pattern. The duration of the average Saros series is close to the number, 1,296, which has a pattern pair of (16/81). The average number of eclipses in the average Saros series is near a pattern number, 72, which has a pattern pair of (8/9).

Defined Value	Pattern Number	Difference Δ	Dev. ± from Pattern
1,298 years (duration)	1,296 (6^4)	2	+.0015
73 (eclipses in Saros)	72 ($2 \cdot 6^2$)	1	+.014

The measure of the duration of the average Saros series in years is a manifestation of the Pattern in nature. The measure of the average number of eclipses in the average Saros series is not a manifestation of the Pattern in nature, because its deviation is outside of the established range.

There are at present some 40 active Saros series producing eclipses. Some are in the process of dying out, some are in full force, producing total eclipses, and some are just beginning their long life. There are an equal number of solar and lunar Saros series, and an equal or near equal number of eclipse members in each series. However, solar eclipses can only be viewed from very limited areas of the Earth's surface, because the Moon moves relatively quickly across the face of the Sun, while lunar eclipses can typically be viewed across vast stretches of the Earth, as the Earth and Moon move together for a long period of time during a lunar eclipse.

A Saros series begins near Earth's polar region, proceeds toward the equatorial regions where total eclipses tend to manifest most frequently, and then dies out in the opposite polar region.

After each Saros eclipse, the next eclipse in the series takes place 120° west of the previous eclipse, and is displaced north or south across the face of the Earth by some 180 miles, depending on whether the series is on the ascending or descending node. When the series completes an exeligmos cycle (three successive Saros eclipses), the series makes a full circle and returns roughly to the same geographic location, but is displaced to the east, because the node has shifted east along the ecliptic on the celestial sphere by approximately 1.44°, and is approximately 540 miles north or south of its previous location (180 miles · 3 Saros cycles).[34] These measures are numbers from the Pattern.

Node shift between Exeligmos eclipses: 1.44°
.48° for each Saros eclipse in series
Eclipse center shift: Exeligmos 540 mi. N or S,
for Saros: 180 mi. N or S

Both of these numbers are identical to numbers from the Pattern. The measure of the node shift between Exeligmos cycles is identical to the pattern number, 144, which has a pattern pair of (16/9), while the displacement in the focal point of the eclipse is identical to the pattern number, 54, which has a pattern pair of (2/27).

Defined Value	Pattern Number	Difference Δ	Dev. ± from Pattern
1.44° (Node shift)	1.44 ($4 \cdot 6^2 \cdot 10^{-1}$)	0	0
540 mi. (Shift of focus)	($9 \cdot 6 \cdot 10$)	0	0

Both of these measures are manifestations of the Pattern in nature.

Other lunar eclipse measures. Two other measures of lunar eclipses are of interest here, because of their relation to the Pattern. First is the fact that the Sun is precisely eclipsed by the Moon, i.e. Sun and Moon have equal angular measures and the center of the Moon passes precisely across the center of the Sun during eclipse, when the Moon is at a distance from Earth equal to 108 X its (Moon's) diameter. And second is the fact that the longest duration lunar eclipses last approximately 106.5 minutes.[35] Both of these measures equal or approximate the number 108 from the Pattern.

Defined Value	Pattern Number	Difference Δ	Dev. ± from Pattern
108 lunar dia. (dist.)	108 ($3 \cdot 6^2$)	0	0
106.5 mins. (duration)	108 ($3 \cdot 6^2$)	1.5	-.014

The first measure is a manifestation of the Pattern in nature, but the second one is not because it lies outside of the range of acceptable deviation.

[34] NASA web site, "Eclipses and the Saros", http://eclipse.gsfc.nasa.gov/SEsaros/SEsaros.html

[35] NASA Lunar Eclipse web site; http://eclipse.gsfc.nasa.gov/SEpubs/5MCLE.html ; Five Millennium Canon of Lunar Eclipses, -1999 to +3000, pages 35 and 36

b. Venus transit.

A transit is an astronomical phenomenon, wherein a substantially smaller celestial body crosses in front of, or 'transits', a larger body. In the solar system, there are only two bodies that transit a larger one that are visible from the Earth; these are the planets Mercury and Venus, which transit the Sun on a regular and recurring basis. Only the Venus transits will be analyzed here.

Venus transits occur on a regular, predictable cycle. They are grouped together in series around a 243-year cycle, that may include occurrences at years 8 + 105.5 and then again at years + 8 + 121.5 of each cycle, although there are numerous instances where there is only one transit after a longer interval, instead of the 8-year double. A series can last for 5,000 years or more. Venus transit series are very analogous to the solar and lunar Saros series, because of their predictability, longevity, and regular recurring pattern, and are one of the longest, recurring astronomical phenomena. At present there are four active series, numbers 3, 4, 5, and 6. [36]

The 243-year cycle of Venus transits is a measure that is identical to a number from the Pattern, 243, which has a pattern pair of (243).

Venus transit cycle: 243 years

Defined Value	Pattern Number	Difference Δ	Dev. ± from Pattern
243 yrs. (Venus transit)	243	0	0

The measure of the Venus transit cycle is a manifestation of the Pattern in nature.

c. The Phoenix cycle.

The Phoenix bird is a mythological creature that supposedly lives for 500 to a 1,000 years and then perishes in a conflagration, only to be reborn from the ashes, heralding a new age. It is a myth that is common to a great many of the peoples of the world.[37] Our interest in this myth stems from its connection with an implied cycle of time, and the fact that the bird perishes in a conflagration, which is suggestive of a transit cycle, and in particular, the Venus-transit cycle, where Venus does indeed appear to 'perish' in the flames of the Sun, only to rise from its ashes and return to life. On the face of it, however, Venus transits, though rare, do occur with much greater frequency than once every 500 to 1,000 years, which would seem to argue against any possible connection between the myth and the Venus-transit cycle. But there are

[36] NASA Eclipse Web Site; "Six Millennium Catalogue of Venus Transits: 2000 BCE to 4000 CE"; http://eclipse.gsfc.nasa.gov/transit/catalog/VenusCatalog.html

[37] Metapedia contributors, "Phoenix (Mythology)", Metapedia, The Alternative Encyclopedia; http://en.metapedia.org/wiki/Phoenix_(mythology) (accessed July 5, 2016)

other time cycles that may be involved, which include both the Saros series and a closely related series called the Inex, or In-Exiligmos series.

The Inex cycle is based on the fact that after a Saros eclipse, an eclipse with very similar characteristics can occur some 29 years later, on the same longitude, but at the opposite latitude of the other hemisphere, ex. if a Saros eclipse occurs at longitude 33° West and latitude 47° North, then 29 years later a similar eclipse will occur at longitude 33° West and latitude 47° **South**. The Saros series and Inex series can be plotted together in a manner that creates a striking panorama.

The panorama is shown below and while it is on far too small of a scale to provide any meaningful information, the important detail that can readily be grasped is that the panorama has a regular, wave-pattern to it, with recurring crests and troughs that originate from an implied central axis and propagate above and below this axis. It is very analogous to the sinusoidal rhythm of a heart. The Saros series seem to beat with a regular, discernible rhythm, with 3-5 longer period series (crests) regularly interspersed with 14 or 15 shorter period ones (troughs). It is this rhythm and its measure that is of immediate importance to this discussion.

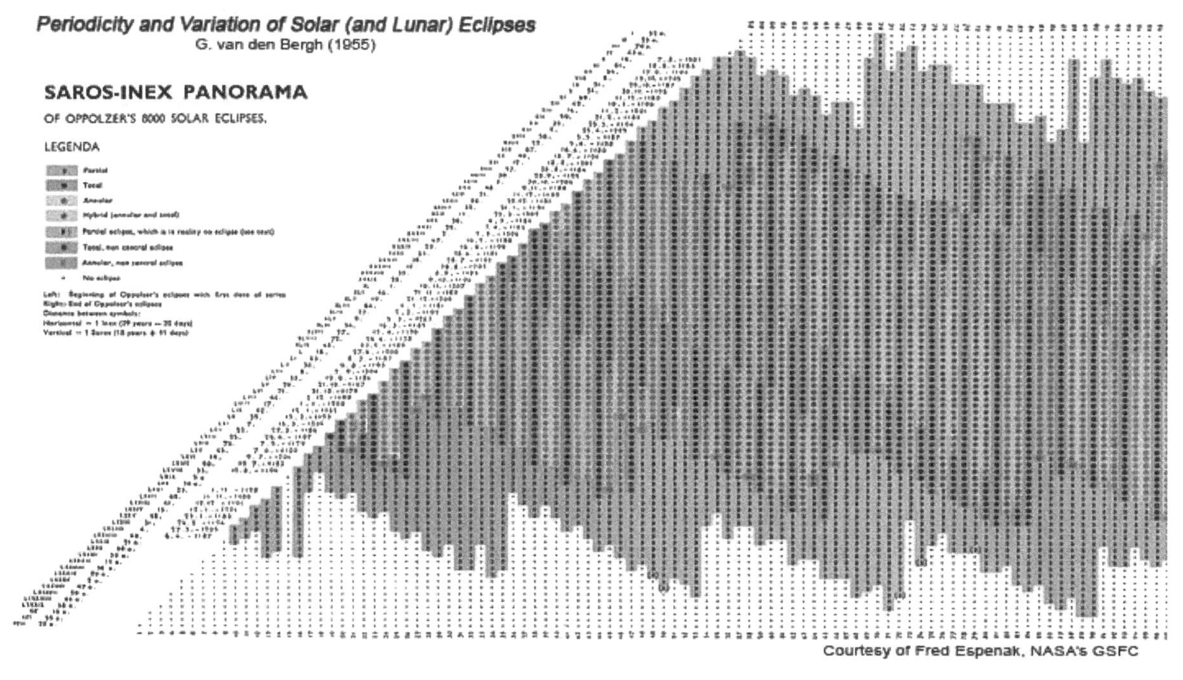

Saros-Inex Panorama

The numbering system used for the Saros series was introduced by the Dutch Astronomer G. van den Bergh in his book *Periodicity and Variation of Solar (and Lunar) Eclipses* (Tjeenk Willink, Haarlem, Netherlands, 1955). He placed all 8,000 solar eclipses in von Oppolzer's *Canon der Finsternisse* (1887) into a large two-dimensional matrix. Each Saros series was arranged as a separate column with the eclipses in chronological order. The Saros series columns were then staggered so that the interval between any two eclipses in adjacent columns was 10571.95 days (= 29 years -20 days). This is another important eclipse cycle called the Inex . The resulting Saros-Inex Panorama proved useful in organizing eclipses. For instance, one step down in the panorama is a change of one Saros period (6585.32 days) later, while one step to the right is a change of one Inex period (10571.95 days) later. The rows and columns were then numbered with the Saros and Inex numbers. [38]

The peaks (longer lasting Saros members) in the sinusoidal pattern of the Saros-Inex Panorama correspond to the timing of transit members of a specific Venus transit series. (See appendix 1 "The Phoenix Cycle" for an in-depth analysis of this issue.) There isn't perfect synchronization between the two cycles, but they are broadly commensurate with one another in their timing. In my opinion, this natural phenomenon is the basis for the Phoenix myth.

The Phoenix cycle is based on a multiple— 2 or 3—of the Venus transit cycle (243 years), which are 486 years (2 cycles) and 729 years (3 cycles). Both of these numbers are identical to numbers from the Pattern: 486, which has a pattern pair of (2/243), and 729, which has a pattern pair of (729).

Phoenix Cycle: 486 years or 729 years

Defined Value	Pattern Number	Difference Δ	Dev. ± from Pattern
486 years	486 ($2 \cdot 3^5$)	0	0
729 years	729 (3^6)	0	0

The measures of the Phoenix cycle are manifestations of the Pattern in nature.

d. The precession of Earth's equinoxes[39]

Precession of the equinoxes proceeds westward along the ecliptic of the celestial sphere is a natural pattern that can be measured and forecasted with great accuracy. Generally, the equinoxes are days with equal daylight and equal nighttime, which occur on March 20 and September 20 of each year. However, strictly speaking the equinoxes mark the specific instant when the apparent movement of the Sun brings the center of the Sun directly above the equator of the Earth, marking the instant when the ecliptic intersects the celestial equator.

Precession is caused by what is commonly referred to as a "wobble" in the Earth's axial rotation. While the Earth rotates around the North celestial pole on a daily basis, which is referred to as axial rotation, it is also slowly rotating around a second point called the North ecliptic pole. This second rotation takes some 26, 000 years to complete one circle of 360°.

[38] NASA Eclipse Web Site. "Eclipses and the Saros", http://eclipse.gsfc.nasa.gov/SEsaros/SEsaros.html ; A much expanded and far more detailed view of the Panorama can be obtained at the web site.

[39] Wikipedia contributors, "Axial Precession". Wikipedia the Free Encyclopedia, https://en.wikipedia.org/wiki/Axial_precession (accessed July 11, 2016)

This is the Platonic Year or Great Year, and it consists of a series of twelve successive great ages based on the zodiac of the ecliptic, with each great age lasting some 2,160 years. The Earth is currently in the Age of Pisces.

The intersection of the ecliptic with the celestial equator on the occasion of the vernal equinox (March 20) is of great significance to astronomers as it marks the new reference point in the celestial spheres for affixing the location of all of the heavenly bodies in relation to Earth. The right ascension and celestial longitude measures for all heavenly bodies are readjusted, using the latest location of the vernal equinox as a reference point—0 right ascension or 0 longitude. The vernal point is also referred to as the "first point of Aries", a term of reference of long standing tradition, even though the vernal point has long been in the sign of Pisces.

The tropical calendar (365.2417 days), with its current provisions for leap years, tracks the vernal point almost perfectly so that the seasons will not be affected by this phenomenon. In earlier times, societies used calendars with a year length of 365 days or 365.25 days, which caused the seasons to gradually drift out of alignment with the calendars, resulting in considerable problems for the societies using them.

The current rate of precession is measured at 50.28 arcseconds of celestial longitude or .838 arcminutes[40] of celestial longitude per year, or 20.4 minutes of mean solar time (clock time) per year. In 72 years, the vernal point will travel westward along the ecliptic by approximately 1° of celestial longitude (exact measure is 60.336 arcminutes of celestial longitude). It takes the vernal point approximately 25,775 years to make a complete circuit around the ecliptic. Several of these numbers are close approximations of numbers from the Pattern; 72 is identical to 72, which has a pattern pair of (8/9), 27,775 is very close to the number 25,920, which has a pattern pair of (32/81), and .838 is very close to the number .839808, which has a pattern pair of (128/6561).

Rate of precession: .838 arcminutes of celestial long./yr. or 20.4 mins. mean solar time/yr.
In 72 years, the vernal point transits approx. 1° of celestial long.
Vernal point completes circuit of ecliptic in 25,775 yrs.

[40] IAU Astronomical Constants; Other Constants: "Rates of Precession" at J2000·0 (IAU 2006), General precession in longitude.

Defined Value	Pattern Number	Difference Δ	Dev. ± from Pattern
.838 arcminutes/yr.	.839808 ($6^7 \cdot 3 \cdot 10^{-6}$)	.001808	-.0022
20.4 minutes/year	20.736 ($6^4 \cdot 16 \cdot 10^{-3}$)	.336	-.016
72 years for 60' long.	72 ($6^2 \cdot 2$)	0	0
25,775 years for circuit	25,920 ($6^4 \cdot 2 \cdot 10$)	145	-.0056

The measures of precession: 1) .838 minutes of celestial longitude/year, 2) 72 years for vernal point to move 60 minutes or 1° of celestial longitude, and 3) 25,775 years for the vernal point to complete a circuit of the ecliptic are all manifestations of the Pattern in nature. The measure of annual precession in mean solar time, 20.4 minutes/year, is not a manifestation of the Pattern in nature.

CHAPTER 11

ELECTROMAGNETIC ENERGY

Electromagnetic energy or radiation is the fundamental energy of the universe. It moves at the speed of light and in measurable waves. The component parts of electromagnetic energy, or: electromagnetic radiation, are distinguished by the magnitude and frequency of the waves that carry it. The electromagnetic energy spectrum consists of: radio radiation, infrared radiation, the visible light spectrum, x-ray radiation, and gamma radiation. Large waves and low frequency characterize radio radiation, while the waves grow smaller and their frequency increases moving toward the opposite side of the spectrum, with gamma radiation characterized by extremely small waves with very high frequency.[41] Energy is inversely proportional to wavelength.

Electromagnetic energy is composed of energy waves and also of individual packets of energy referred to as quanta. The waves consist of separate magnetic and electrical waves that travel in synchrony away from the source, at a ninety-degree angle from one another. The wave and particle forms of energy comprise what is referred to as the wave–particle duality of electromagnetic energy. Many of the same terms apply to both states; however, waves are measured not only by their energy, but also by their wavelength and frequency. Electromagnetic energy can be expressed in terms of energy (joules, electron volts, coulombs, etc.) but also as wavelength, which is measured from crest-to-crest or trough-to-trough.

Electromagnetic energy is a vast field of study, and what follows below is at best only a cursory investigation of it. The purpose of the investigation is strictly limited to an analysis of its broader terms and principals to determine whether or not the Pattern manifests itself. The basic measures of electromagnetic energy and their relationship with one another will be examined first.

Basic values.
Elementary charge, $e = 6.24 \cdot 10^{18}$ electrons
1 coulomb = $1e = 6.24 \cdot 10^{18}$ electrons

[41] Wikipedia contributors, "Electromagnetic radiation", Wikipedia, the Free Encyclopedia, https://en.wikipedia.org/wiki/Electromagnetic_radiation (accessed June 16, 2016)

1 electron volt, eV[42] = joule/coulomb = joule/6.24 · 10^{18} electrons = 1.6 · 10^{-19} joules
1 joule (J) = 6.24 · 10^{18} eV
Speed of light, c = 2.99 · 10^{8} m $sec.^{-1}$ or 9.81 · 10^{8} ft. $sec.^{-1}$

The speed of light has already been analyzed in an earlier chapter and was determined to be a manifestation of the Pattern in nature. Those findings are not repeated here.

a. The Elementary Charge:

The number of electrons in the elementary charge is 6.24 · 10^{18} and the number of eV in one joule is 6.24 · 10^{18}. The number 6.24 is very close to a number from the Pattern, 6.2208, which has a pattern pair of, (256/243).

Elementary charge = 6.24 · 10^{18} electrons
1 joule = 6.24 · 10^{18} eV

Defined Value	Pattern Number	Difference Δ	Dev. ± from Pattern
6.24 · 10^{18} elect and eV	6.2208 (8 · 6^5 · 10^{-4})	.0192	+.003

Both of these measures are considered to be manifestations of the Pattern in nature.

b. Photon Energy, E,[43] wave frequency, and wavelength.

Planck Constant, h = 4.13566 · 10^{-15} eV; also, when multiplied by 1.6 · 10^{-19} joules/eV =
= 6.626070 · 10^{-34} J $sec.^{-1}$

In spectroscopy, the wavelength cm^{-1} = 0.000123986 eV is used to represent energy since energy is inversely proportional to wavelength from the equation E = h · v = h · c/ λ, where h is the Planck constant, c is the speed of light, v is frequency, and λ is wavelength.

$$E = h \cdot v = \frac{h \cdot c}{\lambda \text{ nm}} = 4.13566 \cdot 10^{-15} \text{ eV} \cdot 2.99 \cdot 10^{8} \text{m sec.}^{-1} = \frac{124 \; eV \; nm}{\lambda \; nm} \text{ also,}$$

$$= \frac{h \cdot c}{\lambda \text{ nm}} = 6.626070 \cdot 10^{-34} \text{ J} \cdot \text{sec.}^{-1} \cdot 9.81 \cdot 10^{8} \text{ ft. sec.}^{-1} = \frac{65.17 \text{ J}}{\lambda \text{ nm}} \cdot 10^{-26} \cdot \text{ft. sec.}^{-1}$$

$$E = \frac{124 \; eV \; nm}{\lambda \; nm}$$

[42] Wikipedia contributors: "Electronvolt", Wikipedia the Free Encyclopedia, https://en.wikipedia.org/wiki/Electronvolt (accessed June 17, 2016)

[43] Wikipedia contributors, "Photon", Wikipedia, the Free Encyclopedia, https://en.wikipedia.org/wiki/Photon (accessed June 19, 2016)

$$E = \frac{65.17 \text{ joules}}{\lambda \text{ nm}} \cdot 10^{-26} \cdot \text{ ft. sec.}^{-1}$$

The number 65.17 J is very close to a number from the Pattern, 64.8, which has a pattern pair of (8/81).

Defined Value	Pattern Number	Difference Δ	Dev. ± from Pattern
124 eV nm	120	4	+.033
$65.17 \cdot 10^{-26}$ J · ft. sec.$^{-1}$	64.8 ($3 \cdot 6^3 \cdot 10^{-26}$)	.37	+.006

The measure of E in eV is not a manifestation of the Pattern in nature. The measure of E in joules is a manifestation of the Pattern in nature.

The table below depicts the relationship between frequency, wavelength and energy:[44]

CLASS	FREQUENCY	WAVELENGTH	ENERGY
	300 EHz	1 pm	1.24 MeV
γ	30 EHz	10 pm	124 keV
HX	3 EHz	100 pm	12.4 keV
SX	300 PHz	1 nm	1.24 keV
	30 PHz	10 nm	124 eV
EUV	3 PHz	100 nm	12.4 eV
NUV	300 THz	1 μm	1.24 eV
NIR	30 THz	10 μm	124 meV
MIR	3 THz	100 μm	12.4 meV
FIR	300 GHz	1 mm	1.24 meV
EHF	30 GHz	1 cm	124 μeV
SHF	3 GHz	1 dm	12.4 μeV
UHF	300 MHz	1 m	1.24 μeV
VHF	30 MHz	10 m	124 neV
HF	3 MHz	100 m	12.4 neV
MF	300 kHz	1 km	1.24 neV
LF	30 kHz	10 km	124 peV
VLF	3 kHz	100 km	12.4 peV
VF/ULF	300 Hz	1 Mm	1.24 peV
SLF	30 Hz	10 Mm	124 feV
ELF	3 Hz	100 Mm	12.4 feV

44 Wikipedia contributors, "Electronvolt", Wikipedia, the Free Encyclopedia, https://en.wikipedia.org/wiki/Electronvolt (accessed June 20, 2016)

Photon frequency vs. energy per particle in electronvolts. The energy of a photon varies only with the frequency of the photon, related by speed of light constant. This contrasts with a massive particle of which the energy depends on its velocity and rest mass.[5][6][7] Legend:

γ: Gamma rays
HX: Hard X-rays
SX: Soft X-rays
EUV: Extreme ultraviolet
NUV: Near ultraviolet
Visible light
NIR: Near Infrared
MIR: Mid infrared
FIR: Far infrared
Radio waves
EHF: Extremely high freq.
SHF: Super high freq.
UHF: Ultra high freq.
VHF: Very high freq.
HF: High freq.
MF: Medium freq.
LF: Low freq.
VLF: Very low freq.
VF/ULF: Voice freq.
SLF: Super low freq.
ELF: Extremely low freq.
Freq: Frequency

All of the relationships between energy, wavelength, and frequency within the electromagnetic spectrum are either depicted in this table or can be extrapolated from it. The table also shows the various classifications of electromagnetic radiation and their location on the spectrum.

Radiation energy measured in electronvolts can be directly related to the SI units of mass, momentum, and temperature. By mass-energy equivalence, the electronvolt is also a unit of mass. In high energy physics, the electronvolt is a unit of momentum. In plasma physics, the electronvolt is also a unit of temperature. Additionally, using the reduce Planck-constant, the electronvolt can be used as a unit of time and distance. The mass-energy equivalence is further analyzed below.

c. Energy and Mass.

Energy and mass are related as, $E = M \cdot c^2$ and $M = E/c^2$, where c is the speed of light constant.

$M = eV/c^2 = 1eV / (9.81 \cdot 10^8 \text{ ft.})^2 = .01039069 \cdot 10^{-36}$ eV or kg.
$= 1.6 \cdot 10^{-19} J / (9.81 \cdot 10^8 \text{ ft.})^2 = .016625 \cdot 10^{-36}$ J or kg.

$$M = eV/c^2 = .01039069 \cdot 10^{-36} \text{ eV or kg.}$$
$$= 1.6 \cdot 10^{-19} J / c^2 = .016625 \cdot 10^{-36} \text{ J or kg.}$$

The number .01039069 is very near a number from the Pattern, .010368, which has a pattern pair of (128/81). The number .016625 is near the Pattern number, .01679616, which has a pattern pair of (256/6,561).

Defined Value	Pattern Number	Difference Δ	Dev. ± from Pattern
$.010391 \cdot 10^{-36}$ eV or kg.	$.010368\ (8 \cdot 6^4 \cdot 10^{-6})$	.000023	+.002
$.016625 \cdot 10^{-36}$ J or kg.	$.01679616\ (6^8 \cdot 10^{-36})$	.00017116	-.01

The measure of mass in terms of eV is a manifestation of the Pattern in nature. The measure of mass in terms of joules is not a manifestation of the Pattern in nature.

Using the foregoing figures and converting kg to pounds (avdp.):

$$eV/c^2 = .01039069 \cdot 10^{-36} \cdot 2.2046 \text{ lbs. (avdp.)/kg.}$$
$$= .02290 \cdot 10^{-36} \text{ lbs. (avdp.)}$$
$$J/c^2 = .016625 \cdot 10^{-36} \cdot 2.2046 \text{ lbs. (avdp.)/kg.}$$
$$= .03665148 \cdot 10^{-36} \text{ lbs. (avdp.)}$$

The number .02290 is very close to a number from the Pattern, .02304, which has a pattern pair of (256/9). The number .03665 is near Pattern number, .03600, which has a pattern pair of (4/9)

Defined Value	Pattern Number	Difference Δ	Dev. ± from Pattern
.02290 $\cdot$ 10^{-36} lb.(avdp.)	.02304 ($64 \cdot 6^2 \cdot 10^{-5}$)	.00014	-.006
.03665 $\cdot 10^{-36}$ lb. (avdp.)	.03600 ($6^2 \cdot 10^{-36}$)	.00017116	+.018

The measure of mass in terms of eV, as defined in lbs. (avdp.), is a manifestation of the Pattern in nature. The measure of mass in terms of joules is not.

d. The human eye and the color green.

In the visible light spectrum, within the electromagnetic spectrum, the human eye is most sensitive to the color green with a wavelength of 555 nanometers, which corresponds to a frequency of 540 THz ($540 \cdot 10^{12}$ Hz). The photon energy of this frequency is derived from: $E = h \cdot f$ (the Planck-Einstein relation) where h is the Planck constant and f is frequency.[45]

$$E = 6.626 \cdot 10^{-34} \text{ J} \cdot 540 \cdot 10^{12} \text{ Hz} = 3.58 \cdot 10^{-19} \text{ J}$$

To convert the extremely small measure of $3.58 \cdot 10^{-19}$ J into the typical energy of everyday life, multiply this measure by Avogadro's number to derive a mole of energy at this frequency.

$$E = 3.58 \text{ J} \cdot 10^{-19} \cdot 6.02214 \cdot 10^{23} = 215.6 \text{ kJ/mol.}$$

[45] Wikipedia contributors, "Planck Constant", Wikipedia, the Free Encyclopedia: https://en.wikipedia.org/wiki/Planck_constant . See section, "Significance of the Value" (accessed June 21, 2016)

The number 3.58 is very close to the Pattern number, 36, which has a pattern pair of (4/9). The number 215.6 is very close to the Pattern number 216, which has a pattern pair of (8/27).

Defined Value	Pattern Number	Difference Δ	Dev. ± from Pattern
$3.58 \cdot 10^{-19}$ J	3.6 ($6^2 \cdot 10^{-1}$)	.02	-.0006
215.6 kJ/mol.	216 (6^3)	.4	-.0018

Both of these measures are manifestations of the Pattern in nature

CHAPTER 12

HUMAN VISUAL PERCEPTION AND THE EYE

> "God invented and gave us sight to the end that we might behold the courses of intelligence in the heaven, and apply them to the courses of our own intelligence which are akin to them, the unperturbed to the perturbed; and that we, learning them and partaking of the natural truth of reason, might imitate the absolutely unerring courses of God and regulate our own vagaries." Plato, *Timaeus* 47

a. The structure of the human eye and the Pattern.

It is mainly through the sense of sight that man perceives the world he lives in. Through his visual perception, man interprets and analyzes the perceived world around him, and when perception is reflected onto his reason man learns to recognize and understand the rhythms and patterns of this world. All of this begins with the eye and the relationship of its physical structure to the Pattern. Several instances of this relationship are examined below.

The orb of the eye is almost spherical in shape and is comprised of two major parts. The corneal structure represents approximately 1/6th of its volume, with the remaining 5/6th comprising the chamber of the sclera—the white of the eye.[46] This proportion is a manifestation of the Pattern and derives from the number 6.

Defined Value	Pattern Number	Difference Δ	Dev. ± from Pattern
6 (1/6th to 5/6ths)	6	0	0

The proportion of the main structural parts of the eye derives from the number 6, which is a manifestation of the Pattern in nature.

[46] Wikipedia contributors. "Human Eye", Wikipedia, the free encyclopedia: https://en.wikipedia.org/wiki/Human_eye . Accessed August 13, 2016.

b. The lens of the eye

The lens of the eye is composed of a number of lens fibers that originate at the focal point of the pupil of the eye on the anterior of the lens and end opposite it on the posterior of the lens. These fibers have their centers on the equator of the lens. Examined in horizontal cross section, the fibers appear to be concentrically layered, similar to an onion. If a vertical cross section of the fibers is made at the lens equator, they have the appearance of honey comb, with each fiber having a hexagonally shaped cross section.[47] This hexagonal shape is, of course, a manifestation of the number six, a number from the Pattern.

Defined Value	Pattern Number	Difference Δ	Dev. ± from Pattern
6 (Hexagon)	6	0	0

The hexagonal shape of the cross-section of the lens fibers of the eye is a manifestation of the Pattern in nature.

The eye's sensitivity to the color green is not a physical attribute of the human eye. This sensitivity was dealt with in an earlier chapter; however, the earlier entry regarding it is repeated here in its entirety because of its relevance to the relationship of the eye to the Pattern and its significance to discussions that will follow concerning the eye.

c. The human eye and the color green.

In the visible light spectrum, within the electromagnetic spectrum, the human eye is most sensitive to the color green with a wavelength of 555 nanometers, which corresponds to a frequency of 540 THz ($540 \cdot 10^{12}$ Hz).[48] The photon energy of this frequency is derived from: $E = h \cdot f$ (the Planck-Einstein relation) where h is the Planck constant and f is frequency.[49]

$$E = 6.626 \cdot 10^{-34}\ \text{J} \cdot 540 \cdot 10^{12}\ \text{Hz} = 3.58 \cdot 10^{-19}\ \text{J}$$

To convert the extremely small measure of $3.58 \cdot 10^{-19}$ J into the typical energy of everyday life, multiply this measure by Avogadro's number to derive a mole of energy at this frequency.

$$E = 3.58\ \text{J} \cdot 10^{-19} \cdot 6.02214 \cdot 10^{23} = 215.6\ \text{kJ/mol.}$$

[47] Wikipedia contributors. "Lens (anatomy)", Wikipedia the free encyclopedia: https://en.wikipedia.org/wiki/Lens_(anatomy) . See section headed, lens fibers. Accessed August 13, 2016.

[48] Wikipedia contributors. "Color Vision", Wikipedia, the free encyclopedia: https://en.wikipedia.org/wiki/Color_vision . See section, "Wavelength and hue detection." Accessed August 13, 2016.

[49] Wikipedia contributors, "Planck Constant", Wikipedia, the Free Encyclopedia: https://en.wikipedia.org/wiki/Planck_constant . See section, "Significance of the Value" (accessed June 21, 2016)

The number 3.58 is very close to the Pattern number, 36, which has a pattern pair of (4/9). The number 215.6 is very close to the Pattern number 216, which has a pattern pair of (8/27).

Defined Value	Pattern Number	Difference Δ	Dev. ± from Pattern
$3.58 \cdot 10^{-19}$ J	3.6 ($6^2 \cdot 10^{-1}$)	.02	-.0006
215.6 kJ/mol.	216 (6^3)	.4	-.0018

Both of these measures are manifestations of the Pattern in nature.

CHAPTER 13

THE STRUCTURE OF THE HUMAN LIVER AND THE PATTERN

The human liver would seem to have no obvious connection with vision, but Plato in *Timaeus* 71-72 does make a connection between that organ and the vision that is associated with dreaming. This is hardly a definitive reference that can be confidently relied upon to make such a connection, and there is every justification for dismissing it out of hand as a curiosity and nothing more. However, the microscopic structure of the liver is of more than passing interest as it has a very definite connection with the Pattern, in a manner that is strikingly similar to that of the human eye.

The individual hepatic lobules that make up the structure of the liver's lobes are hexagonal shaped. At each of the six corners of a hepatic lobule are the portal triad, which consists of a branch of the hepatic artery, a branch of the hepatic portal vein, a branch of the bile duct, as well as lymphatic vessels and branches of the vegas nerve.[50] This hexagonal shape is, of course, a manifestation of the number six, a number from the Pattern.

Defined Value	Pattern Number	Difference Δ	Dev. ± from Pattern
6 (Hexagon)	6	0	0

The hexagonal shape of the cross-section of liver lobules is a manifestation of the Pattern in nature.

It should also be noted in passing that the color of the bile produced by the liver is dark green to yellowish green-brown. An interesting fact in light of the eye's aforementioned sensitivity to the color green, but probably one of no significance.

[50] Wikipedia contributors, "Liver", Wikipedia, the free encyclopedia, https://en.wikipedia.org/wiki/Liver (accessed September 23, 2016), see section on Microscopic Anatomy

PART II

EARLY EVIDENCE OF THE PATTERN IN CIVILIZATION

CHAPTER 14

THE EGYPTIAN EYE OF HORUS

In many respects this is the end of the first part of this book and the beginning of the second. It is the eye, as symbol—specifically, the ancient Egyptian symbol for the Eye of Horus— that will mark the divide. Part of this discussion is in Part I and the remainder should be, but is not, in Part II. There is a reason for this as will become apparent, because the two narrative lines gradually merge and then separate.

a. The Eye of Horus

The eye has frequently appeared as an important symbol in many cultures around the world and throughout human history, though few if any command the importance that the Eye of Horus did in ancient Egypt. The Eye of Horus was the left eye of a falcon, while the falcon's right eye was considered to be the Eye of Re. The left and right eye correspond to the Moon and the Sun, respectively, in this myth, and this association is of the utmost importance to what follows.

The Eye of Horus is called the *wedjet*[51] eye, which, interestingly, is believed to derive from the word *wedj*, the Egyptian word for **green. Also, Wedjet is the name for the Cobra goddess, Edjo; or, for a colonnade, particularly of green papyrus columns.**[52] Wedjet is translated as the 'complete eye' or the 'sound eye.' This puzzling formulation is explained by the fact that the Eye of Horus was torn apart and thrown away during a fight between Horus and Seth, his arch rival. The parts of the eye were later recovered and the eye was restored to its original vigor by the god, Thoth. In its damaged state, the parts of the Eye are identified with a series of fractions, which are all multiples of two and comprise a geometric progression.

[51] Wikipedia contributors. "Eye of Horus", Wikipedia, the free encyclopedia: https://en.wikipedia.org/wiki/Eye_of_Horus . Accessed August 13, 2016.

[52] Gardiner, Sir Alan, Egyptian Grammar, Third Edition (2005), Griffith Institute, Oxford; page 560, upper left corner

See also: Wikipedia contributors, "Gardiner's sign list", Wikipedia, the free encyclopedia, https://en.wikipedia.org/wiki/Gardiner%27s_sign_list (accessed September 27, 2016)

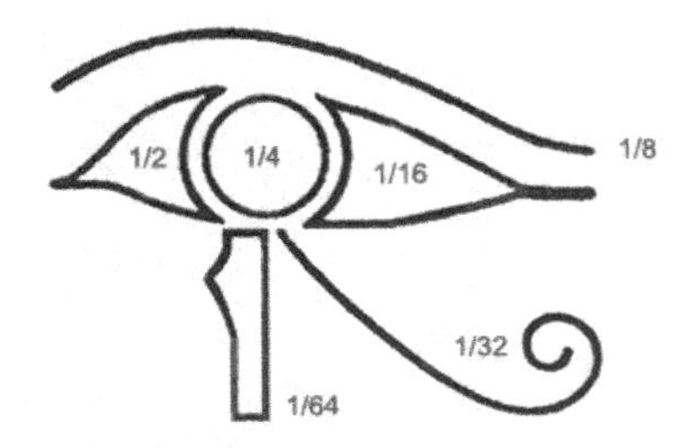

The progression, then, is: 1/2, 1/4, 1/8, 1/16, 1/32, and 1/64—all of which are the inverse of numbers from the Pattern—which when added together, total: 63/64. This is a most important fraction for our purposes. Also, the ancient Egyptians quite clearly and unequivocally considered the *wedjet* eye to be a 'completed eye.' But, it is obvious from the fraction that there is 1/64th missing. Modern Egyptologists usually explain that the eye is restored when the god Thoth, who is often associated with the Moon and cycles of time[53], magically or miraculously adds the missing 1/64th to make it complete again.[54] This would imply that the ancient Egyptians were erroneous or careless in their formulation of this tale. They were not. In fact, they were profoundly correct, as will be demonstrated in the following section.

b. The Astronomical Significance of the Fraction 63/64.

Recall that the right eye of the falcon was referred to as the Eye of Re and was identified with the Sun, and that the falcon's left eye was referred to as the Eye of Horus and was identified with the Moon. Modern astronomy established that the angular, or apparent, size of the Sun ranges between 31.51 and 32.55 arcminutes on the celestial sphere, while the Moon's angular size ranges between 29.33 and 34.1 arcminutes—both of which measures vary, depending on the relative position and distance of the Sun and Moon from the Earth, as determined by the respective elliptical orbits of the Earth about the Sun and the Moon about the Earth.[55] Using these parameters, the average angular size of the Sun is 32.03 arcminutes, and the average angular size of the Moon is 31.7 arcminutes. The average angular size of the Moon (31.7) is almost exactly 63/64ths of the average angular size of the Sun (32.03). 63/64 X 32.03 = 31.5 arcminutes, as opposed to the best modern estimate of 31.7, a difference of .2 arcminutes or .006. The ancient Egyptians, then, were quite correct when they stated that the fraction 63/64ths represents a completed Eye of Horus. There is nothing missing, nothing to be added.

[53] Wikipedia contributors, "Thoth" Wikipedia, the free encyclopedia: https://en.wikipedia.org/wiki/Thoth . Accessed August 22, 2016.

[54] Ibid, page 197. See, also, the Wikipedia entry cited in footnote 50.

[55] Wikipedia contributors. "Angular Diameter", Wikipedia the free encyclopedia: https://en.wikipedia.org/wiki/Angular_diameter . Accessed August 20, 2016.

Moon's Average Angular Diameter is 63/64 of that of Sun's

Furthermore, if the Egyptian year of 360 days is multiplied by 63/64, the result is 354.375, which the almost the exact measure of the lunar year of 354.367 days. Also, 63/64 of the 30-day, Egyptian month is 29.53125 days, which is almost the exact measure of the average, lunar synodic month of 29.53058 days.[56] These differences are too small to matter for most practical purposes.

c. The Measure of the Egyptian Foot and the Fraction 63/64.

As was discussed in Chapter on 'Earth and its Dimensions' on the derivation of the nautical and geographic measures of the foot, there are two arithmetic progressions that underlie these measures of the foot: the first is based on the number 8, and consists of: 64, 72, and 80; while the second is based on the number 9, and consists of: 63, 73, 81. The nautical and geographic measures of the foot have already been addressed in aforementioned earlier chapter and will not be repeated here. The third measure of the foot that was alluded to in this earlier chapter is based on the fraction, 63/64, which comprises the numerator 63 and denominator 64, both numbers from the two arithmetic series. The fraction 63/64 can be used to generate the measure of the ancient Egyptian foot of 300 mm[57] from the international foot of 304.8 mm. Of course the inverse of this fraction, 64/63, can be used to generate the international foot of 304.8 mm: 64/63 X 300 mm = 304.8 mm.

The Egyptian foot is: 63/64 X 304.8 mm = 300 mm

It may be noteworthy that if the Eye of Horus myth is prelude to the derivation of the measure of the ancient Egyptian foot, then it would suggest that the ancient Egyptian foot of 300 mm is associated with the Moon, while the foot of 304.8 mm (International foot) is associated with the Eye of Re and the Sun. This is an interesting possibility, but unfortunately there is nothing more that can be offered to substantiate it.

A number of questions regarding the ancient Egyptian foot arise, not the least of which, is: Why was it created if the International foot was the primordial measure of distance? No entirely satisfactory answer presents itself. However, it may be noteworthy that at latitude 14° the cosine measure of the geometric foot (309.22 mm at the equator) is: cos 14° X 309.22 mm = 300 mm. The geographic location is unimportant to our analysis, but the measure of 14° is. The limits of the band of the ecliptic, or zodiac, is established by the planet Mercury, which has the highest orbital inclination of any planet at 7°. Mercury's orbital inclination establishes

---

[56] Wikipedia contributors, "Lunar Calendar", Wikipedia, the free encyclopedia, https://en.wikipedia.org/wiki/Lunar_calendar . Accessed August 29, 2016

[57] Wikipedia contributors. "Ancient Egyptian units of measurement", Wikipedia, the free encyclopedia: https://en.wikipedia.org/wiki/Ancient_Egyptian_units_of_measurement . Accessed August 21, 2016.

the limits of the band of the ecliptic at 7° above and 7° below the ecliptic (apparent path of the Sun), giving the band a width of 14°.

Egypt is sometimes considered to be equivalent on Earth of the northern 7° of this band, which directly overlays it when the Sun reaches the Tropic of Cancer—the upper limit of the Sun's apparent travel across the northern hemisphere, which currently crosses Egypt at 23°-27' N and historically was believed to have been located at 23°- 51' N. This is close to the historic, southern boundary of ancient Upper (southern) Egypt that was located at the Nile's first cataract near Aswan, which is at latitude 24°-06' N.[58] The northern limit of ancient Lower (northern) Egypt was located at the tip of the Nile delta, at latitude 31°-30; N. This gives ancient Egypt a range of some 7°-24' of latitude, a range nominally consistent with the upper band of the ecliptic when the Sun reaches the Tropic of Cancer. Ideally, the limits of Egypt should range from 24N to 31N, but this does not accord with Egypt's actual geography, although it closely approximates the ideal. If this association with the band of the ecliptic is correct, then the ancient Egyptian foot of 300 mm is reflective of this association and, because of this association, likely the source of all measurement in ancient Egypt. There is a geographic association with the measure of the ancient Egyptian foot, but it will be discussed in a later chapter and is unimportant to our analysis here.

d. Select Right Triangles and the Fraction 63/64

Assume the following triangle, where angle **A** is 30°, angle **B** is 60°, and angle **C** is 90°, and where side **a** is 1, side **b** is √3 and side **c** is 2. This triangle is exactly half of an equilateral triangle, which is usually considered to be a perfect triangle. It is also one of the tringles discussed in Plato's Timaeus 53-55.

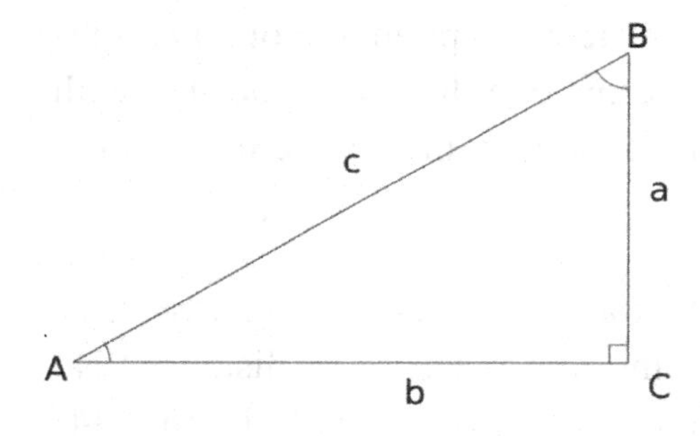

If each of the respective sides are multiplied by 36, then side **a** becomes 36, side **b** becomes 62.354, and side **c** becomes 72. The measure of side b (62.354) is close to the number 63. If this triangle is minimally modified so that side **b** is 63, then angle **A** becomes 29° 52' 16", and side **a** becomes 36.18 and side **c** becomes 72.66. The significance of these measures for our immediate purposes is that the number 72.66 squared is 5,279.5 or approximately 5,280—the

[58] Wikipedia contributors. "Aswan", Wikipedia, the free encyclopedia: https://en.wikipedia.org/wiki/Aswan > Accessed August 22, 2016.

measure of the international mile in feet—and a number within 0.5 feet of the actual measure or an error of .000095.

Further, assume a second triangle, where angle **A** is 26° 33' 54", angle **B** is 63° 26' 06", and angle **C** is 90°, and where side **a** is 1, side **b** is 2 and side **c** is √5. (This triangle is a close relation to the previous triangle, in that in each of them two of their three sides are 1 and 2.) A close approximation to this 1–2–√5 triangle is one where side **a** becomes 4, side **b** becomes 8, and side **c** becomes 9. When each of these sides is squared, the results are: side **a** becomes 16, side **b** becomes 64, and side **c** becomes 81. There is an obvious distortion in this triangle, as side **a** squared (16) plus side **b** squared (64), should equal side c squared (81). Sides **a** and **b** total 80, not 81. (This discrepancy and its significance was discussed earlier in the Chapter on "Earth and Its Measure.")

The significance of these two related triangles (1–√3–2 and 1–2–√5) is that when their modified versions are compared, their respective measure for side **b** relate to one another as 63:64, or 63/64. This is not an insignificant relationship; it is one that underlies much if not all of measurement, as was demonstrated here and elsewhere in this work, it relates to the measure of both the international mile of 5,280 feet and the international foot of 304.8 mm. One final fact will serve to further emphasize this point.

As was noted in the foregoing discussion of the modified 1–√3–2 triangle, side **b** is 63 and angle **A** is 29° 52' 16". This measure for angle A (29° 52' 16") is important. If this measure is assumed to be a latitude measure, then the circumference of the Earth at this latitude, in international miles, is: 24,902 miles X cosine 29.87° (equal to 29° 52' 16") = 24,902 miles X .867157 = 21,593.9 miles, of 5,280 feet of 304.8 mm each, which is within 6.1 miles of the Pattern number 216 X 10^2 that has a Pattern pair of (8/27). (Significantly, WGS 84 measures the circumference of the Earth at this latitude at: 21,598 international miles.)

Diameter of Earth at Latitude 29° 52' 16" N = 21,593.9 miles

Defined Value	Pattern Number	Difference Δ	Dev. ± from Pattern
21,593.9 miles*	21,600 ($6^3 \cdot 10^2$)	6.1	-.00028

*This measure is a manifestation of the Pattern in nature.

One final note on latitude 29° 52' 16". This is the precise latitude location of the Step Pyramid complex of Saqqara in Egypt, which is widely believed to be the earliest, large-scale stone structure built in the world; it was constructed around 2,700 BCE, only several hundred years after the stone age ended and civilization began.[59] In light of its history, then, how can there possibly be a connection between this structure and the aforementioned fact? Early man,

[59] Wikipedia contributors. "Saqqara" Wikipedia the free encyclopedia: https://en.wikipedia.org/wiki/Saqqara . Accessed August 22, 2016.

as we understand his development, was clearly ill-equipped to master the several sciences required for any of this to happen! He would have had to know the precise measure of the Earth's circumference, in international miles, the numbers from the Pattern, and the ability to accurately survey the location for the Step Pyramid complex before building it. A tall order for certain and the odds are clearly against this happening. However, if there is a connection, then this is just one more of the many extreme anachronisms encountered in this work that needs to be addressed.

e. Other Measures that derive from the 63/64 Ratio

Recall from our earlier discussions that the apparent mean diameter of the Moon (31' 30") and apparent mean diameter of the Sun (32') in terms of arc were related to one another as 63/64. If we apply this relationship to the circle of 360°, from which was derived the ancient Egyptian year of 360 days, we arrive at: 63/64 X 360 = 354.375, which is a very close approximation for the lunar year of, 354.36 days. Of course, if we multiply the ancient Egyptian month of 30 days by 63/64, we obtain, 29.531 days, which is almost an exact match for the length of the synodic month, 29.530 days. There is yet another number that can be derived from this ratio and it comes from the *temenos* walls of the Step Pyramid complex.

The *temenos* wall surrounding the Step Pyramid complex has a nominal perimeter of 3,140 royal cubits of 524 mm each, which is equal to, 5398.163 international feet, based on the international foot's measure of 304.8 mm. If the span of the gateway to the complex, which is approximately 9.5 meters (derived from Lauer's scaled drawing of the entryway) and equivalent to 18.13 royal cubits or 31.17 international feet, is subtracted from this number, 5398.163, the result is, 5367 international feet. If this number is multiplied by the 63/64 ratio, we obtain, 5283.14 feet, a very close approximation for the international mile of 5,280 international feet, which might be improved upon with a finer measure of the gateway.

f. Further Significance of the Appendages of the Eye of Horus

The Eye of Horus symbol discussed earlier has two strange appendages attached to the lower rim of the eye. One is the curly-cue appendage that is identified with the fraction, 1/32. The other is an odd-shaped appendage that is identified with the fraction, 1/64.

Eye of Horus appendages

The curly-cue appendage often appears on the crowns of Egyptian kings, while the second is not attested anywhere else in the records of ancient Egypt. The key to the meaning of these two symbols is to be found in the so-called, *nsw-bty* title of the ancient Egyptian kings, which has the literal meaning of, 'He of the Sedge and Bee', but it is often translated for modern purposes as, "King of Upper and Lower Egypt." [60] The hieroglyph for this title is shown below.

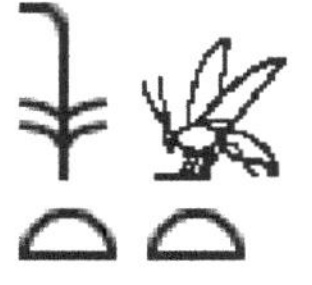

Nsw-bty sign

The second image in the hieroglyph depicts an insect that more closely resembles a wasp than a bee, but the meaning accorded it by the ancient Egyptians was, 'bee', which is attested by the fact that the same hieroglyph appears in the hieroglyph for the word for, honey.[61] Given this, the curly-cue appendage likely is the proboscis of a bee, which symbolizes the bee. But the meaning goes deeper: the bee is a metaphor for the hive, with its hexagonal-shaped cells, which evokes the number 6, a key and fundamental number from the Pattern. The curly-cue appendage, then, is a metaphor for the number 6, which symbolizes all of the numbers and measures from the Pattern.

The first image in the *nsw-bty* hieroglyph is related to the second, but is profoundly different. There are four leaves or branches (stamens?) below the 'pistil' or female reproductive organ of the plant, which collectively evoke the number 5, but is 5 of 6 if the stem of the plant is taken into account, which directly relates it to the second symbol. The second appendage, then, is symbolic of the number 5, and the hieroglyph likely depicts a measuring rod with a side marking to denote 5 of 6. But what is the significance of the number 5? It is a basic number from a different pattern, one that in many ways defines the planet Venus, which also factors in the measure of time.

Several of the time measures defined by Venus have already been addressed, including the Venus transit cycle and the phoenix cycle, but there is yet another: the Venus inferior conjunction cycle, which will be described in the following chapter.

---

[60] Wikipedia contributors, "Ancient Egyptian royal titular", https://en.wikipedia.org/wiki/Ancient_Egyptian_royal_titulary Wikipedia, the Free Encyclopedia. Accessed September 14, 1016

[61] Gardiner, Sir Alan, *Egyptian Grammar*, (Griffith Institute, Oxford, 3rd Edition), page 477, Section L 2
See also: Wikipedia contributors, "Gardiner's sign list", Wikipedia, the free encyclopedia, https://en.wikipedia.org/wiki/Gardiner%27s_sign_list (accessed September 27, 2016)

CHAPTER 15

VENUS AND THE FIVE-POINTED STAR

She is Earth's closest neighboring planet in the solar system, but in many respects she is the most mysterious. Perpetually wrapped in a dense, heavy cloud cover, she hides her secrets even from modern science to a large degree. She has factored in numerous myths and legends in every culture in every age, and is a source of awe and wonder even today. Her appearance in the heavens is always spectacular, but the beauty of some of her astronomical displays is simply stunning, even to today's somewhat jaded audiences. And none of these displays is more stunning than when she appears besides the crescent Moon, as she nears her closest point of approach to Earth, an event that regularly repeats in the Venus inferior conjunction cycle.

a. The Octaeteris Cycle

The Octaeteris cycle is based on the fact that after eight solar (tropical) years, the Moon will appear in the same phase and on the exact same calendar date, within a day or day and a half. This cycle is in very good synchronicity with five Venusian synodic cycles, where Venus enters inferior conjunction with the Earth and Sun after each cycle. On these occasions, Venus will appear beside the Moon on the same date as a previous appearance, and because Venus lies comparatively close to the Sun, with an elongation never greater than 48°, the Moon is always in crescent phase when the two appear together. These appearances are based on a predictable Sun-Earth-Venus inferior conjunction cycle, as described in the following paragraph.

b. The Venus Inferior Conjunction Cycle

There are thirteen Venusian sidereal revolutions, i.e. returns to the same location with respect to the Sun, when measured against the backdrop of the celestial sphere.[62] During these thirteen Venusian sidereal periods, (equal to eight Earth solar (tropical) years) the Earth and Venus align in inferior conjunction (i.e. Venus is located between the Earth and the Sun, in direct alignment), five times. This Earth-Venusian cyclical pattern is: 5 inferior conjunctions with the Earth and Sun for every eight Earth solar (tropical) years and for every thirteen Venusian

[62] Wikipedia contributors, "Octaeteris", https://en.wikipedia.org/wiki/Octaeteris Wikipedia the Free Encyclopedia, accessed September 14, 2016.

sidereal years, or a number pattern of: 5—8—13.[63] This pattern paints the five, evenly-spaced (72°) points (cusps) along the circular orbit of Venus, from which the five-pointed star—the pentagram—is drawn.[64]

Today, this historically important figure is often associated with the occult, but in the world of astronomy it obviously has a far different association. However, its odious association with the occult is not undeserved.

The five-pointed star pattern is the first and foremost manifestation of the number five by the planet Venus. Without the circle, the five-pointed star is part of the Egyptian hieroglyphic name for *duat*, the planet Venus.[65] With the circle, the five-pointed star is a part of the Egyptian hieroglyphic name for the *duat*, which translates as the sky at morning twilight, or the underworld. However, complicating the definitive identification of the five-pointed star in Egyptian hieroglyphs with the planet Venus, is the fact that the five-pointed star symbol is also used in conjunction with stars in general as well as for adoration.[66] This may have been an evolutionary expansion in the use of the symbol in hieroglyphic writing, as no star can be linked to the number five. In some of the earliest appearances of five-pointed stars in ancient Egypt, though, there is no mistaking what they refer to—the planet Venus.

Venus is not a star, but many cultures have referred to the planet as the 'evening star' as its orbital flight approaches the Earth, and as the 'morning star' as it re-emerges from the daylight after reaching inferior conjunction with the Earth and Sun and resumes its orbital flight away from the Earth. The Greeks called the evening star Hesperos and the morning star Phosperous; the Romans called the evening star, Hesperus and the morning star, Lucifer. The identification of Lucifer with the devil came about in later ages. However, all of the imagery necessary for this connection was already in place long before this identification was made: A star, Hesperus, appearing to fall from the heavens and enter the underworld, from which it later emerges at dawn as a new entity, Lucifer. The similarities of this imagery

[63] Wikipedia contributors, "Venus", https://en.wikipedia.org/wiki/Venus Wikipedia the Free Encyclopedia, accessed September 14, 2016; see, in particular the section, Orbit and Rotation, wherein a dynamic graphic of Earth-Venus movements, and inferior conjunctions are depicted.

[64] Wikipedia contributors, "pentagram", https://en.wikipedia.org/wiki/Pentagram Wikipedia the Free Encyclopedia, accessed September 14, 2016.

[65] Budge, E. A. Wallis, *An Egyptian Hieroglyphic Dictionary*, Parts 1 and 2, (Dover Publications, New York, 1978), see, in particular, references on pages 403 (*neter ṭuaut*) and 870 (*ṭuat*).

[66] Gardiner, Sir Alan, *Egyptian Grammar* (Griffith Institute, Oxford, 3rd Edition 1957), page 487, items N-14 and N-15; Gardiner mentions the 'morning twilight' and the 'underworld' in the context of these five-pointed star hieroglyphs, but does not mention Venus by name in these references; however, he does mention its use in context with stars in general as well as adoration.

with the phoenix myth are obvious. The identification of the pentagram with the devil readily followed this connection. This connection, while understandable, was not the understanding of the ancient Egyptians. To them, the *duat* or underworld was a place of transformation and rebirth, not eternal damnation, but it was a place of immense danger and the journey through it was extremely difficult; rebirth was by no means a sure thing.[67] This whole issue, though, is very complicated and almost impossible to sort out, and, while it may be of interest to some readers, is a digression and one that will not be followed further.

When the Step Pyramid complex of Saqqara was first subjected to extensive clearing and exploration in the 1920's, the archaeologists involved found a number of cut stone blocks with five-pointed stars carved on them in relief. The blocks were all of similar dimension, L .540 m x H.270 m x D. .370 m. Each block had two equal-sized stars symmetrically cut on one long side, each star being .30 m in diameter. These blocks were found in the lose fill around the central burial chamber under the pyramid. As to what they were, no one knew, but the speculation was that they were the roofing blocks from an earlier structure that had been dismantled. This theory seemed plausible until blocks with stars cut on two long sides opposite to one another were found, with two stars on each side for a total of four. The mystery only deepened when more such blocks were found around the complex's south tomb. The blocks were found nowhere else in the complex, only around the two tomb structures, and always as lose fill around them.[68] If the five-pointed star is accepted as the symbol of Venus, then the meaning of these blocks becomes obvious: they represent the 243-year Venus transit cycle. The blocks with two stars represent the cycle when there are only two transits, each separated by 121.5 years. The blocks with four stars represent cycles where there are three or four transits in the cycle, with intervals of 105.5, 8, 121.5 and 8 years, or 8, 113.5, 121.5 years, or 105.5, 129.5, and 8 years. In any case, the 243-year duration of the cycle is its determining dimension, and the intervals between the individual transits within the cycle always total 243 years. Why these Venus transit blocks were placed around the two tomb structures and nowhere else is a mystery that cannot be answered at present. However, as was noted earlier, the Step Pyramid complex of Saqqara is one of the oldest, carved stone structures on Earth and is at least 4,700 years old. The knowledge necessary to construct these blocks is incompatible with the structure's place in time, and, to be certain, their presence in these ruins is yet one more example of a number of glaring anachronisms that have been encountered so far in this work.

---

[67] Budge, E. A. Wallis, *The Egyptian Heaven and Hell*, (Open Court Publishing, Chicago, Illinois, 4th printing 1997).

[68] Firth, Cecil M. and Quibell, J.E., *Excavations at Saqqara, The Step Pyramid Complex*, (Service Des Antiquities de L'Égypte, 1935, Cairo; Re-published by Martino Publishing, Mansfield Conn. 2007), pages 46 and 47

c. The Henti Period in Ancient Egypt

Henti period sign.

There will be other evidence from the Step Pyramid complex that will be presented later that further supports the association of the star blocks with the 243-year Venus transit cycle. Before moving on from this issue, though, the existence of a time cycle in ancient Egypt that closely approximates ½ of the 243-year transit cycle, or 121.5 years, needs to be discussed. This cycle is the 120-year 'ḥenti period' the hieroglyphic sign for which is depicted above.[69] If this cycle was associated with the Venus transit cycle, then the fact that it was called a '120-year' period may reflect a public reference to the celestial event and one that was broadly evocative of it, but not exactly accurate. There is also a possibility that the ancient Egyptians may have recognized a double ḥenti period, of 240 years, which the double sign seems to imply, but this is not proven.[70] So, does the ḥenti period have a connection with the Venus transit cycle? There is no definitive proof that it does or doesn't, so the issue must remain an open one at present. If it was ever used in a hieroglyphic context with a five-pointed star, that might decide the argument for such an association, but there is none, so far as I know.

There may be other instances of the number five occurring in relationship to Venus, but we will leave the subject by noting in conclusion that there are five Venus synodic days (mid-day-to-mid-day) of 116.75 days each between successive inferior conjunctions, for a total of 584 days (5 X 116.75 = 583.75).

d. Re-defining an Ancient Sign

One can search the available references in vain for an Egyptian hieroglyph that specifically stands for the 5—8—13 Venus inferior conjunction cycle. However, there is a sign that has long befuddled Egyptologists, who have either left the sign undefined or given it the meaning of, "half-month festival."[71]

---

[69] Budge, Sir E.A. Wallis, *An Egyptian Hieroglyphic Dictionary*, (Dover Publications, Inc. New York, 1978, a republication of an original work published by John Murray in London in 1920) page 488, ḥenti. See also, Wikipedia contributors, 'List of Egyptian Hieroglyphs by alphabetization' https://en.wikipedia.org/wiki/List_of_Egyptian_hieroglyphs_by_alphabetization#H1_and_H2 Wikipedia, the Free Encyclopedia (accessed September 19, 2016), H1 and H2

[70] Gardiner, Sir Alan, Egyptian Grammar, (Griffith Institute, Oxford, Third Edition, 2005 printing) sect. § 77, page 60

See also: Wikipedia contributors, "Gardiner's sign list", Wikipedia, the free encyclopedia, https://en.wikipedia.org/wiki/Gardiner%27s_sign_list (accessed September 27, 2016)

[71] Gardiner, Sir Alan, *Egyptian Grammar* (Griffith Institute, Oxford, 3rd Edition 1957), page 486, sign N-13

'Half-month festival" sign

At each of the points or cusps of the star, a Venus inferior conjunction event occurs, where the planet Venus will appear near the crescent Moon. However, unlike the octaeteris cycle, where Venus will appear by the crescent Moon at the same cusp eight years later, on exactly the same calendar date, within a day or day and a half, the Venus inferior conjunction cycle occurs, cusp-to-cusp, after approximately 18.5 months or 584 days. In other words, the time between successive Venus inferior conjunctions is 18.5 months or 584 days. The half-month festival sign, then, as depicted above, should be accorded the meaning of "Venus inferior conjunction cycle" or "584-day festival."

e. Venus and the Golden Mean:

As was discussed earlier, Venus makes five inferior conjunctions with the Earth and Sun in eight Earth or tropical years, or in thirteen Venusian sidereal years. This creates an 8:13 orbital resonance between Earth and Venus. Each of the five inferior conjunctions is removed from the previous one by 144° of Venusian sidereal movement.[72] To better explain this, we should envision a circle divided into 360° and the orbit of Venus tracked by a fast-moving conjunction indicator and the orbit of Earth tracked by a slower moving conjunction indicator. Both conjunction indicators start at 360° in the same sidereal location. Then, as noted earlier, they both come into alignment again at 144° from 360°, but measured in a counter clock wise direction in accordance with their prograde orbital motion. Accordingly, the inferior conjunctions occur as follows:

Start	360°	(Both indicators in alignment.)
1st conjunction	216°	
2nd conjunction	72°	
	0°	(The indicators arrive at 0° (360°) and begin a second rotation.)
3rd conjunction	288°	
4th conjunction	144°	
5th conjunction	0°	(The indicators align again at 0° (360°), or after two rotations)

Thus, the conjunction indicators must turn through two complete sidereal circles in order to achieve five inferior junctions in eight Earth years or thirteen Venusian sidereal years, but this may not be apparent, until the five inferior conjunctions and their angular locations are placed in clock-wise, normalized position. When viewed in this manner it can readily be seen that each conjunction advances by 72° from its predecessor, though, they do not occur successively.

[72] Wikipedia contributors, "Venus", Wikipedia the free encyclopedia, https://en.wikipedia.org/wiki/Venus (accessed September 20, 2016)

2nd conjunction	72°
4th conjunction	144°
1st conjunction	216°
3rd conjunction	288°
5th conjunction	360°

This is significant because the inferior conjunctions occur in a repeating resonance of 2:3, with two conjunctions occurring in the first sidereal circle and three occurring in the second for a total of 5. In summary, there are 2 inferior conjunctions in the first sidereal circle, 3 in the second, for a total of 5, which occur in 8 Earth years or in 13 Venusian sidereal years. The Earth, Venus, Sun inferior conjunction resonance then can be stated as: 2:3:5:8:13. Excluding the number 1, these are the first five numbers from the Fibonacci series, where each number in the series, beginning with 5, is the sum of its two predecessors (e.g. 5 + 8 = 13). The series runs on ad infinitum, but importantly each succeeding additive pair, when divided into one another (e.g. 8/5 = 1.6), produces ever closer approximations to *phi* or the golden mean, ≈ 1.6180339, which is often represented by *phi*. Mathematically, *phi* can be derived from the equation: $(1 + \sqrt{5})/2$.[73] This equation contains the exact dimensions of the 1: 2: $\sqrt{5}$ triangle, from which the golden mean can also be generated.

The association of Venus with the golden mean is also important, as the five-pointed star that is used to represent Venus also produces the golden mean at each of the five intersections of the star's five defining lines, where each whole line stands in relation to its longer portion as *phi* and each longer portion stands in relation to the shorter portion as *phi*. Furthermore, the length of the Venusian year in Earth days is 224.7 days, which when divided into the Earth year of 365.24 days equals 1.625, a close approximation of *phi*.[74] There may be more such relationships between Venus and the golden mean, but that there is such a connection seems to be beyond question. As to the significance of this connection, we will address this issue in a later chapter.

f. The Appendages of the Eye of Horus Revisited

The previous chapter, "The Eye of Horus as Symbol and Measure" included a section on the significance of the two appendages of the Eye. The section concluded with an assertion that the *bty* appendage, 1/32, is a metaphor for the number 6, which is a symbol for all of the numbers from the Pattern. The significance of the *nsw* appendage, 1/64, was interpreted as a possible metaphor for the number 5 and Venus. That interpretation can now be affirmed, but this metaphor can be further understood as symbolic of the entire Fibonacci series and of the golden mean. However, there is more to be said on the appendages. Taken together,

[73] Wikipedia contributors, "Golden ratio" Wikipedia, the free encyclopedia, https://en.wikipedia.org/wiki/Golden_ratio (accessed September 20, 2016)

[74] Wikipedia contributors, "Venus" Wikipedia, the free encyclopedia, https://en.wikipedia.org/wiki/Venus (accessed September 20, 2016)

their literal meaning in ancient Egyptian is *tĭt*, which is also the identical word for the "knot of Isis", a term that is represented by the hieroglyph shown below—which itself is a variation of the *ankh* sign of life.

Knot or *tĭt* of Isis.

However, no matter which hieroglyph sign is used for it, the *tĭt* is literally interpreted as, "the image" or "the figure", but of what is not known for certain, though the hieroglyph sign shown above does strongly suggest that "the image" is human.[75] Several questions arise here. Does the *tĭt* sign beneath the Eye of Horus imply that the images or figures that can be perceived by man arise from the golden mean and from the measures from the Pattern? Do the golden mean and the Pattern, when taken together, define reality, at least in so far as it can be perceived? There is more to be said on this issue, but for now it must be concluded, though, not before revisiting the Pattern.

g. The Pattern and the Number Five

It is clear from the underlying theme of the two foregoing chapters on, "The Eye of Horus as Symbol and Measure" and "Venus" that the number five, with its connection to the transcendental number *phi*, the golden mean, and the Fibonacci series, is the basis for a second pattern. This is true, and it is well documented that the golden mean plays a significant role in the designs of nature and has a demonstrable effect on man's concepts of beauty and harmony.[76] However, the number five can also generate the numbers of the Pattern, which, in effect, brings us back to where we began.

If the progression of the defining number three from the Pattern is taken as a given, where each succeeding number is the product of its predecessor and the number three, as was shown on the top row of the table in the Chapter on "The Pattern of Dimension and Measure", we have:

1—3—9—27—81—243—729—2,187

Then, if each number in this progression is divided by the number five, all of the other numbers from the Pattern can be derived. (However, as was the case earlier, only the root numbers are considered.) For example, we will start with the number 27 from the foregoing progression:

[75] Gardiner, Sir Alan, *Egyptian Grammar,* (Oxford Institute, Oxford, reprinted 2005) see sign D-17 on page 452 and signs S-34 on page 508 and, in particular, the following parenthetical entry for (V-39) also on page 508 See also: Wikipedia contributors, "Gardiner's sign list", Wikipedia, the free encyclopedia, https://en.wikipedia.org/wiki/Gardiner%27s_sign_list (accessed September 27, 2016)

[76] Wikipedia contributors, "Golden Ratio", Wikipedia, the Free Encyclopedia, https://en.wikipedia.org/wiki/Golden_ratio (accessed October 10, 2016)

$27 \div 5 = 54 \times 10^{-1}$
$54 \div 5 = 108 \times 10^{-2}$
$108 \div 5 = 216 \times 10^{-3}$
$216 \div 5 = 432 \times 10^{-4}$
$432 \div 5 = 864 \times 10^{-5}$, etc.

Other examples can be used, such as 3, 81, etc. but the results are all the same. Each defining number from the progression of three when divided by five generates the succeeding number below it, as shown in the table of the Pattern, which, in turn, when divided by the number five, generates the number below it, etc.

So, does the number five introduce a second pattern, one based on the golden mean, or does it just define the Pattern in a different manner? The answer is, it does both, but it is a matter of no small interest that both of the numbers five and six are key factors of the Pattern. This is not a coincidence.

CHAPTER 16

ANCIENT ARCHITECTURE AND MEASURES OF THE PATTERN

Throughout much of the 1st Part of this book the question of how early civilized man could learn and utilize many of the measures discussed has been a lingering and disconcerting one, sometimes to the point of distraction. These issues are addressed in the following chapter.

There is a broad consensus within the scientific community that mankind progressed to civilization sometime around 3,000 BCE in several areas of the world, including Mesopotamia, the Indus River delta region in India, The Yellow River Valley in China, Caral in Peru, and the Nile Valley in Egypt. All of these early civilized cultures seem to have emerged simultaneously, and to have possessed a remarkably complete development across a broad spectrum of disciplines and fields that comprise our understanding of the word 'civilized.' These include: language, writing, mathematics, measures of time and distance and weight/volume, religion, administrative organization, architecture and building, transportation, farming, animal husbandry, etc. The complexity and relative sophistication of these early civilizations is one of the most remarkable and startling things about them. There seems to have been little or no evolution in their development, and it is almost as if they suddenly and rather dramatically appear in their entirety at their very beginning. How this came to be is difficult to comprehend and even more difficult to explain, given the steady development and evolution of modern cultures from these ancient ones. However, by addressing the narrow issue of the origin of measure, including mankind's understanding and use of it, a plausible explanation may be attained, which may have ramifications for these broader issues.

Early mankind seems to have mastered several broad areas of measure that have specific measurements, which are either exact or close approximations to numbers from the Pattern. The circle and its arc measures, all related to numbers from the Pattern, seem to have been adopted early and universally.[77] The broad outline for tracking time and the adaptation and use

[77] Klein, Herbert Arthur, *The Science of Measurement, A Historical Survey* (New York, NY, Dover Publications 1974), p 100-101. While Klein, with some justification, assigns the origin of arc measure and the divisions of the day to the early Mesopotamian people, I believe it was universally adopted early on. The Mesopotamian civilizations wrote their records of such matters on relatively long-lasting clay tablets, whereas fewer records

of specific measures for this purpose seem to been similarly adopted, and remarkably much of this has an astronomical basis of some sophistication, such as the several lunar cycles and the cycles of the planet Venus. The division of the day into smaller units of hour and minute, and possibly second, were also early developments.[78] Measures for length and distance, while typically dissimilar when compared with one another across different cultures, were also early developments, as were the measures for weight and volume.

With regards to the measures of physical dimension and distance, all of these early civilizations developed units of measurement from an early date, and some such as Sumer, tried to standardize these measures across a broad expanse of territory beyond their immediate control that they influenced through trade and other contacts. The exact measures of many of these units is either unknown or often subject to divergent opinions, with one exception: the measures of Ancient Egypt, which are fairly well attested in written records and actual measuring rods, as well as the substantial remains of contemporary structures against which these measures can be applied for verification. This fact will prove of immense value for our purposes.

With respect to certain astronomical phenomena and their measure it is a matter of considerable controversy as to whether or not these ancient cultures were aware of the band of the zodiac and its arc dimensions and subdivisions, and the related phenomenon of precession of the equinoxes. However, if they were aware of these matters, it raises the question of how they learned of such arcane matters and what instruments they used to measure them? More problematic, yet, is the issue of whether or not they were aware of even more arcane astronomical phenomena, including, the: Saros eclipse cycle, Venus transit cycle, and Phoenix cycle, all of which have long-duration cycles and are clearly well beyond the known technical capabilities and sophistication, and documentation and record keeping capacities of these early civilizations.

The derivation and adoption of the Sumerian sexagesimal system for measuring the circle and tracking time may be a key issue, particularly because it eventually was adopted world-wide. Interestingly the equivalents of several Sumerian measures of mass were known to be in use in medieval England—thousands of years after the disappearance of the Sumerian city-states. Also, weights based on the almost exact measure of the English ounce avdp. of 28.35 grams and a measuring rod graduated in increments of 1/16th of an inch have been found in early Indus River ruins.[79]

---

are preserved from these other civilizations. Egypt has a large quantity of early records, but much of it is of a religious or funerary nature.

[78] Ibid, pp 97-104. See, also, Whitelaw, Ian; *A Measure of All Things, The Story of Measurement Throughout the Ages*, (David & Charles, Cleveland, Ohio; 2007) p 10, where he asserts that the Sumerians, an early Mesopotamian people, were the source of these measures. (Note: It is generally believed that the Sumerian culture flourished between 3,500-2,000 BCE.)

[79] Whitelaw, Ian, *A Measure of All Things, The Story of Measurement Through the Ages*, (David & Charles, Cincinnati, Ohio, 2007), page 14.

Extensive cultural contact through conquest and trade may explain the dispersal of the sexagesimal system and these measures of mass and length across seemingly insurmountable distances. If the Sumerians were the originators of this system, then it would seem to have evolved in different directions and measures after its dispersal.

However, the fact that the numbers of the sexagesimal system that lie at the heart of the Sumerian measures of the circle and time are identical to the numbers from the Pattern could also suggest that the Sumerians may have learned the Pattern and then drew numbers from it to apply to their measures. This is possible, but it doesn't explain how the many different subsequent measurements, as documented in Part I, which were apparently derived long after the passing of the Sumerians also were comprised of numbers from the Pattern, unless their originators also used the Pattern for their source for numbers to define their measures. Not impossible, but if it happened, such a multi-generational, cross-cultural, unity of purpose and effort by the human race would be unique in human history, with its constant rivalry, strive, and bitter divisions. A prospect which hardly seems realistic!

The all-inclusive nature of the Pattern and its numbers, and the fact that so many measures are related to it, would seem to strongly suggest that these measures were all derived at the same time. Also, the fact that these measures seem to accord so well with so many of the natural phenomena that these measures are applied to also suggests that these measures were derived at the same time. Furthermore, as was noted in the beginning of this book, the only thing that unites many of these disparate measures and their underlying natural phenomena is the Pattern, which would seem to be far too harmonized for these measures to have gradually evolved and then been adopted piecemeal over thousands of years and across widely dispersed cultures. The most likely answers, then, are that: 1) <u>all</u> of these disparate measures—speed-of-light, diameter of Moon, average duration of a Saros eclipse series, diameter of Earth's magnetic core, etc.—were specifically derived and adopted to reflect their underlying phenomena <u>at the same instant</u>, and 2) all of these phenomena seem to obey the imperative and order of the Pattern. What drives the Pattern is a topic that will be taken up later in this book. The issue at hand, though, is how man learned of the Pattern and its measures?

The most direct and concise answer to this question is: He was taught. But, by whom? The classical explanation is that God formed his mind and instructed him at the beginning. The current explanation is that advanced beings from some other galaxy taught him. Before analyzing these possibilities further, it should be recalled that the Pattern and its numbers are apparently reflective of much, if not all, underlying natural phenomena in the solar system, possibly beyond, and that man's knowledge of these measures requires that he be able to perceive and comprehend them before he can be taught their specifics. In other words, man is a part of this system, and whoever created it also made man a part of it, in that he was given the ability to perceive and comprehend the Pattern and its manifestations in nature.

A fortuitous visit to Earth by a race of beings from another galaxy does not explain how all of these harmonized natural phenomena came into existence or how man could be a part of it all, or how the Pattern underlies so much natural phenomena, unless of course they possess god-like qualities and created all of this. Admittedly, this is a possibility, but it seems unnecessarily complicated as an explanation. If these mysterious, extraterrestrial beings created all, then how did they get here, how did they do it, and why did they leave? We could pursue this line of inquiry much further, but there is no need; all of it is an unnecessary distraction and pointless digression. Plain and simple, God is the author of all; there is no need to give Him an address from another galaxy. Furthermore, if man was given this knowledge and taught its meaning and application, then this would explain how ancient man could possess this knowledge.

Obviously, not all of mankind was taught, nor could those select few that were be taught the full range and ramifications of the Pattern's measures. As was seen in Part I of this book, many of these ramifications have only recently been explored, such as those for electromagnetic energy and the measures applicable to the Earth's interior, including the diameter of its electromagnetic core and the location of the Earth/Moon barycenter with the Moon at perigee. However, enough of the Pattern and its manifestation in nature's measures was taught to give ancient man a grounding in the science and to build on it with the measures that he was taught.

In addition to their use for early calendrical purposes, there are also widespread references to many of the numbers from the Pattern in both the myths as well as the religious beliefs of many ancient peoples, frequently to the numbers 12, 30, 72, 360, 2160, and 4320. Often, these same numbers, or derivations thereof (e.g. 54, 108, etc.), also turn up in the measures of the monumental architecture of these peoples. These societies span the ages and the globe and include the Sumerians, the ancient Egyptians, the Greeks, the Indian, the Chinese, and the Maya. A number of authors have explored this fact in considerable detail, but invariably they have concluded that all of these references are attributable to the phenomenon of precession of the Earth's equinoxes—an important measure to be certain, but, as has been widely demonstrated in Part I of this book, it is only one of the many, many measures found in nature that are manifestations of the Pattern.[80] For example: 2,160 is the number of years in one zodiac sign of the precession calendar, but it is also a number found in the following measures: 1) the diameter of the Moon: 2,160 international miles, 2) the equatorial circumference of the Earth: 21,600 nautical miles, of 6,000 nautical feet (309.22 mm/foot) per mile, 3) the distance between the Earth and the Moon, measured surface-to-surface with the Moon at extreme perigee: 216,000 international miles, and 4) the circumference of the circle: 21,600 arcminutes. While it can be concluded that ancient man was aware of the precession of Earth's equinoxes, he

---

[80] A good summary of this entire issue, as well as of the efforts of several authors in this regard can be found in: Hancock, Graham, *Fingerprints of the Gods, The Evidence of Earth's Lost Civilizations* (Three Rivers Press, New York, 1995); see, in particular, Chapter 31, "The Osiris Numbers", where specific reference is made to the works of: Santillana and von Dechend, *Hamlet's Mill*, and Jane B. Sellers, *The Death of the Gods in Ancient Egypt*.

was also aware of many more measures than just this one and knew that all of these measures originated from the Pattern.

So, when, where, and how was ancient man taught? The available records of the myths and religious tenets of the ancient peoples provide clues, but these are frequently difficult to reconcile with one another or summarize, as they often involve differing versions with variations in the writing styles over the often widely separated time periods in which they were compiled, and their narrative context also often evolved and changed over the years. Gods were gods with portfolios of certain characteristics and powers at one time, and then in later years their portfolios changed. Gods with established associations with fellow gods in one age were given new associations in later years.

At a meeting of ways I was ware of a bronze god lying prone at my feet,
And I knew him the offspring of Zeus, whom we prayed to of old, as was meet.
"Lord of the triple moon" I cried, "avatar of woe,
Ever a lord hast thou been, but behold, in the dust thou art low."
That night, with a smile on his lips, the god stood by me sublime,
And said, "A god though I be I serve, and my master is time."[81]

However, the numbers referenced to in these myths and religious tenets often remain stable and unchanging. The myths and religious tenets were useful in preserving and perpetuating these numbers, but they were not the source for ancient man's first instruction or for his continuing exploration of the subject matter and his further education. That role likely was played by monumental architecture.

Ancient monumental architecture was far more enduring, providing ancient peoples with both a pattern and a record to emulate over the ages. Such structures likely were constructed around the world, involving every culture. While there are many remains of ancient monumental architecture that could be examined, only two will be considered. One has already been mentioned earlier: The Step Pyramid complex of Saqqara, Egypt, which as remarked upon earlier, is the oldest, large-scale stone structure on Earth, dating to approximately 2,700 BCE. The second is the Great Pyramid of Giza, one of the most celebrated structures in history and the sole remaining member of the seven wonders of the ancient world. Much has been written about the Great Pyramid, unfortunately a lot of which is wild, unsubstantiated speculation, which has tended to taint even reasoned inquiry into its history. However, the reader may rest assured that almost all of the sections concerning this pre-eminent structure are rooted in the hard sciences of geodesy, cartography, metrology, geometry and astronomy. Most of this material is entirely original.

---

[81] Poet Palladas of Alexandria. Bury, J.B. *History of the Later Roman Empire*, (Dover publications, New York 1958) author's version on pages 374 and 375

It should also be pointed out that ancient monumental architecture often does not perfectly reflect the Pattern; instead, it mimics the imperfections of the manifestations of the Pattern as found in nature. As there are no perfect reflections of the Pattern in nature, so there are no perfect reflections of the Pattern in ancient monumental architecture. This does not mean, nor should it imply, that the measurements of these ancient structures as determined by modern methods are treated lightly; they are not.

a. Analysis Methods and Key Numbers

The analysis of the Step Pyramid complex and the Great Pyramid is based on actual measurements of the various components of these structures, their geographic location and their orientation with respect to the cardinal directions. The measurements used are derived from a measure of the royal cubit found in the Step Pyramid complex, which is 524mm (see Appendix 2, page 1), and the corresponding Egyptian foot of 299.42mm. Both measures were also used in the Great Pyramid. These structures are also analyzed with respect to counts of major architectural features, particularly so in the case of the Step Pyramid.

Where appropriate and necessary to the context of the discussion, relevant religious texts and hieroglyphic symbols are analyzed. It cannot be overemphasized that there is little to nothing in the way of contemporaneous written records concerning either of these structures, which was likely by design. The knowledge imbedded in these structures was meant to endure, and to enable those in possession of such knowledge to train and instruct younger initiates. Written records can be lost, destroyed or corrupted; memories can be corrupted or diminish with time. Stone defies the ages, and it is a testament to the designers and builders of these structures that their purposes have been admirably achieved.

There are, however, several contemporaneous written records, in the Step Pyramid in the form of six relief panels or stelae; three under the Step Pyramid and three under the South Tomb. These are very important records, and their meaning and significance will be explored in Part V, and, of necessity, there will be extensive reference to ancient Egyptian religious tenets and imagery.

Throughout the following sections on these two structures, there are frequent references to the Pattern, as presented in Part 1, as well as several key measures from geometry and trigonometry. There also a number of references to factors that have been introduced earlier, including 5:6, and 63/64, both of which derive from the "Eye of Horus." Further, several triangles that were presented earlier, are also frequently cited, including: the equilateral triangle, the isosceles triangle, and the triangle with sides 1 and 2, whose diagonal is √5. The key fractions associated with these three triangles also are frequently cited, including: √3 or 1.732, the height of the equilateral triangle; √2 or 1.414, the hypotenuse of the isosceles triangle; and √5 or 2.23, the hypotenuse of the 1:2 triangle.

There are also frequent references to the circle, the sphere, and their key factor, π (pi) or 3.1416. Above all, however, there are frequent references to *phi* or the golden mean, 1.618. These two transcendental numbers are the keys to much of the knowledge imbedded within these structures and particularly to the larger cosmic order that they memorialize. The cognates of *phi* are also frequently cited, including: its square, 2.618; its square root, 1.272, and its reciprocal or 1/*phi*, .618. *Phi* is an underlying factor in both structures and is frequently found in their measures, and in their relationships to one another and other nearby structures. In many respects, it is *the* fundamental factor in the architecture of these two monuments.

Several geometric figures are also frequently cited, including the five-pointed star in a circle, which has been introduced earlier; and the overlapping circles of the Vesica Pisces, which will be introduced later.

PART III

THE STEP PYRAMID COMPLEX OF SAQQARA

CHAPTER 17

PRELIMINARY COMMENTS ON PART III

This part assumes a certain, rather in-depth level of understanding of astronomy, particularly of lunar eclipse and Venus transit cycles and their measurement; therefore, a glossary is not provided as the specialized terms used should be familiar to those equipped to understand this book. However, eclipses are discussed, often in considerable detail, for the reader to fully grasp the significance of the various features of the Step Pyramid and its complex. Links to various on-line sites— particularly to the U.S. National Aeronautics and Space Administration's very informative web site on eclipses— are provided throughout, should a reader desire more information in this regard than is provided in the text. Egyptologists who read this book will most likely struggle with the astronomy and associated numbers found in it. To them I can only say that there was simply no other way to present this information.

The level of detail provided on the Step Pyramid and its complex is only that necessary to allow the reader to fully appreciate its significance in respect of the eclipse cycles. This part is not intended as a work exclusively on Egyptology, although a review of a significant amount of this material is provided throughout the part for context. Toward the end of the part, considerably more such detail is provided for further elaboration on certain points considered critical to the thrust of the part and on the formulation of conclusions, but it shouldn't prove overly demanding. An in-depth understanding of Egyptology is not a prerequisite for readers otherwise equipped to read and understand this part. Readers desiring more information on the Step Pyramid and the ancient Egyptians associated with its construction can consult the various references cited throughout the book.

I have used modern astronomical terms throughout, recognizing of course that these are anachronisms. However, there is no viable alternative as the terms that the ancient Egyptians may have used are either unknown, obscure, or ambiguous. However, this should not detract from the text, as the astronomical phenomenon described are the same today as they were millennia earlier, regardless of the terms applied to them.

Similarly, I have used the terms commonly applied by Egyptologists to the various structures and architectural features of the Step Pyramid complex. In certain areas, particularly for the surrounding wall, I have used the terms employed by the architect and Egyptologist, Jean-Philippe Lauer, who spent a lifetime working on the complex, starting in 1927. This is not only desirable in order to minimize the possibility of confusion from arising, but there is simply no viable alternative, as the ancient Egyptian terms for the complex's various features are unknown, with very few exceptions.

Many of the aspects of the Step Pyramid and its complex that are related to eclipse cycles are based on numerical counts of specific features of the complex, such as the numbers of panels and recesses in the wall that surrounds it and in the reeds in the entryway columns. Others are based on the linear dimensions of these features and these rely on measurements. The scaled drawings of the complex drawn by Jean-Philippe Lauer are an invaluable resource and are depicted and cited throughout. Lauer joined with the archaeologists Cecil M. Firth and J. E. Quibell, who had been engaged in an extensive and detailed excavation of the complex since 1924, to assist them in interpreting the monument and with its reconstruction. The structures that are visible today are largely due to the patient and faithful reconstructions done by Lauer.

Many, but not all, measures are given in meters followed by the equivalent figure in Egyptian royal cubits in parenthesis, e.g. 12 m (22.92 cu.). The royal cubit is equal to 20.63" or .524 meters. 1 meter is equal to 1.91 cubits.[82] Where there is some doubt about the measurement intended in the structure, the cubit measure is considered the more accurate, as it was the measure used by the ancient Egyptians and it specifically was the measure used in the construction of the Step Pyramid complex.

A final word about number and the Step Pyramid complex before beginning: a wealth of number—for dimensions, counts of features, etc.—will be presented throughout the part and I believe that all of these numbers are very significant. Many of them I can identify and relate to measures of time and to various astronomical cycles and principals, but there are others that I have no idea what they mean. The whole structure is a mathematical treatise and it usually leaves no room for doubt as to what the intent was of its designers, as they often repeat important numbers over and over again, sometimes with quite startling effect. I believe that the designers left nothing to chance and there are probably few if any numbers in the structure that aren't there by design to serve some very specific purpose. They left nothing to chance.

There is another aspect of the structure that will become more and more apparent as we analyze and review it. It seems that its designers were well aware of the Pattern that defines

[82] Firth, Cecil M. and Quibell, J. E. with Plans by Lauer, Jean-Philippe; *Excavations at Saqqara, The Step Pyramid*; Vols. I (text) and II (plates) (Martino Publishing; originally published by L'Institut Francais D'Archaeologie Orientale, Le Caire); 1935; vol. I, Description of Plates, Plate 1; the length of the cubit used in the construction of the Step Pyramid complex, as measured from a red line datum found inscribed on a wall of the court of the south princess, is 524mm.

most if not all measures of time and distance, and they designed the Step Pyramid complex to reflect and mimic the Pattern as closely as possible. As a result, the complex embodies a startling wealth of information on these measures, particularly of lunar eclipse cycles, but to the extent that it mimics the Pattern it also includes much more information than this.

CHAPTER 18

PRESENTATION AND ORGANIZATION OF PART III

I have tried to introduce the most basic aspects of lunar cycles and eclipses in their relation to the Step Pyramid complex first and then proceeding to the more complex ones, so as to follow a logical progression in the order of presentation of the information that is memorialized in the various structures of the complex, but there are a number of deviations from this intent. The reason for this is because an irreconcilable conflict arises when following this progression with the equally important need to maintain an order in the presentation and discussion of the various structures of the complex and their significance. In other words, the structures can and often do include information of the most basic nature along with that which is far more complex as well. This conflict and the compromises that it necessitated in this part's order of presentation will become apparent at the very outset.

Similarly, I have tried to limit the amount of material presented on the Step Pyramid complex to only that which is relevant to the discussions of lunar cycles and eclipses, and Venus transit cycles. However, in some cases there may appear to be an excessive amount of such detail, I can assure the reader it is all relevant and it will all be cited and used at one point or another. The numbers presented, whether of measurements of things or structures or to the counts there of, are of particular importance.

CHAPTER 19

AN ANCIENT TALE

There is a story told by Manetho—a 3rd century BCE Ptolemaic Priest who composed the *Aegyptiaca*, a history of Egypt that is now largely lost—about a lunar eclipse which supposedly occurred during the reign of Necherophes, a late second or early third dynasty Pharaoh of the early 3rd millennium BCE. "In his reign the Libyans revolted against Egypt, and when the moon waxed beyond reckoning, they surrendered in terror." [83] Although the existence of this Pharaoh is considered doubtful,[84] which would tend to call into question the accuracy of this story, nevertheless it is an interesting story that raises several intriguing questions.

How is it that the Egyptian soldiers seem to have been able to maintain their composure in the face of such a singular event as a total lunar eclipse, while the enemy soldiers were terrified of it and panicked? Of course, this could be attributable to the Egyptian soldiers' superior discipline, or it could be that they were aware of the eclipse because they had been told ahead of time what to expect. And if they were aware of the eclipse, then how did they come by such information? And how were the Egyptians able to predict both its timing and nature in order to turn it to their tactical advantage on a battle field? The answer to these questions could have enormous ramifications to the generally accepted understanding as to the level of scientific knowledge possessed by the ancient Egyptians of the early third millennium BCE. In order to accurately forecast an eclipse, they would have to have had: 1) a long-term and disciplined eclipse observation program, 2) a means for documenting observations and analyzing them, and finally, 3) accurate and reliable forecasting tools. And all of this would have had to have been sustained over many, many years, or a period of time roughly comparable to what it took for modern eclipse science to develop. All interesting, yes, but the tale is not true and the ancient Egyptians were not capable of such scientific feats. Or, were they?

Eclipses were amongst the most awe inspiring, yet feared, natural phenomenon in the ancient world. They were widely believed to portend great evil and people went to great lengths to propitiate the gods to spare them. Fortunately for the ancient peoples, though, eclipses occur

[83] Waddell, W.G. (translator). *Manetho*. (Loeb Classical Library, Harvard University Press; Cambridge, Massachusetts) 1997; p 41

[84] Gardener, Sir Alan: *Egypt of the Pharaohs*. (Oxford University Press, New York, N.Y.) 1961: p. 431

so infrequently that most witnessed only half a dozen or so in their lifetime, whether lunar or solar. This is just as true today as it was in ancient times, although today their timing, location and characteristics can be forecast with great accuracy, and this information can be provided to the public well in advance, which has tended to diminish much of their mystery.

There are scattered references to eclipses throughout history, including the one described above, and many of these have been verified using modern computers, which have been used to compute thousands of years of records for both lunar and solar eclipses that have occurred in the past. These records have been invaluable to historians in establishing an exact chronology for these ancient references.[85] However, it has been considered doubtful that ancient peoples were capable of accurately forecasting their occurrence over long periods of time, using such reliably repetitious cycles as the Saros. This view may be changing as recent analysis of an artifact found in ancient ship wreck in the Mediterranean Sea, which has been dated to around the 2nd century BCE and that is referred to as the "Antikythera Mechanism", has been shown to have a geared mechanism of surprising sophistication and accuracy for using the Saros cycle to predict eclipses.[86]

While this cycle is called Saros, after an ancient Babylonian term "sar" that was used by them to describe an interval of time of 3600 years, it was not identified and employed as an eclipse period until the English astronomer Edmund Halley adopted it in 1691. "According to R.H. van Gent, Halley '…extracted it from the lexicon of the 11th-century Byzantine scholar Suidas who in turn erroneously linked it to an (unnamed) 223-month Babylonian eclipse period mentioned by Pliny the Elder in his *Naturalis Historia II*.' "[87]

Although there is some evidence that the ancient Babylonians were familiar with eclipses and their cycles, there is currently considerably less that the ancient Egyptians were familiar with them. However, as this book will show there is an abundance of evidence available from the ancient Egyptian Step Pyramid complex of Saqqara that this view may not only be wrong, but that the ancient Egyptians may have had a very sophisticated and detailed understanding of them. This complex is one of the most iconic monuments of the ancient world, and it is composed of some of the most extensive remains of early ancient Egyptian civilization. The evidence of eclipse cycles, in particular the Saros, from this complex is not focused on one or two architectural details or structures, but pervades almost every aspect of it. It is a scientific instrument of the first order, and demonstrates a level of knowledge and understanding that is breathtaking both in its range and scope.

[85] Espenak, Fred and Meeus, Jean; NASA/Technical Publication-2009-214172; "Five Millennium Canon of Lunar Eclipses: -1999 to +3000 (2000 BCE to 3000 CE)"; National Aeronautics and Space Administration, Goddard Space and Flight Center, Greenbelt, Maryland 20771; January 2009; (Also available on line: http://eclipse.gsfc.nasa.gov/LEcat5/LEcatalog.html; see Acknowledgements Section for link to publication.)

[86] Freeth,Tony. "Decoding an Ancient Computer" *Scientific American*, Volume 301, Number 8 (December 2009): pp 76-83.

[87] Espenak, Fred. "Periodicity of Lunar Eclipses" NASA Eclipse web site: (http://eclipse.gsfc.nasa.gov/LESaros/LEperiodicity.html), dated 23 October 2009; pp. 4 of 10.

Why the focus on eclipses? Why would anyone attach so much importance to them that they would build a whole complex of structures to document and memorialize their occurrence? Simply put, they are naturally occurring phenomena that reliably repeat over predictable periods of time, thus they are a reflection of time. As such, they are a manifestation of the Pattern that underlies all existence. The ancient Egyptians were fascinated with such manifestations and referred to them collectively as *Mayet*, or world order, a concept that lay at the very center of their religious beliefs.[88] The relative infrequency of eclipses served to shroud them in deep mystery, which inspired an endless fascination for them by the Ancient Egyptians.

But there is much more to the Step Pyramid complex than just eclipses. There is also extensive focus on Venus transit cycles and on the legendary and mythical Phoenix Cycle, as well as substantial information on several patterns and harmonies occurring in nature that are related to the measures of time and distance. There is more, yet. The information provided in the stones of the complex suggests that the Ancient Egyptians pushed the frontiers of science much further than anyone could ever have imagined and that they achieved results that would rank with the very finest of modern science. All of this is based on hard numbers and cycles of time, particularly long cycles such as the Saros and Venus transit cycles.

Perhaps, then, the soldiers of Necherophes knew exactly what to expect when they faced the Libyans on the battlefield some five thousand years ago and the Moon suddenly "waxed beyond reckoning." The eclipse had been accurately forecasted well in advance for them, possibly from the Step Pyramid complex.

---

[88] Clark, R. T. Rundle; *Myth and Symbol in Ancient Egypt* (Thames and Hudson, New York, New York) 1959; p. 143.

CHAPTER 20

LOCATION, HISTORY, ORIENTATION, AND DIMENSIONS OF THE COMPLEX

a. Location, Historical Context, and Orientation

The Step Pyramid complex of Saqqara is located near the village of Saqqara, Egypt, on the west side of the Nile River, just south of the city of Cairo. It is in the high desert beyond the cultivated areas of the Nile River valley. The geographic coordinates of the structure are: 29° 52' 16" North, 31° 12' 59" West. [89] It was located immediately adjacent to the ancient city of Memphis.

The location of Step Pyramid complex marks the parallel of latitude where the measure of the Earth's circumference is very close to being equal to 21,600 international miles of 5,280 international feet each of 304.8mm each; it measures 21,592 international miles to be exact (cosign 29° 52' 16" X 24,900 miles = 21,592 miles). This factor may been a principal consideration in the selection of the Step Pyramid's location, because the Earth's circumference at this latitude, 21,600 international miles when divided by the 1,440 minutes in a day, makes the measure of a minute of time at the Pyramid exactly equal to 15 international miles. It is also noteworthy that the number 216 (the root number of 21,600) is a number from the pattern, and is one that is encountered many times in nature, especially in measures of the Moon and its motions, which as will be shown later on is very relevant to the purpose and function of the Step Pyramid complex.

The complex is composed of several major surface structures, including: a *temenos* (royal or sacred enclosure) wall, the Step Pyramid that is the focal point of the complex, a colonnaded entry way, and many smaller structures and courtyards. All were found in a heavily ruined state, although there were limited renovations during the extensive explorations conducted by

[89] Wikipedia contributors, "Pyramid of Djoser", Wikipedia, the Free Encyclopedia, https://en.wikipedia.org/wiki/Pyramid_of_Djoser (accessed November 7, 2016). See also, Mark Lehner, *The Complete Pyramids* (Thames and Hudson Ltd, London, 1997) pages 84-93. Both are good general references.

Cecil M. Firth, J.E. Quibell and Jean-Phillipe Lauer during the 1920's and 1930's.[90] The entire complex is believed to date to approximately 2, 600 BC, and to be the work of Pharaoh Djoser (Zoser), the first king of the ancient Egyptian third dynasty. There are no known contemporary records of its construction.

The orientation of the complex is not known for certain, but the orientation of the pyramid itself is given as 4° 30' east of true north.[91] However, Lauer's plan number 1 depicts the compass orientation of the complex, which appears to be the same. [92]

b. Dimensions and Measures

The measures of the complex and its various structures were first taken by J.P. Lauer, a trained architect. Several of measures of the complex's major dimensions are cited in Firth and Quibel, but these appear to be rounded and are even contradictory in places. For example: the dimensions of the enclosure wall are given as 550 X 300 meters, and later as 545 X 280 meters.[93] These measures are not considered to be reliable, and are probably the authors broad interpretation of the actual measures performed by Lauer, which can be found in appendix two of this book, and indicates that they are: 544.8 X 276.85 meters, or 1039.7 X 528.34 royal cubits of 524 mm each. (Using the rounded figures of 1,040 X 530 royal cubits for the overall dimensions is not unreasonable.) These latter measures are considered both accurate and reliable, and most are likely to be the best currently available. However, even these measures are not without their problems, as will be discussed below.

The ancient Egyptians measured in terms of the foot and cubit, which is 1½ feet. However, there was also what is referred to as the "royal cubit", which contains an extra hand for a total of seven, whereas, the Egyptian foot contains four hands and the ordinary Egyptian cubit contains six hands. The measure of the Egyptian foot was earlier given as 300mm, which provides for an ordinary cubit of 450mm and a royal cubit of 525mm. However, a royal cubit, as used in the construction of the Step Pyramid complex, was found in the ruins of the complex and was measured as 524mm (see page 1 of appendix 2). This may not seem like a meaningful difference, but the magnitude of the dimensions involved in the construction of the complex makes such a slight difference very meaningful, especially considering the symbolism that I believe attaches to these measures. The cubit of 524mm and the foot of 299.43mm, therefore, are the measures that will be used in all further analysis of the dimensions of the complex.

---

[90] Cecil M. Firth, J.E. Quibel and J.P Lauer, *Excavations at Saqqara, The Step Pyramid,* (Service des Antiquitiés de L'Égypte, Le Caire Imperimerie de Institute Français D'Archaeological Orientale, 1935; republished by Martino Publishing, Mansfield Centre, Connecticut, USA, 2007); This is a very authoritative reference, but one that, unfortunately, is not readily available. Firth and Quibel are given as authors; Lauer is credited only for the plans.

[91] Firth and Quibel, page 2.

[92] Firth and Quibel, plan number 1: "The Step Pyramid. General Plan", J. P. Lauer, which shows cardinal orientation of complex

[93] Firth and Quibel, pages 1 and 9.

Lauer's measures, as shown in Appendix 2 are not without their issues.

- His measures do not provide for a range of uncertainty, which seems a questionable practice when measuring ruins.
- His measures are cited to within a hundredth of a meter, not an unacceptable precision, but it omits for finer measures by the builders of the structures.
- He translates from meter measures to cubits of 524mm each, and then rounds up or down to give even numbers of cubits for the measures. This is not a wholly unacceptable practice, but it could compound errors in measure by the original architects and builders of the complex—errors that are not known for certain.

Regardless, Lauer's measures, as cited in Appendix 2, are the most reliable available and will be used in all further analysis, recognizing the shortcomings cited above.

CHAPTER 21

FUNDAMENTAL MECHANICS AND TERMINOLOGY OF ECLIPSES[94]

In simplest terms, eclipses occur when the Sun, Earth and Moon are in conjunction or in line with one another on the ecliptic (Sun's apparent path). Solar eclipses occur when a new Moon on the ecliptic passes between the Earth and Sun. Lunar eclipses occur when the Earth passes between a full Moon on the ecliptic and the Sun.

The points on the ecliptic where the Moon intersects it in its orbit about the Earth are referred to as the nodes. The ascending node is where the Moon intersects the ecliptic as it climbs above it and the descending node is where it intersects it as it heads below. Eclipses can only occur with both the Sun and Moon at the same or opposite nodes. If on the same node, the eclipse will be a solar; if at opposite nodes, the eclipse will be lunar. Total solar eclipses are only possible if the Moon is in its new phase as it passes through a node. Total lunar eclipses are only possible if the Moon is in its full phase, as it passes through a node. In either case, the Sun must be at or near a node at the same time.

Forecasting eclipses requires knowledge of several basic factors, including:

1) The Location of the Sun (the eclipse year): Forecasting eclipses, whether solar or lunar, requires precise knowledge of where the Sun is on the ecliptic and when it will arrive at a node. This does not occur at the same time each year, as the nodes are in constant motion due to lunar precession. This precession causes the nodes to gradually regress westward along the ecliptic, returning to the same point with respect to the celestial sphere once every 18.6 years. **The length of time that it takes for the Sun to return to the ascending node is 346.62 days, which is referred to as the eclipse year.** The interval for its passage between the ascending and descending nodes is 173.31 days.

[94] A good succinct summary of eclipses and the principals behind them can be found at: Wikipedia Contributors, "Eclipse Cycle", Wikipedia, the Free Encyclopedia, http://en.wikipedia.org/wiki/Eclipse_cycle (accessed April 18, 2017. This article is very well sourced. Much more information in far greater detail can be found at NASA's web site, http://eclipse.gsfc.nasa.gov

2) Favorable eclipse geometry (the eclipse season): When the Sun is within approximately 17days of arriving at a node, the eclipse season begins; it ends some 17 days later after it passes through the node. Thus, the eclipse season lasts some 34 days at each node. On the celestial sphere, this season describes an arc of some 17° to either side of the node, for a total of 34°. Anytime that the Sun is in the eclipse season and a Full or New Moon is at a node, an eclipse will be visible somewhere on Earth. If the Sun is within 4° 30', to either side of a node, then the eclipse will be total. There are two eclipse seasons each year, one at the ascending and one at the descending node.

3) The location of the Moon (the draconic month): The Moon takes approximately 27.2 days to travel from and return to the ascending node, or 13.6 days to travel from one node to the other. This is called the draconic month and its name derives from the fact that this period is associated with a dragon that was believed by some cultures to be responsible for causing eclipses by swallowing the Sun or the Moon. For an eclipse to occur, the Moon must be at a node sometime during the eclipse season.

4) New or Full Moon (the synodic month): The Moon must be either in its new phase for a solar eclipse or full phase for a lunar eclipse, when it arrives at or near one of the lunar nodes, for an eclipse to occur. The period of time from full-Moon-to-full-Moon, or new- Moon-to-new-Moon is approximately 29.5 days, which is called the Moon's synodic period; it is also called a lunar month or lunation. The time that it takes for the Moon to progress from full to new phase is approximately 14.75 days, which is called a fortnight.

There are many other important factors that also enter into forecasting eclipses and the foregoing are just the barest fundamentals. However, these few factors will permit us to begin our journey to trace out the information on eclipses that is memorialized in the structures of the Step Pyramid complex. There are many more, which will be presented later in this part as we delve ever deeper into the information in these structures.

CHAPTER 22

THE PRINCIPAL COMPONENTS OF THE COMPLEX DESCRIBED

Unlike its more famous cousin to the north of it—the Great Pyramid of Giza—the Step Pyramid of Saqqara was virtually unknown to the outside world, until the savants who accompanied Napoleon's army into Egypt visited it during their explorations of the country and wrote of it in their monumental *Description de L'Egypte*. The Pyramid was subsequently entered and described in some detail by a number of western explorers of the early nineteenth century, including: Minutoli and Segato in 1821, Vyse and Perring in 1837, and Lepsius in 1846. However, there was no organized effort to explore the ruins surrounding it, until Cecil Firth and J. E. Quibell, who were later joined by the Jean-Philippe Lauer, began a systematic clearance of the entire complex, which revealed an unimaginable wealth of architectural remains. Their project began in 1924 and continued on for years thereafter. Lauer, an architect by training, was brought into the effort in 1927, to survey and document these remains. His careful attention to detail enabled him to partially reconstruct many of these remains, and those structures that are visible in the complex today are largely the result of his efforts. Indeed, Lauer spent almost his entire life engaged in this enterprise, a remarkable level of commitment and dedication, which has earned him wide-spread praise and recognition.[95]

Early explorers attributed the Step Pyramid to Pharaoh Netjerikhet, based on a number of stelae of the king found in the subterranean chambers and passages under it. This name was not widely attested elsewhere in the remains of ancient Egypt and was not associated with Pharaoh Zoser, 1st king of the 3rd Dynasty, until late in the Egyptian nineteenth dynasty. Aside from the Step Pyramid complex, little is known of Zoser or his life. Contemporary records are almost none existent.[96] Indeed, his memory, even today, would likely have been as shadowy, as that of most of the kings of this early age, were it not for the Step Pyramid. The Turin Canon records a reign of 19 years for Zoser.[97]However, there are some ruins bearing his name from

---

[95] Drioton, Etienne and Lauer, Jean-Philippe; *Sakkarah; The Monuments of Zoser* (Le Caire Imprimerie De L'Institut Francais D'Archeologie Orientale) 1939; p. 7.

[96] For further reading on Zoser's life, see: Wilkinson, Toby A. H.; *Early Dynastic Egypt* (Routledge, 11 New Fetters Lane, London, EC4P 4EE) 1999 pp 95-98.

[97] Gardiner, Sir Alan; *Egypt of the Pharaohs*; (Oxford University Press) 1961; page 433.

the great temple of Atum-Re, now almost totally destroyed, at Heliopolis, which is on the other side of the Nile and not too far north of Saqqara. I mention this because this temple was famous from time immemorial as a great center of learning. It was here that the 'heaven-watchers', the astronomers, of ancient Egypt were educated and trained. And it was in Heliopolis that the Egyptians believed that time was first measured and the calendar was created.

Credit for the design and construction of the Step Pyramid and its complex has traditionally been assigned to Imhotep, Zoser's vizier and architect and arguably the greatest of the legendary seers and wise men of ancient Egypt. Although Imhotep's knowledge and achievements were widely celebrated throughout the long history of ancient Egypt—he was even deified and worshiped as a god in later times—his actual existence was considered very doubtful by Egyptologists, until his name was found on the pedestal of a statue (the statue itself is missing) of Zoser found in the Step Pyramid complex.[98]

The Step Pyramid and its surrounding complex are located at: 29°—52.28' N, 31°—13' W. It is situated in the high desert to the west of the Nile River, and the farms as well as the nearby palm groves of the modern Egyptian village of Saqqara. The name of the village is believed to have derived from Sokar, an ancient Egypt god worshiped from the earliest of times. The area surrounding the complex is occupied by the extensive remains of one of ancient Egypt's largest cemeteries, which was in almost continuous use throughout Egypt's long history, down to and including Roman times. It is a wild and haunting place, even today with tourists all around; the silence of the desert is never far off. Memphis, one of the greatest of ancient Egypt's cities and its capital during several eras, was located nearby, but little of it remains today. [99]

Speaking of tourists, there are a number of graffiti left behind by some earlier ones—Egyptians from the time of Ramses the Great and earlier. This creates a daunting perspective on the great age of the Step Pyramid and its complex, as we pause to reflect that in their time it was already over two thousand years old and a place of legend and mystery. These people are long gone, but the Pyramid and its surrounding buildings, which are now approaching five thousand years of age, are still here. We are just mere 'tourists' passing briefly as the complex sails through time.

The Step Pyramid complex consists of the following primary features, as depicted on the following figure.[100] More specific detail will be provided later when they are examined in light of the specific information memorialized in them:

---

[98] Drioton and Lauer, p. 6.

[99] For further general reading on the Saqqara necropolis, see: Lauer, Jean-Philippe, *Saqqara; The Royal Cemetery of Memphis; Excavations and Discoveries Since 1850.* (Charles Scribner's Sons, New York, New York; Copyright: Thames & Hudson Ltd) 1976.

[100] 'Drawn by Philip Winton, © Thames & Hudson Ltd., London from *The Complete Pyramids* by Mark Lehner, Thames & Hudson.'

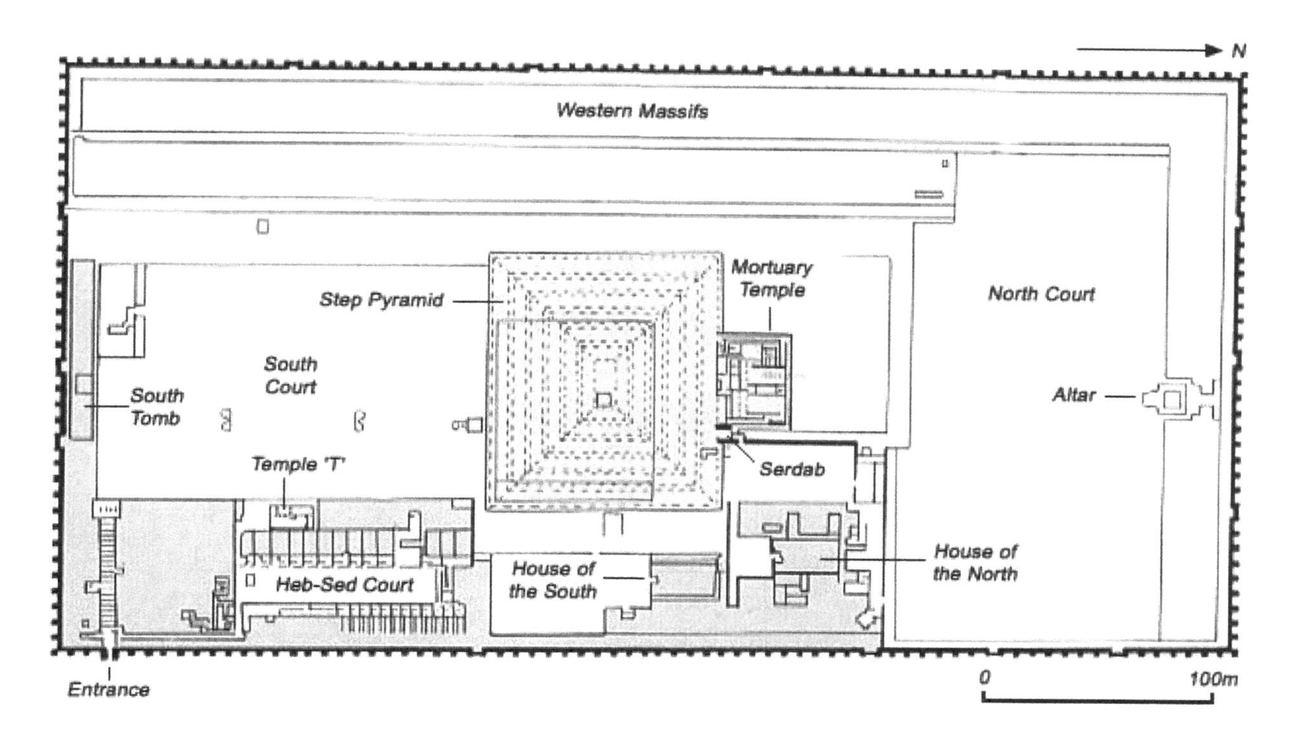

a. <u>The *Temenos* Wall</u>

The wall surrounding the complex is often called a *temenos* wall because it surrounds an area that is considered forbidden or sacred. The wall is approximately 544.8 m. (≈ 1040 cu.) in length and 276.85 m. (≈ 530 cu.) in width. It is deeply niched all along its outer face, and along its inner face as well, and has regularly spaced projections and insets on the outer face, which Lauer referred to as *redans* and *courtines*, respectively, terms that I will also use when referring to these structures. Lauer referred to the niches in the wall as *panneauxs* and I will use this term as well when referring to them. At each of the four corners there is a bastion, for a total of four. There are fourteen *faux portes*, or false doors, found at various locations along the wall, and one real doorway located at the southeast corner that provides the only access to the complex. The niches in the wall are all well shaped and evenly spaced, and, amazingly, all were carved into the stone after the wall was built, an enormous expenditure in manpower and one that defies easy explanation. The wall is composed of fill enclosed by roughly cut local stone blocks and faced with finely cut limestone blocks. Originally the wall stood approximately 10 m. (19 cu.) high. It was crowned with a rounded coping with eight regularly spaced recessed squares set below the crown all along the wall's length. The wall encloses some 15 hectares or 37 acres, the area of a small city of its time, and has an overall appearance of a fortification.[101] (See Appendix 3, Plates 1,34, 47, 48 and 54)

---

[101] Firth and Quibell, vol. I; pages 9-10 and 53-54. See also, Stecchini, Livio Catullo; Notes on the Relation of Ancient Measures to the Great Pyramid; appended in: Tompkins, Peter; *Secrets of the Great Pyramid* (Harper and Row; New York, NY) 1971; page 376; remarks on Lauer's figures for the dimensions of the wall.

Details of the *Temenos* Wall:

Number of *panneauxs*:

14 *faux portes* with 24 *panneauxs* in each, for a total of:	336
1 *porte* with 32 *panneauxs*, for a total of:	32
4 bastions with 20 *paneauxs* in each, for a total of:	80
192 *redans* with 15 *panneauxs* each, for a total of:	2,880
211 *courtines* with 5 *panneauxs* each, for a total of:	1,055
Total number of *panneauxs*:	4,383

Number of *redans* and *redan* equivalents*:

192 regular *redans*:	192
15 *portes* with 2 *redan* equivalents each:	30
4 bastions with 1 *redan* equivalent each:	4
Total number of *redans*[102] *and equivalents*:	226

Number of *courtines* and *courtine* equivalents*:

211 regular *courtines*:	211
15 *portes* with 1 equivalent each:	15
Total number of courtines and equivalentrs:	226

***Redan* and *courtine* 'equivalents' convert the entire wall into a continuous 'full/hollow' form, following its *redan-courtine-redan-courtine* pattern.**

b. The Entryway Colonnade

The only doorway to the complex is approximately 1 m. (1.9 cu.) in width and the door to it is made of stone, permanently set in the open position. Inside the doorway, there is a colonnade of parallel columns that has an overall length of 64.4 m (122.9 cu. ≈ 123 cu. or ≈ 216 Egyptian feet of 299.42mm each; the number 216 is a number from the pattern and the root number in the measure of the circumference of the Earth: 21,600 geographic miles). These columns appear to be fluted in the manner of Greek columns, but are in fact fasciculated or ribbed in a manner that is suggestive of bundled reeds. None of the columns are free standing; rather, they are engaged to the surrounding structure. The colonnade runs roughly from east to west. At its western end, it intersects with a transept or crossing structure that is composed of four pairs of similarly fasciculated columns oriented in a north-south direction. Each pair of the transept's columns is engaged structurally, but they are otherwise free standing. The colonnade is divided by a low gate, into two sections; the first has twelve pairs of columns and the second section, to the west of it has eight pairs, for a total of forty columns in the

[102] For reference purposes from this point forward, the term 'total *redans*' will be used to refer to the total count of all regular and equivalent *redans*. This same usage also applies to *courtines,* where the term 'total *courtine's* will be used.

main colonnade. The transept is also set off from the main colonnade by a low stone gate. A similar gate on the far side of the transept, leads into the south court of the complex.[103] (See Appendix 3, Plates 52 and 54.)

The details of the colonnade are as follows[104]:

i. Section I—columns in the eastern section of the main hallway, closest to the entrance. There are 12 pairs of columns in this section (24 total), each of which has 17 reeds, for a total of 408 reeds.

ii. Section II—columns in the western section of the main hallway. There are 8 pairs of columns in this section (16 total). There are four columns on the south side of the western-most part of the section that have 19 reeds each. All of the rest of the columns in this section have 17 reeds. There is a total of 280 reeds in this section: 4, 19-reed columns, for a total of 76 reeds, and 8, 17-reed columns, for a total of 204 reeds.

iii. Section III—the transept. There are 4 pairs of columns in this section (8 total), each of which has 19 reeds, for a total of 152 reeds. The transept is intimately connected to several Venus cycles, including the Venus inferior conjunction cycle and the Venus transit cycle.

c. The South Tomb

Just past the colonnade transept and to the left lies a structure that is embedded in the southern section of the *temenos* wall. Immediately in front of it lies a structure that is referred to as the south tomb chapel. The outer structure of the south tomb is usually described as a *mastaba*, an Arabic word meaning bench, which is commonly used to describe the rectangular tombs of Egypt's earliest pharaohs. While the use of *mastabas* was discontinued for pharonic burials after the 3rd dynasty, they remained in widespread use for many generations later for the burial of ancient Egypt's prominent citizens. The south tomb is approximately 85 m (162 cu.) in length. A deep pit, some 28 m. (53.5 cu.) in depth and some 7 m. (13.3 cu.)/side square, was dug under the south tomb. Access to the bottom of the pit is provided by a stairway accessed from the western side of the *mastaba* at the top of the *temenos* wall. The south tomb chapel is noteworthy for the decorative frieze of *uraei*, or rearing cobras, surrounding it.[105] (See Appendix 3, Plates 47 and 48)

[103] Cecil M. Firth, J.E. Quibel and J.P Lauer, *Excavations at Saqqara, The Step Pyramid,* (Service des Antiquitiés de L'Égypte, Le Caire Imperimerie de Institute Français D'Archaeological Orientale, 1935; republished by Martino Publishing, Mansfield Centre, Connecticut, USA, 2007); pages 13-15 and 64-67

[104] Ibid. Pages 13-15 and 65-67. For reed count, see, in particular, page 67. See also, vol. II, plate 54.

[105] Ibid. Pages 18-19 and 54-64.

d. The Step Pyramid

The Step Pyramid is clearly the dominant structure of the complex and provides its focal point. In simplest terms, it is a six-stepped, rectangular-shaped structure, 60 m. (115 cu.) in height, 125 m. (239 cu.) in length and 109 m. (208 cu.) in width[106]. The center of the Pyramid is slightly offset from the center of the complex, toward the southeast quarter. The lengthwise or longer axis of the Pyramid is oriented roughly east-to-west, some 4° 30' east of true north. Most of the fine limestone outer surface was stripped away centuries ago, so that all that is now visible is the inner packing stones of rough cut, local stone, set in limestone-clay mortar. The 6-stepped Pyramid encompasses several earlier structures, beginning with a square *mastaba* built above a great pit similar in dimensions to the pit under the south tomb. The square *mastaba* was extended all around and later extended again, but this time only toward the east. Subsequently, the *mastaba* was transformed into a four-stepped pyramid and then finally into the six-stepped pyramid that is visible today, by extending the four steps of the earlier pyramid to the west and north and by adding two additional steps. A great granite lined pit, 28 m. (53.5 cu.) I depth and 7 m. (13.3 cu.)/side square, identical to the pit under the south tomb, lies under the Pyramid.[107] Interestingly, the pit is not at the center of either the four-stepped or the six-stepped pyramid, both of which have different center points. However, it is believed to have been central to the earliest form of the *mastaba.*[108]

e. The *Serdab* and its Court

The Temple to the north of the Pyramid is largely ruined, with little more than the ground plan of its various components remaining. However, the *serdab*, Arabic for cellar, contains a statue of the king in a good state of preservation. The statue is completely enclosed in a box, which is equipped with two peep-holes, presumably placed so that the statue could look out. It rests against the slope of the Pyramid's bottom step, which gives the statue an incline of some 16°-17° above the horizontal.[109] Thus, the statue faces 4°30' east of north, which is the orientation of the Pyramid, and at an angle of elevation of 16°-17° above the horizon; these are important facts to the discussions of eclipses that will follow later in the book.

Lauer measured the dimensions of the *serdab* court as: 50 cubits X 117 cubits, or 87.5 X 204.75 Egyptian feet of 299.42mm each. The perimeter of the court in Egyptian feet is 584.5 ≈ 584—a significant number in astronomy, which will be discussed later.

---

[106] Lehner, Mark; *The Complete Pyramids; Solving the Ancient Mysteries*; (Thames and Hudson, London) 1997; page 85. Lehner cites a figure of 121 m. (229 cu.) for the Pyramid's length. The difference may be do to the fact that the Pyramid's west side sits atop the western galleries, and the location of its edge must be estimated.

[107] Firth and Quibell, vol. I; Pages 2-9.

[108] Ibid, vol. I page 7 and vol. II, plate 23.

[109] Bauval, Robert; *The Egypt Code*; (Disinformation Co. New York, NY) 2008; pages 29-30; Bauval provides a thorough discussion of this angle and of the measurements obtained by previous Egyptologists, as well as a detailed recounting of how he measured it himself.

f. The *Heb-Sed* Court

The *Heb-Sed* or *Sed*-festival is probably the most preeminent festival of kingship in ancient Egypt, yet even today little is known of it. Exactly what its purpose was, and when and how often it was celebrated are all matters that are vaguely understood and are matters of considerable debate within Egyptology. In the Late Period in ancient Egypt, it was commonly referred to as the 'thirty-year festival.' However, in practice there was no set interval to it, as some kings celebrated it more often than every thirty years, while others waited considerably longer or did not celebrate it at all during their reign, even kings who ruled for more than thirty years and would have been expected to celebrate it at least once.[110] The *Heb-Sed* Court in the Step Pyramid complex is situated on the east side and just above the entryway colonnade. It is one of the most intricate parts of the complex, but in broad terms is composed of a series of 'chapels', thirteen arrayed along the west side of the court and twelve along the east side. All are dummy structures with solid interiors, and although they have an outer structure and many are provided with an elaborate entryway, and in some cases vestibules, they do not have interior spaces or passageways. Several have niches which are believed to have held statues, though this is not known for certain. In the southern part of the court, there is a large stone dais or platform, with twin stairways on its eastern face that give access to the top. It is a low-profile structure. One of the stairways has four steps, while the other has five. This structure is the hieroglyph used to symbolize the *Sed*-festival, and it was for this reason that the court was named the *Heb-Sed* Court in modern times.[111] (See Appendix 3, Plate 67.)

g. Temple "T"

Off of the *Heb-Sed* court, but connected directly to it via a narrow passageway, is a small structure referred to as Temple "T." The floor dimensions of the Temple, as measured by Lauer, are: 38 cubits X 17 cubits, or in Egyptian feet of 299.42mm, 29.75 X 66.5 Egyptian feet. The perimeter I Egyptian feet is 192.5, a number which is reflected in the number of regular redans in the complex, 192—a number that is astronomically significant, which will be discussed later. The dimensions of the floorplan have further significance: 38/17 = 2.235, which is the square root of 5 and approximately the number of synodic months in the Saros cycle, which will be discussed in detail later. Furthermore, the diagonal of the floor plan creates a triangle with sides: 38 length, 17 width, and 41.6 hypotenuse. The interior angles of this triangle are 24.1° and 65.9°, which may represent, respectively, the approximate latitude of the tropic and the approximate latitude of the polar circle, which are important measure that define the orientation of the Earth. This factor is reinforced by the presence of the *djed* columns in the lintels of the doorways in the structure. The *djed* columns are believed to represent the Earth, with its two tropic lines and two polar circle lines, which will be discussed in greater detail later.

---

[110] Wilkinson; pages 212-215

[111] Firth and Quibell, vol. I pages 10-13 and 67-69; also see vol. II, plates 63 and 67.

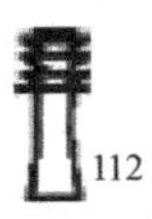[112]

h. The Tombs of the Royal Princesses, Intkaes and Hetep-hernebti

These two structures are found on the eastern side of the complex, immediately above the *Heb-Sed* Court. In outward appearance, the facades of the two structures are similar, but not identical. Like the chapels in the *Heb-Sed*, these two structures are largely dummies, with mostly solid-fill interiors, although each has a short passageway that leads to several small rooms or shrines inside. More recent Egyptologists refer to these two structures as the Chapel of the North and Chapel of the South, respectively. No evidence of a burial was found in either of them. Inkaes means, "She who brings her Ka" and Hetep-hernebti means, "The faces of the two goddesses are at peace." These two names, together with the name of the king, Netjerikhet, were found on a number of boundary stellae or markers found in the complex. However, on none of them are the two princesses referred to as daughters. (See Appendix 3, Plates 77 and 79.)

i. The Underground Galleries

To the west and north of the complex, there are extensive underground corridors with countless numbers of storerooms branching off of them. Many were empty, but many others contain the broken remains of vast numbers of stone pots, vases, and plates. The numbers of such vessels found under the step pyramid is upwards of 30,000; tens of thousands more are estimated to lie in these underground galleries. These galleries are the least explored and least understood areas of the Step Pyramid complex, and are very difficult and dangerous to access, because many of them are structurally unstable or have collapsed.[113]

---

[112] Wikipedia contributors, "List of Egyptian Hieroglyphs by English Names", Wikipedia the Free Encyclopedia, https://en.wikipedia.org/wiki/List_of_Egyptian_hieroglyphs_by_English_name/A%E2%80%93L Item under D. (Accessed May 16 2017)

[113] Cecil M. Firth, J.E. Quibel and J.P Lauer, *Excavations at Saqqara, The Step Pyramid,* (Service des Antiquitiés de L'Égypte, Le Caire Imperimerie de Institute Français D'Archaeological Orientale, 1935; republished by Martino Publishing, Mansfield Centre, Connecticut, USA, 2007); vol. I, pages 71-72.

CHAPTER 23

LUNAR ECLIPSE FACTORS MEMORIALIZED IN THE STEP PYRAMID COMPLEX

a. Factors from the *Temenos* Wall, (*panneauxs*)

Tropical Year: If the total number of *panneauxs* in the *temenos* wall (4,383) is divided by 12, the result is **365.25 days**, which is a fairly close approximation of the tropical year. (The modern figure is **365.242 days**, while **365.25 days** is the length of the year of the Julian Calendar; nevertheless, for the purposes of this book I will use the latter year of 365.25 days, when referring to the tropical year, unless otherwise noted.) This implies that the wall is designed to either: 1) track the tropical year for 12 years, or 2) track the tropical year for one year, with each 12 *panneauxs* representing one day. If it tracks one tropical year, then each of the 12 *panneauxs* in a day represents two hours, and each of the 8 insets atop each *panneaux* represents 15 minutes. Using the one-year format allows for the derivation of several additional important factors, as detailed in the following sections.

Venus Transit Period: There are 1,458 *panneauxs* along the eastern side of the *temenos* wall, which is equal to 6 Venus transit cycles of 243 years each. There are also 243 earth days in the time that it takes Venus to rotate through 360°.

Eclipse Year: The total number of *panneauxs* along the *temenos* wall adjacent to the length-wise axis of the south tomb is 223.5. If this figure is divided by the number of *panneauxs* in a single day (12), the result is 18.63 days. If this figure is subtracted from the tropical year (365.25), the result is **346.62 days**, which is the exact length of the eclipse year, to two decimal positions. There are, of course, two nodes, and the period of time that it takes for the Sun to travel from one to the other is the eclipse half-year of 173.31 days.

Saros Cycle in Synodic Months: If the length of the south tomb mastaba (162 cubits ≈ 161.8 cubits or *phi* · 100) is divided by the perimeter of the complex (3,140 cu.) and multiplied by the total number of *panneauxs* in the wall (4,383), the result is 226.13, which should be the number of *panneauxs* along the length-wise axis. However, this assumes that all of the *panneauxs* are of equal dimensions; they are not. From Lauer's several drawings, the *panneauxs* in the *courtines* are larger than those in the *redan*s and *faux portes*. Therefore, the number of *panneauxs* along the south tomb must be counted, and not calculated based on the measure of its length. As stated in the foregoing paragraph, the actual count is 223–223.5, which is equal to the number of synodic months in the Saros cycle.

Lunar Precession Period: If the eclipse year (346.62 days) is divided by 18.63, the result is 18.6, which is the length of the lunar precession cycle in tropical years. This is the length of time that it takes for the nodes to regress 360° or one complete circle along the celestial sphere. Thus, the location of the nodes can be tracked on the wall, by starting out at one end of the south tomb and ending when the opposite end is reached, 346.62 days later, from which point the next eclipse year would begin and then end 346.62 days later, and so on, until the node returns to its original starting point on the south tomb, some 18.6 tropical years later.

b. Factors from the *Temenos* Wall (*redans/courtines*):

Synodic Month: The familiar and widely followed, full-hollow (30 day-29 day) scheme used in pure lunar calendars for measuring and tracking the length of the synodic month (29.53 days as measured from full-moon-to-full-moon, or new-moon-to-new-moon) and for tracking the lunar year (354.53 days) is readily accomplished with the wall's total *redans-courtines*. This calendar is readily corrected, using visual observations of the moon. Accordingly, there are 226 full lunar months and 226 hollow lunar months represented on the wall. However, if the half lunar-month, or fortnight of 14 days, is used instead, where each two *redans* represents a full month and each two *courtines* represents a hollow month, then there are 226 lunar or synodic months represented. This latter use was most probably the builder's intent, as the fortnight is an important measure for short term eclipse predictions. However, the number 226 would seem to be an odd choice for the designers, when the near adjacent number for the length of the important Saros cycle of 223 is of far more importance for eclipse forecasting and tracking. In fact, the wall was designed with the Saros cycle of 223 lunar months in mind and the extra 3 'units' are not superfluous, but extremely useful for tracking the life of a Saros series with the wall. All of this, though, anticipates later discussions on these points, so I will defer further remarks on this subject for the time being.

Venus Orbital Period: The 192 *redans* in the *temenos* wall, appear to represent the number of Venus days—sunrise-to-sunrise, measured as 116.8 Earth days—for the planet to orbit the Sun: 1.92 Venus days X 116.8 = 224.256 days ≈ 225 Earth days for Venus to complete one orbit, which is the length of the Venus year.

Jupiter and Saturn: The twelve-year measure of the *temenos* wall also lends itself to tracking the sidereal period of Jupiter (11.86 years) and Saturn (29.64 years), but there is no evidence of this.

c. Factors from the Step Pyramid *Serdab*:

Eclipse Season: The statue of Zoser that is in the *serdab* on the north face of the Pyramid is elevated 16°-17° above the horizon and faces 4°30' east of north, as does the Pyramid. These two figures are the controlling geometric limits for the eclipse season for lunar eclipses. The first figure is the maximum angular distance of the Sun from a node at which a lunar eclipse is possible (The actual figure varies because of the eccentricity of the Earth's orbit as well as the Moon's orbit, and range between: 15.3°–17.1°). The second is the maximum angular distance of the Sun from a node at which a total lunar eclipse is possible, currently this angular measure is approximately 4°35'. These angles vary depending on the Moon's distance from the earth at the onset of the eclipse season. As a practical matter, since the Sun's apparent angular velocity is 0.99°/day, the eclipse season typically begins and ends some 17 degrees to either side of the node, or 34 degrees in total. This is the 'eclipse season' and generally there are several solar eclipses and one or more lunar eclipses each season. There are two eclipse seasons each year, one at each node.[114] On the wall, the onset of the eclipse season would be marked some 204 *panneauxs* (17 (days) X 12 (*panneauxs*/day) before and after the end point for the eclipse year.

It bears remarking that the statue does face to the north and, because of the Pyramid's geographic location, it views an area of the sky where the Moon never appears. Hence, it might be expected to more appropriately be positioned on the Pyramid's southern face, so as to view an area of the sky where the Moon is visible. However, if the statue were positioned on the southern face, its designers' intent may not have been properly understood, as it would suggest that the statue might mark something in the southern heavens at an angle of 184°30', instead of memorializing an angle of 4°30'. Similarly, if the Pyramid's orientation were 17° east of north and the statue inclined at an angle of 4°30', instead of the other way around, the inclination of the statue may have gone unnoticed entirely or unremarked.

d. Factors from the *Heb-Sed* Court and the Tombs of the Two Princesses:

Draconic Month. Tracking and timing the location of the Moon in relation to the ecliptic is a critical measure for eclipse forecasting. As Jean Meeus observed, "...the place of the eclipse depends mainly on the Sun, while the instant of opposition depends principally on the Moon."[115] The length of the draconic month, measured from successive appearances of the Moon at the ascending node, is 27.2 days. The length of time that it takes for the Moon to move between the ascending and descending nodes is 13.6 days. It is the half-draconic month that is tracked in the *Heb-Sed* court. Each of the thirteen chapels along the western side of the *Heb-Sed* court counts one day of the half-draconic month. The final 6/10th of a day is measured

[114] Espenak, Fred; *Periodicity of Lunar Eclipses*; 'Eclipse Seasons', at NASA Eclipse Web Site: http://eclipse.gsfc.nasa.gov/LESaros/LEperiodicity.html Also, see definition of 'ecliptic limit', found in the International Astronomical Union's 'Small Eclipse Dictionary', at: http://www.eclipse-chasers.com/iaueclipsedictionary

[115] Meeus, Jean; Mathematical Astronomy Morsels III; (William-Bell, Richmond, VA.) 2004; page 136.

at the appropriate Princess Tomb, with the northern tomb representing the ascending node and the south tomb representing the descending node. The façade of each tomb has five sections, marked by engaged columns, with an offset doorway located in the third section, counting from the right. Each section represents $2/10^{th}$ of a day or 4.8 hours, with the middle section representing an elapsed time of $6/10^{th}$, or 14.4 hours total. The doorway marks the arrival of the Moon at the node. The count for the Moon's transition to the next node begins again in the *Heb-Sed* court chapels, as before. These same western chapels also figure in the Saros series cycle, as defined in draconic months, but, I am anticipating myself.

CHAPTER 24

LUNAR ECLIPSE PERIODICITY AND THE SAROS CYCLE

The most useful and reliable eclipse cycle for predicting lunar eclipses and forecasting their form and duration is the Saros cycle. The Saros arises from the beat period of synodic and draconic months that result in lunar eclipses repeating in a regular and predictable fashion. For the synodic period the number is 223 synodic months. For the draconic period the number is 242 draconic months. A third period is also involved: the anomalistic month of 27.55 days. The anomalistic month tracks the Moon's location along its elliptical orbit, and this period is important for determining the relative size of the Moon's orb in relation to the orbs of the Earth and Sun. If the Moon is at apogee, it appears to be smaller than it is at perigee, and it is important for calculating the nature and duration of an occultation. Ideally all three components should be whole integers or nearly so. To summarize, an eclipse of a Saros series occurs after the following time periods have elapsed from a prior eclipse in that series:

223 synodic months of 29.53 days
242 draconic months of 27.2 days
239 anomalistic months of 27.55 days

Also, there are approximately 6,585 days between successive eclipses in a Saros series, which is equal to 18 years, 11 days, and 8 hours. Each eclipse member of a Saros series would repeat near the same geographic location, except for the fact of the 8-hour difference, which means that the next eclipse member in the series will take place some 120° west of the previous member. This means that every third eclipse member of a series will return to approximately the same geographic location. This is a related eclipse cycle, called the Exeligmos, or 'turn of the wheel.' The Exeligmos cycle is 54 years and 34 days.

a. Saros Period in Synodic Months and the Temenos Wall.

As discussed in the previous chapter, there are 226 total *redans* in the wall, which equates to 223 synodic months, plus three additional 'units' that are used for tracking the progress and life of a Saros series. By way of background, the various members of a Saros series, starting with their first appearance, move progressively eastward across the 34 days, or 34°, of its eclipse season until it passes beyond this season and the series ends or dies. The reason for this is the slow, eastward shift of the lunar nodes. The modern figure for this factor is 0.48°, meaning that each Saros series will move progressively eastward by 0.48° (1.44° between successive appearances of a Saros series at the same geographic location—the Exeligmos cycle). Using the 3 'units' from the *temenos* wall, the figure for this factor as used by the ancient Egyptians is derived as follows: 3/223 X 34° = 0.46°, which compares very favorably to the modern figure. Accordingly, if the cycle of a Saros series is tracked with the total *redans* in the wall and ends after 223 total *redans* (synodic months), and the next cycle is begun at the end point of the previous cycle, the series will return to the starting point on the wall some 74 Saros cycles later, which is very close to the modern figure for the average life of a Saros series: 73 cycles or members. The life of the Saros series, using the *temenos* wall to track its progress, will end at that point.[116] That it was the intent of the complex's designer for the *temenos* wall to serve this dual purpose seems evident if the count is begun from the doorway to the complex and then proceeds north and around the wall from that point, counting synodic months as described earlier. The 223 synodic-month period of the Saros cycle would then terminate on the southeast bastion of the complex, ending exactly 3 total *redan* 'units' short of the doorway.

b. Saros Period in Days and the *Temenos* Wall.

The eastern edge of the south tomb chapel, with its uraeus frieze, marks the east-west centerline of the complex. The western edge of the south tomb's shaft is approximately 11 *panneauxs* from this centerline. There are 4,383 *panneauxs* in the wall, which as previously noted represents 12 tropical years of 365.25 days. Counting 18 years takes 1.5 circuits of the wall or 1.5 X 4,383 = 6,574.5 *panneauxs*. If this count is started on the wall at a point adjacent to the western edge of the shaft of the south tomb, and ends on the complex's centerline immediately after the 18-year count is satisfied, the total number of *panneauxs* would be 6,585.5, which is very close to the modern estimate for the Saros period in days: 6,585.3 days.[117] Starting this count at the western wall of the south tomb seems somewhat arbitrary, but as will become evident later on, this wall has a very profound significance and is a most appropriate place to begin the count.

Interestingly, the Saros Period can also be tracked on the *temenos* wall, starting at the wall immediately adjacent to the west wall of the shaft of the south tomb and ending directly at the wall adjacent to the eastern side of the south tomb chapel. If 243 (the measure of the Venus transit cycle period) is substituted for the tropical year of 365.25 days, then there are exactly 18.03 tropical years from the measurements starting point at the shaft of the south tomb to the ending point at the south tomb chapel. (4,383/243 = 18.037, which is almost exactly equal to the

[116] Espenak, Fred; *Periodicity of Lunar Eclipses*; NASA Eclipse web site, at: http://eclipse.gsfc.nasa.gov/LESaros/LEperiodicity.html; See in particular, Section 1.4 Saros.

[117] Ibid.

Saros period of 18.03 days. The difference between 18.037 – 18 = .037, which when multiplied by 243 = 9 days, vs. the expected 11.33 days in the actual measure of the Saros period.)

c. Saros Period in Years and the *Temenos* Wall.

The western side of the *temenos* wall contains 1,458 *panneauxs,* which when divided by 6 is equal to 243, which, as we have seen, is a number closely associated with Venus. There are 243 earth days in the Venus sidereal period and 243 years in the Venus transit cycle. If the total number of *panneauxs* in the temenos wall, 4,383, is divided by 243, the result is 18.037, which is almost exactly equal to the Saros period of 18.031 tropical years (18 years, 11 days, and 8 hours). If the number 18.037 is multiplied by 3, the result is 54.111, which is almost exactly equal to the Exeligmos period of 54.1 tropical years, which is the period for a particular Saros series to return to approximately the same geographic area.

The use of the number 243 may seem to be arbitrary, but it is not. The number 243 is approximately 2/3rd of 365.25 (243.5 actual), which is the length of the tropical year and the basic measurement factor of the *temenos* wall ($12 \cdot 365.25 = 4{,}383$). As stated earlier, the number is also closely associated with the planet Venus, and Venus has a close connection with the Moon through the Octaeteris cycle, where Venus and the crescent-phase Moon come into proximity with one-another once every 584 days. (This is also the Venus inferior conjunction cycle with the Earth, which was discussed earlier.) 7.5 X 584.6 = 4,384.5 ≈ 4,383, which is the number of *panneauxs* in the temenos wall. This is not a coincidence. I will have more to say about the Moon's association with Venus later.

d. Saros Period in Synodic and Draconic Months; the *Heb-Sed* Court and South Tomb.

From Chapter 3, it was noted under the section for the eclipse year, that there were 223.5 *panneauxs* in the *temenos* wall adjacent to the south tomb, and that this number divided by the 12 *panneauxs* in a day was equal to 18.63 days, or the difference between the tropical year (365.25 days) and the eclipse year (346.62 days). Now this figure of 18.63 relates the synodic month period of the Saros cycle to the draconic month period in a way that was apparently known to the ancient Egyptians, based on evidence found in the *Heb-Sed* court. Recall from the description in the Introduction that there are 12 chapels along the eastern side of the *Heb-Sed* court, adjacent to the *temenos* wall, and 13 along the western side. Now 12 X 18.63 = 223.5, as we've already seen and this number is a very close approximation of the period in synodic months—223, But 13 X 18.63 = 242.19, a number that is a very close approximation of the length of the Saros cycle period in draconic months—242. This relationship arises from the fact that the lunar year of 12 synodic months (354.36 days) is close in length to 13 draconic months (353.6 days). Thus, the period of the Saros cycle in synodic months (223) relates to the cycle's period in draconic months (242), as 12 and 13, when multiplied by the common factor of 18.63 (the difference between the tropical and eclipse years). Using the *panneauxs* adjacent to the south tomb, both the synodic and draconic month periods can easily be tracked and related to one another, as 12 to 13, through the common factor of 18.63. Similarly, close approximations can be had by using the number 18.6 as the common factor, which yields 223.2 synodic months and 241.8 draconic months, respectively. The number 18.6 is the length of the lunar precession cycle in years, also known as the lunar node regression cycle.

e. Precession of the Lunar Line of Apsides and the *Temenos* Wall.

I could not find any evidence in the Step Pyramid complex that the ancient Egyptians used the anomalistic month (27.5 days) for tracking the Saros cycle. The anomalistic month measures how long it takes for the Moon to return to the same point along its elliptical orbit, and is usually measured between successive appearances of the Moon at its point of perigee or apogee. At perigee, the orb of the Moon is larger in appearance than it is elsewhere along its orbit, while it is smallest at apogee. For the Saros cycle, the Moon should be of the same apparent size at each lunar eclipse, which allows for comparisons of the magnitude and duration of the occultation between a Saros series' various members and permits reliable forecasting of future eclipse members in the series. This is the reason why the period between Saros cycle eclipses should be integers of anomalistic months or close to it.

While the ancient Egyptians didn't use the anomalistic month, there is evidence that they employed a related measure as a rough proxy for this purpose and that is the gradual precession of the lunar line of apsides of the Moon's elliptical orbit. The lunar line of apsides moves in the same sense as the Moon and it takes 2,190 days, or almost exactly 6 years, for it to return to its same position with respect to the lunar nodes. This means that every 6 years the Moon will be in the same location along its orbit in relation to the lunar line of apsides as it nears a particular node and the lunar orb will have roughly the same apparent size as it did 6 years earlier. On the *temenos* wall, there are 4,383/2 or 2,191 *panneauxs* for each 6-year period, which is within 1 day of the 6-year rotational period of the lunar line of apsides. More importantly, though, after 18 years—roughly the period of the Saros cycle—the lunar line of apsides returns to its same position with respect to the lunar nodes, thus the lunar orb at the time of eclipse is roughly the same size in appearance as it was at the previous eclipse in the series.[118] However, the period for the return of the lunar line of apsides in relation to the Saros cycle is 3 X 2,190 = 6,570 days, whereas the period for the Saros cycle in whole integers of anomalistic month is 242, which when multiplied by the number of days in an anomalistic months (27.5 days) is equal to 6,585 days, a 15 day difference. The Saros cycle is 18 years and 11 days 8 hours long, so 11 days should be added, which would more closely synchronize the lunar line of apsides figure to the Saros cycle, yielding a figure of 6,581—still short by a little over 4 days, but a negligible factor, as there would be little discernible change in the appearance of the lunar orb from a Saros

[118] Meeus, Jean; *Mathematical Astronomy Morsels III*; page 18.

series member eclipse when compared to those immediately succeeding it, although their characteristics do change throughout the series. In any case, the appearance of the lunar orb, while significant, is not as essential for eclipse forecasting as are the measures of the period of the Saros cycle in synodic and draconic months.

CHAPTER 25

THE SAROS CYCLE DESCRIBED; LIFE AND DEATH OF A COBRA

To the Chinese, they were dragons but to the ancient Egyptians the Saros series were cobras. Pick your favorite metaphor, but the cobra seems particularly apt, because of the way that the cobra rears up and flares its hood as it prepares to strike.

This is one of those Chapters that in some respects should have appeared earlier in the book, but the significance of several important points would not have been as apparent earlier. However, as I remarked earlier in the Prologue, there is an unavoidable tension between the orderly presentation of the material memorialized in the Step Pyramid complex with relating it to the science of lunar eclipses in a logical and progressive fashion. Also, a brief review of some fundamental concepts of the Saros is presented here so as to set the stage for a further analysis of the Step pyramid complex in regard to these concepts.

Lunar Saros series eclipses begin their life near one of Earth's poles and then move progressively toward the opposite pole over the course of their life, where they die off some thirteen centuries later. The reason the series does not last indefinitely is that as it moves toward the opposite pole it enters and then moves progressively across the 34° window of the eclipse season until it reaches the opposite end. As we've seen earlier each member eclipse of a Saros series moves across the window of the eclipse season at a rate of 0.48° (recall that the rate used by the ancient Egyptians was 0.46°) for each eclipse member in a series. This movement is always eastward. Each member of a Saros series also moves progressively north or south at the rate of approximately 300 km (180 international miles) for each eclipse member in the series. Saros series lunar eclipses at the ascending node originate in the Antarctic polar region and then move progressively north until they die off in the Arctic polar region. Lunar eclipses at the descending node move in the opposite direction. They originate in the Arctic, then move progressively toward the south, and die off in the Antarctic polar region. The following describes the life and death of a typical Saros series.

The series begins at the node in the Antarctic polar region with a small penumbral eclipse on the Moons lower limb. The next 9 or so eclipse members in the series will move ever deeper into the penumbra of Earth's shadow, but all of them will be almost imperceptible to the naked eye. Then the Moon will enter the umbra of Earth's shadow and the first perceptible partial eclipse will occur on the Moon's lower limb. For the next several centuries each eclipse member in the series will move progressively deeper into the umbra, until, after some 20 partial eclipses, a total eclipse will occur of short duration. Thereafter, each eclipse member will be a full eclipse that moves progressively deeper into the umbra, until the series nears the ecliptic, where it will produce the deepest eclipses of the longest duration in the series. There will be around 12-13 total eclipses in the series. As the series continues its northward movement, the series will begin producing partial eclipses again as the Moon gradually moves out of the umbra of Earth's shadow. Progressively decreasing partial eclipses will occur on the Moon's upper limb with each eclipse member, until the Moon transits entirely into the shadow of the penumbra again. After another group of penumbra eclipses near its upper limb, the Moon will pass entirely beyond the penumbral shadow and the series will die off in the Arctic polar region, some 13 centuries after it began. In all, the series will have produced some 73 eclipses: 20 penumbral eclipses, 40 partial eclipses, and 13 total eclipses. Recall, though, that each member eclipse in the series is displaced some 120° to west of its predecessor, because the period of the Saros is not a whole number of days, but includes an 8-hour difference. Thus, only a third of the member eclipses of Saros series are visible at one of the three particular areas of the Earth where the series manifests itself. The other 2/3rds of its members are equally distributed at areas located 120° and 240° to the west of it, respectively. This, of course, is the Exeligmos cycle, which is three Saros periods long.[119]

a. Eclipse Forecasting and the Step Pyramid.

The Step Pyramid has long been regarded as a crude effort at pyramid building by the ancient Egyptians. It is anything but that, as it is a complex design that incorporates physical aspects of both the Earth and the Moon, as well as marking the location of the Sun. However, for the moment only those aspects of its design that relate to the Earth are pertinent; the others will be presented and discussed later on.

Almost all descriptions of the Pyramid describe it as a six-step pyramid, as I did in introducing the structure earlier. This description is not incorrect, but it overlooks a structural detail about the structure that is readily apparent on closer review. That is, the pyramid's rise isn't stepped at all, but consists of regularly interspersed breaks between sharp angles of ascent and softer ones. The sharp angles are 17° from the vertical, while the soft ones are some 20°-25° from the horizontal. However, even this is overly simplistic as is demonstrated by the following quote summarizing Lauer's incredulity over what he found when examining the remains of the stones from the Pyramid's outer casing:

[119] Espenak, Fred; *Eclipses and the Saros* and *Periodicity of Lunar Eclipses*; NASA Eclipse Web Site; http://eclipse.gsfc.nasa.gov/SESaros/SESaros.html; and, http://eclipse.gsfc.nasa.gov/LESaros/LEperiodicity.html; See in particular, Section 1.4 of the second reference, "Saros." See also, Meeus, Jean; Morsels III, Chapter 18 'About Saros and Inex Series.'

> "...Lauer noticed several blocks, each with two weathered sides forming an angle varying from 127° to 132° 30', and recognized that they must have come from the edge of the pyramid steps. The slope of the upright faces being 17° from the vertical, an angle of 20° to 25° is left for the inclination of the steps to the horizontal. Later he saw reason to suspect that the variation of 5° 30' in this angle was not due to any lack of care in workmanship, but that the angle changed from step to step, increasing upwards!"[120]

Lauer's scaled drawing of the Pyramid's elevation clearly shows that the first three 'steps' of the Pyramid are similar in their vertical rise, but the final three noticeably and progressively diminish, which accounts for the change in the angle of the casing stones that Lauer found. The conclusion seems inescapable; this was all by design and the decreasing angles toward the top of the Pyramid mimic a similar change that occurs in a sphere toward its top.

In considering the significance of the eclipse material that we have found memorialized in the complex so far, the necessity at this point for a three-dimensional scale model of the Earth for further analysis and interpretation is rather obvious. The Pyramid provides this in the form of a scale model of Earth's northern hemisphere, which is divided by the angular breaks in the its vertical rise into 12 distinct steps, which diminish in their rise toward the top. There's more.

The 4-step pyramid that is embedded under the 6-step structure may not have been the remains of an earlier structure, as has long been assumed, but an intentional design feature. Recall from our introduction description, that the 6-step pyramid was built atop the 4-step structure, by first extending its four steps toward the north and west, and then adding two additional steps to finish the 6-step pyramid. This preserves the 4-step structure's southern and eastern faces, and incorporates them into the 6-step structure's corresponding faces. This creates an interesting and extremely useful tool for three-dimensional modeling purposes. Using the eastern aspect of the Pyramid, as shown in the following drawing, the angle between the Pyramid's center and the outer edge of the top of the 4-step pyramid is approximately 24°, which is the approximate angle of obliquity of Earth's rotational axis from the plane of the ecliptic.[121] In simpler terms, of course, this is the tilt angle of the Earth, which produces our seasons. This enabled the ancient astronomers to project the path of an upcoming eclipse onto the model and to forecast both its range of visibility and the location of its 'greatest eclipse.'[122]

If the edge of the top of the 4-step pyramid is assumed to be the north pole of the equator or the geographic north pole, then the base of the pyramid would be the ecliptic, the path of the Sun. The Tropic of Cancer would then be at the approximate level of the edge of the 2nd step

---

[120] Firth and Quibell; vol. I, page 50.

[121] The obliquity of Earth's rotational axis is not constant and is believed to vary from approximately 22°45' to 24°30' over very long periods of time.

[122] Espenak, Fred and Meeus, Jean; NASA/Technical Publication-2009-214172; "Five Millennium Canon of Lunar Eclipses; Page 4, Section 1.2.4.

of the 6-step pyramid, on the northern face. The ecliptic is of critical importance, because the Sun travels along this path and the lunar nodes intersect it, which is where the Earth-Moon-Sun conjunction takes place. The mid-point of the base of the Pyramid's eastern face is the location of the 'first point of Aries', an archaic term for the moment of the vernal equinox, when the Sun stands on the equator on the observer's meridian. This point it the beginning for all measures of time and for all calendars that are based on astronomical ephemerides, and affixing its position is essential for accuracy in eclipse forecasting.

As noted earlier in the introduction of the Pyramid, the dimension for its width (shorter side) is 109m (208.19 cu.). This is the dimension for the base of the Pyramid's eastern and western faces. The cubit used here as elsewhere is known as the 'royal cubit', which is based on the addition of a seventh 'hand to the Egyptian cubit of six hands, which is 1.5 times the Egyptian foot. 208.19 royal cubits X 7 hands/r.cu. = 1,457.33 hands, which when divided by 4 hands/foot = 364.33 Egyptian feet.[123] Thus, the width of the Pyramid is 208.19 r. cu. or 364.33 e. ft. The perimeter of a square, whose side is equal to the Pyramid's shorter side, is 364.33 X 4 = 1,457.33 e. ft. And if this figure is divided by 4 again, the result is 364.33, a close approximation of the tropical year of 365.25 days. Accordingly, each four Egyptian feet in the square base of the Pyramid is equal to approximately 1 day. That this was the intent of the Pyramid's designers seems obvious, but the numbers do not match up with sufficient accuracy to state so with certainty. Lauer's numbers are probably sound, but there are two factors that need to be taken into consideration when using them: 1) The Pyramid is heavily ruined, making exact survey's somewhat problematic, and 2) Lauer does not state whether or not he rounded his figures or that they may have been subject to some degree of error. In any case, the figures derived for the base of the Pyramid and their relationship to the length of the year seems too close to be mere coincidence.

At this point, the ancient astronomers could mark the date of the Sun's conjunction with the two lunar nodes, based on the time record for the eclipse year from the *temenos* wall. Once the two dates, one for each node were marked, then the eclipse season of 34 days, 17 days to either side, were marked off. After this, the corrections for the movement of the location of the point of 'greatest eclipse' were marked. These were based on the records from prior eclipse member in the Saros series for this location, and include the 0.46° easterly drift per Saros series member, or 1.38° for each member at this location (the Exeligmos cycle). Next a correction for the Series change in Latitude is entered. This is where the significance of the Pyramid's 12 steps in its vertical rise come into play.

The point of greatest eclipse of each Saros series member moves north (ascending node) or south (descending node) by some 300 km (180 international miles, or 156.75 nautical miles of 6,000 feet of 308mm each) from the previous member. For an Exeligmos cycle, this factor is 3 X 180 international miles = 540 international miles, or 470 nautical miles or roughly 450

[123] The Egyptian foot of the Step Pyramid is calculated at 299.42 mm, and is composed of 4 hands of 4 digits each.

nautical miles. If the pole–to–pole circumference of the earth is taken as 23,000 international miles, or 21,600 nautical miles, then the quadrant measure for the northern hemisphere would be 1/4th of 21,600 or 5,400 nautical miles, which when divided by 12 is approximately 450 nautical miles. Thus, each of the twelve steps of the Pyramid represents a north/south correction factor of 450 nautical miles from the location of the maximum point of occultation of the previous 3 Saros members of the series, or the Exeligmos cycle, in the northern hemisphere. (This assumes that there are some 24 Exeligmos cycle members from a 'typical' Saros series, with approximately 73 members.) But where are the records for the previous Exeligmos cycle members? The answer is with the cobras—literally. But before proceeding, there is one other eclipse cycle that is memorialized in the Step Pyramid.

b. Tetrads

There is another purpose served by the gradual increase in the angle of the Step Pyramid's steps from 127°-132.5°, and this is to memorialize the geometry and time period of the tetrad eclipse cycle. The tetrad is an eclipse cycle where four successive total lunar eclipses occur after intervals of six lunations (six synodic lunar cycles) each. Tetrads occur when the angular distance of the Sun from the node opposite the node of a full Moon is approximately ± 4.8°. (There are other factors that come into play in the occurrence of tetrads, but this geometry is a threshold requirement for them to occur.) Tetrads are fairly common during an approximate 300-year period, followed by a period of some 300-years where there are none. This approximate six-hundred cycle is a recognized eclipse cycle.[124]

The angle of the rise of the Step Pyramid's steps is 17° from vertical, while the angle of the run or step is 17° from the horizontal plus an additional 3°at the lowest step increasing to 8.5°at the highest step. The angle of each step is thus 127° (17° + 90° + (17° + 3°)) to 132.5° (17° + 90° + (17° + 8.5°)). The range of the additional angle is thus 3°-8.5°. If the additional angle at the first step is 3°, then there are 5.5° remaining for the upper five steps, or an incremental increase estimated of approximately 1.1°/step. If these incremental increases are accurate, then the additional angle at each step would be:

1st step: 3°
2nd step: 4.1°
3rd step: 5.2°
4th step: 6.3°
5th step: 7.4°
6th step: 8.5°

The 2nd and 3rd steps have additional angle values that are close to the 4.8° value. However, there are significant additional calculations that must be made in order to forecast tetrads,

[124] Jean Meeus, *Mathematical Astronomy Morsels III*, (Willmann-Bell, Inc., Richmond, VA, USA, 2004), Chapter 21, pages 123-140.

such as the influence of the Earth's elliptical orbit. Therefore, all that can be said on this point, without going into a great deal of additional detail, is that some of the additional angle values of the Step Pyramid's steps are roughly consonant with the Sun's angle at lunar eclipse necessary for tetrads to occur. Also, the number of steps in the Step Pyramid, six, is reflective of the number of centuries in a tetrad cycle, which is approximately six hundred years.

c. Records of Eclipses.

Recall the description of the temple of the south tomb, which was described in the introductions, wherein the decorative frieze of cobras was mentioned. Notice from the drawing that there is a cobra (uraeus) above each 'full' and 'hollow' section in the wall, and that there are 24 courses of blocks below each of them, with 18 or 19 in the hollow sections. How many of these cobras were in the frieze, is impossible to say, but from Lauer's drawings there may have been over a hundred—fifty 'full' and fifty 'hollow.' I believe that this is where the records for the Saros series were kept. It is only informed conjecture, but the numbers work. There are 24 Exeligmos members in a typical Saros series—24 blocks below the cobras, and approximately 18-19 members of the series are umbral eclipses, hence visible—18-19 blocks in the hollow sections. The only other requirement is to distinguish between Saros series at the ascending node and those at the descending node. Equating the 'full' sections to those at the ascending node and the 'hollow' sections to those at the descending node seems more than appropriate to this purpose. This is where the records for the correction factors would have come from.

There is one more critical factor to address in predicting eclipses and that is 'gamma' or the nature and extent of the shadow of the eclipse. This is the subject of the next chapter.

CHAPTER 26

GAMMA AND THE FLARE OF THE COBRA'S HOOD

The measure of the shadow cast on the Moon by the Earth during a lunar eclipse is called 'gamma' and in astronomical texts is simply referred to by the Greek letter, γ. Gamma is measured from the central point of the axis of Earth's shadow cone to the center of the lunar orb, in terms of radii of Earth's umbral shadow, at the point of greatest eclipse. The shadow itself is composed of two components, the umbra or central shadow, and the penumbra or outer shadow that surrounds the umbra. There is no definite boundary between the umbra and penumbra. The penumbra results from the fact that not all of the light from the Sun is blocked by the Earth during a lunar eclipse, thus creating a zone with a more diffuse shadow than that present in the umbra.

The mean diameter of the Earth's umbral shadow at the mean lunar distance is 1°23'; its radius is 42'. The Moon's apparent diameter at mean lunar distance is 31'30"; its radius is 15'45". The Sun's apparent diameter is 32'; its radius is 16". The diameter of the umbral shadow is approximately equal to 2.66 X the Moon's apparent diameter. The mean width of the penumbral shadow is approximately equal to the apparent diameter of the Moon, or 31'30".

For shadows north of the axis of the shadow cone, gamma is a positive number; for shadows south of the axis of the shadow cone, gamma is negative. If the point of greatest eclipse is the center point of the axis of the shadow cone, gamma is zero; e.g. for an eclipse with a gamma factor of 1.16, the point of greatest eclipse is 1.16 earth radii north of the center point of the axis of the shadow cone, and the eclipse will be a partial one.[125]

a. Gamma and the Step Pyramid Complex

The Diameter of the Lunar Orb and the Step Pyramid. As mentioned in the previous chapter, the Step Pyramid is a complex structure that represents at various points, depending on the purpose, the Sun, the Earth, and the Moon. For the purpose of measuring gamma, it represents

[125] Espenak and Meeus; *Five Millennium Canon of Lunar Eclipses*; Section 1, Figures and Predictions.

the Moon, with an apparent diameter equal to the Pyramid's shorter side, or 109 m. (208 cu.). (The Pyramid's longer side is also relevant to the Moon, as it depicts the Moon's extremes of latitudinal and longitudinal libration, which will be discussed later.) The Pyramid's circular footprint, as described, is relevant to what follows.

The Diameter of the Shadow of the Earth and the *Temenos* Wall. Recall from the introductions that the dimensions of the *temenos* wall are: length: 1,040 cu., and width: 530 cu. Thus, the perimeter is 3,140 cu., which is roughly equal to the circumference of a circle with a diameter of 1,000 cu. If this circle is superimposed on the complex such that its center point is the center point of the complex, and another circle is then marked of from the center point of the complex with a diameter equal to the width of the complex (530 cu.), an interesting and very useful image is created. And if the Pyramid's circular footprint, with its diameter of 208 cu., is sketched in, all the necessary elements for measuring gamma are present.

The inner circle, with diameter equal to the width of the complex or 530 cu., is the shadow of the umbra and the center point of the cone of the shadow axis is the center of this circle. The outer circle is then the limit of the penumbra. And the Moon's orb represents the footprint of the Pyramid's circle, with a diameter of 208 cu. A comparison of these numbers with the current ones used for the measurements of the umbra and penumbral shadows follows:

i. The Diameter of the Umbral Shadow: 208 (lunar diameter from Pyramid footprint) x 2.66 (current factor used for the diameter of umbra in relation to the lunar diameter) = 553, which is roughly comparable to the value memorialized for the diameter of the umbral shadow from the complex: 530. Thus, the actual figure (553) is off from the expected figure (530) by some .043%.

ii. The Diameter of the Penumbral Shadow: Modern estimates for the width of the band of the penumbra generally assign it a value equal to one lunar diameter. The width of the penumbral shadow that is memorialized in the complex is calculated by subtracting the diameter of the umbral shadow from the diameter of the outer circle: 1,000 (diameter of the outer circle) – 553 (diameter of the inner circle) = 437 and then dividing by 2: 437/2 = 218. This is the value from the complex (218) whereas the current figure is one lunar diameter or 208. Thus, the actual figure (218) is off from the expected figure (208) by some .048%.

b. Measuring Gamma and the *Temenos* Wall

Recall from the introductions that the *temenos* wall has some 226 total *redans*. If the four bastions (*redan* equivalents) are subtracted out, for reasons that will become apparent from what follows, then there are 222 total redans along the *temenos* wall: 37 each in the north and south segments of the wall, and 74 each in the east and west segments. A grid can then be created by drawing a line between each redan on the north wall to its counterpart on the south wall, and from those on the east wall to their counterparts on the west wall. (The corners of the complex are irrelevant to this purpose, which is why the bastions were subtracted out.) The grid can then be numbered from the north-south centerline and east-west centerline, which bisect one another at the center of the inner circle of the complex, or the umbral shadow. The ancient Egyptians probably calculated gamma with this grid, by equating thedistance of the lunar orb from the center of the umbral shadow in terms of *redans* or 'fingers' east-west of the centerline and 'fingers' north-south of the centerline, instead of using the modern factor of earth radii.

The 37 total *redans* along the northern and southern sides of the *temenos* walls would track the slow movement of the lunar nodes toward the east at the rate of 0.46°/eclipse member in a Saros series (the modern figure is 0.48°). The 74 total *redans* along the eastern and western sides of the *temenos* walls, with those outside of the outer circle—3 to the south and 3 to the north—subtracted out, leaving some 68 inside the larger circle and some 37 in the inner circle. The lunar eclipses occurring in the ring between the inner and the outer circles were penumbral eclipses, and their gamma factor may not have been of interest to the ancient Egyptians. The north-south *redans* along the eastern and western walls would be roughly equivalent to the slow shift north (ascending node) or south (descending node) of the point of greatest eclipse, at the rate of 300 km between eclipses.[126] With each eclipse in a Saros series, the gamma measure moves one *redan* or 'finger' south or north, depending on whether the Saros series is occurring at the ascending or descending node, and one *redan* or 'finger' east. Note that while the point of greatest eclipse of a Saros series at the ascending node moves north, the Moon's orb is moving <u>south</u> through the umbra with each member eclipse in the series. The opposite circumstances apply to a Saros series at the descending node, with the Moon's orb moving north through the umbra. Thus, the Egyptians would have followed the movement of the Moon's orb southward across the gird for a Saros series at the ascending node, as its point of greatest eclipse moved northward. Following is an example of the use of this system:

A Saros series member eclipse for a series at the ascending node produces a total eclipse just above the ecliptic and is the 5th of some 12-13 total eclipses that will be produced by this series. Its gamma factor is +0.25. On the grid of the Step Pyramid complex, this eclipse would also be a total eclipse and the center point of the lunar orb would be largely in the northeastern quadrant of the inner circle (umbra), 1 finger to the east of the north-south centerline and 8 fingers above the east-west centerline.

The accuracy of this system does not begin to compare to the accuracy achieved with modern measurements; however, it does provide reasonably good results that can be used for record keeping purposes and for making predictions.

---

[126] Espenak, Fred. NASA web site articles: *Periodicity of Lunar Eclipses*, sections 1.4 and 1.5; and *Eclipses and Saros*.

Also, it does not provide an identifiable mechanism for adjusting the shadow plot for the obliquity of the Moon's axis of rotation to the ecliptic, which has a mean value of 5.1°. That the ancient Egyptian astronomers were aware of this factor seems beyond doubt, in light of the extent of the level of their knowledge and understanding of eclipses. Nevertheless, I don't know the mechanism that was used to incorporate this important factor.

c. Flare of the Cobra's Hood

With the knowledge of the predicted point of greatest eclipse available from the *temenos* wall shadow plot, the Step Pyramid projection can be completed and an accurate forecast of the location and nature of a lunar eclipse made.

The metaphor of a cobra as eclipse seems particularly apt. The cobra gradually rises from its place of birth in the polar region, climbing steadily toward the opposite pole, and flaring its hood with each lunar eclipse member in a Saros series, gradually covering the Moon in a total eclipse. With the Moon obscured by the cobra's hood, the imagery is complete.

CHAPTER 27

LUNAR LIBRATION AND THE STEP PYRAMID

One of the subtler aspects of lunar eclipse observations is measuring the eclipse shadow's progression in relation to the larger craters and other visible features of the Moon. This is a fairly accurate means for measurement, but it must take into account lunar libration. The Moon's rotation period is synchronous with its orbital period, which is why its visible side remains constant from the perspective of an observer on Earth. However, because of its elliptical orbit and the obliquity of its orbit, the Moon actually oscillates or rocks back and forth and up and down, depending where it is in its orbit. The up and down rocking is termed latitudinal libration and increases the amount of the Moon's face that is visible by as much as 7° of lunar latitude at each of the Moon's poles, or 14° in total. Similarly, the back and forth rocking is termed longitudinal libration and it can increase the amount of the Moon's face that is visible by as much as 8° of lunar longitude to the east or west, or 16° in total. It should be noted, however, that these are upper end values for libration. Lunar longitudinal libration for example, ranges from 3° to 8°.[127] These motions are slow and are not directly observable.

a. Latitudinal libration:

Accounting for latitudinal libration on the face of the Pyramid is a bit involved; it requires visualization of the Pyramid 'rocking' up and down. If we use the top of the first step of the Pyramid, where it breaks from its sharp vertical rise and moderates to a more gentle angle, as the lunar equator, then the first step would be the portion of the Moon that would be made visible if the Moon rocked upwards by 7°. If the Moon is rocked downward by 7°, then this step would be below the lunar equator and the very top of the Pyramid would be just visible above the final or 12th visible step, making it, in effect, a 13th step. Dividing the 90° of the lunar quadrant that is represented by the Pyramid's face by 13 steps and taking each step to be of the same magnitude, then each step would represent approximately 7° of lunar latitude. This value is at the upper end of its range, but this was probably considered necessary by the designers in order to accommodate the entire range.

---

[127] Meeus, Jean; *Mathematical Astronomy Morsels*, (William Bell; Richmond, VA.) 1997; page 31.

b. Longitudinal libration:

As noted in the introduction section, the Step Pyramid is not a square; rather it is a rectangle, with dimensions: length: 125 m. (239 cu.), and width: 109 m. (208 cu.). Using the length-wise face of the Pyramid as the northern hemisphere of the Moon, longitudinal libration is approximately accounted for by the difference between the dimensions of Pyramid's length-wise face and its width, with the dimension for the width being equivalent to the lunar diameter at the lunar equator, unadjusted for longitudinal libration. However, the length-wise dimension (239 cu.) is for the base of the Pyramid, whereas as was the case with latitudinal libration, the level that was probably used for affixing the lunar equator on the face of the Pyramid lies at the top of the first step (see foregoing). From Lauer's scaled drawings, the length of this step is estimated at 228 cu., which is close to the expected value for the lunar equator adjusted for longitudinal libration: (16°/180° X 208 cu.) + 208 cu. = 226 cu. Thus, the length-wise face of the Pyramid would have a distance of 208 cu. centered on the centerline of the Pyramid marked off on it to represent the lunar equator, with 9 cu. to either side to account for lunar longitudinal libration. Again, as was the case with latitudinal libration, this value was at the upper end of its range.

It may seem somewhat arbitrary to use the top of the first step of the Pyramid as the lunar equator, but this is not the case. The southern edge of the funerary temple built on the Pyramid's north side actually abuts directly on its first step and has a roof support at a level that is located approximately half way up the face of this step. What if any structure or platform may have been built above this level cannot be ascertained. The level of the roof was estimated to be 6 m (11.5 cu.). [128] The roof is centered on the face of the Pyramid, but does not cover its entire reach. This may be significant as the measure of libration is generally made from a central point on the face of the Moon. Thus, the edges of the Moon are less important to this function than the central areas. From the standpoint of the Pyramid, then, the values for libration that are implied in its outer dimensions merely serve the narrow purpose of quantifying them. The actual measurement of libration and adjusting the plot of the lunar face to account for them is done from the center of the Pyramid's northern face.

From the vantage point of the roof of the temple, observers could look to the south and observe the Moon, and compare it against the plot of its surface features, adjusted for librations, on the face of the Pyramid. This distinguishes the 'funerary temple' from all of the other structures that we have so far examined, which by and large are timing and tracking devices. This suggests that the Step Pyramid complex may have functioned as an active observatory as well, which should not be surprising. But there is more to the "funerary temple" than this.

The funerary temple is a curious structure that unfortunately is largely in ruins, with little other than its foundations available from which its basic layout was reconstructed by Firth and Quibell and Lauer. Several quotes from Firth and Quibell will serve to give the reader

[128] Firth and Quibell; page 9.

an appreciation for the impressions they formed of it. "It is identified as the funerary temple because the great temple to the south-east side of the Pyramid [the *Heb-Sed* Court] is of another type and there is no [funerary] temple in the usual position on the east side." "The building seems to be on the model of a palace; there are long passages which have to be traversed before entering…." "This funerary temple with its long passages, courts and chambers may well be the prototype of those buildings to which the Greeks gave the name Labyrinths from their supposed resemblance to the Cretan palaces."[129] And finally, "…we find ourselves in the North Temple, the maze-like plan of which is unlike any other known." [130] I relate all of this not to set the stage for some dramatic purpose later on, but to convey to the reader the sense of mystery and wonder that this whole structure aroused in the minds of the archaeologists. Its exact purpose remains a mystery to this day.

[129] Ibid, pp 8 and 9.

[130] Ibid, page 48

CHAPTER 28

THE EYE OF HORUS

There is no known Egyptian record that would argue that the ancient Egyptians possessed the capability to perform fine measurements of the Moon, save one; it is indirect, but compelling. The record is preserved in the mythological story of the fight between Horus and Seth for the throne of Egypt. During this fight, Seth tore out the left eye of Horus and threw it away, beyond the edge of the Earth. Thoth, the Egyptian god who was "...intimately connected with a variety of lunar functions relating to cycles of time, measure and movement"[131] went to look for it and found it lying in pieces in the darkness. He brought the pieces back and reassembled them into the full Moon, but he did so in a descending geometric progression: 1/2 + 1/4 + 1/8 + 1/16 + 1/32 +1/64, which added up to 63/64. Afterwards, Thoth declared:

> "I came seeking the Eye of Horus, that I might bring it back and count it. I found it [and now it is] complete, counted and sound, so that it can flame up into the sky..." [132]

Egyptologists generally are of the opinion that the myth and the geometric progression used for the reconstitution of the Eye of Horus describe the fortnight period between new and full Moon, and that the fraction 63/64 approximates unity. This argument is flawed in several respects: 1) The total of the fractions, 63/64, may approximate, but is not equal to 1 or unity; therefore, it cannot represent the full Moon, at least not by using this explanation; and 2) There are only 6 days represented by the 6 fractions of the progression, which is not representative of the actual number of days between new and full Moon, which is of course 14+. There is a simpler and better explanation, one that demonstrates the sophistication and capabilities of the ancient Egyptian astronomers.

If we take the apparent diameter of the Moon, as seen from the Earth, 31'30", and divide it by the apparent diameter of the Sun, 32', the result is .984375, which is 63/64. Accordingly, if unity is the Sun, then the Moon is 63/64ths of unity. The Moon as 63/64ths of the Sun is complete

---

[131] West, John Anthony; *The Traveler's Key to Ancient Egypt*; (The Theosophical Publishing House; Weaton, Il.) 1995; page 69.
[132] Clark, R. T. Rundle; pp. 224 and 255.

or full, but it is not and can never be equal to unity, even when it is full. This explanation is in complete accord with the ancient Egyptian religion, which throughout recognizes the Sun and the Sun god Ra as superior to all others.

Nice little story, that, but what does this have to do with astronomy? Just this: in order to arrive at the fraction 63/64 and then to relate this measurement to the Moon, the ancient Egyptian astronomers had to have been capable of conducting precise observations and making detailed measurements of both the Sun and the Moon, with accuracy to within at least 2 decimal positions, possibly better.

One final observation: In their written notations of fractions the ancient Egyptians expressed all of them in terms of unity, i.e. the numerator was always 1, even if it required them to go through such contortions as expressing 63/64 as the sum of six smaller fractions, all with 1 as the numerator. As cumbersome as this rule was, it was steadfastly adhered to throughout the long history of ancient Egypt, with one exception: the fraction 2/3. This singular exception is very relevant to our purposes, but I will make no further comment on the issue for the time being.[133]

[133] West, John Anthony; *Serpent in the Sky* (Quest Books; Weaton, Il) 1993; pp, 114-115.

CHAPTER 29

A MOMENT IN TIME?

The center of the six-step Pyramid is not in the center of the complex, nor is the center of the circle that is formed using the Pyramid's width (208 cu.) as its diameter, which was used earlier as the foot print of the lunar orb, whether on the Pyramid's east or west side or anywhere in between. All of the Pyramid's possible center points, including those for the 4-step pyramid and the earlier *mastaba* in its various iterations, are to the south of the complex's center point, with most being to the south east of it. There is no simple explanation for this, as it would have been well within the ancient Egyptian's capabilities to precisely center one or more of these structures or their dimensions on the center point of the complex. Why they chose not to defies easy explanation, but it does raise the issue of whether they intended to locate the Pyramid, or more to the point the foot print of the lunar orb, so as to indicate a particular moment in time. If so, its location in the complex's umbral/penumbral shadow zones is inconclusive at best, as it could represent a total lunar eclipse from any one of a number of different Saros series that were active at the time of the complex's construction, including those centered on the ascending and descending nodes. It is an interesting possibility to consider, though, but for the moment there is insufficient information available with which to make a definitive statement in this regard, one way or the other.

CHAPTER 30

ORIGIN AND NATURE OF THE SAROS ECLIPSE CYCLE

a. Earth's Elliptical Orbit

The Earth is in an elliptical orbit about the Sun, which occupies one of the two focal points of this ellipse, with the other being empty. The measure of how much this orbit deviates from a circle is termed eccentricity, and it is by, definition, the ratio of the distances CS/CP. See the following figure:

A circle would have an eccentricity value of 0. The more pronounced the ellipse, the closer the eccentricity value approaches to 1.[134] In a heliocentric orbital system, such as the Earth's, the closet orbital approach to the Sun is called the perihelion and it greatest distance from the Sun is called the aphelion.

The Earth's present orbital eccentricity value (*e*) is: .01671. In the year 0, it was .01750. It is currently decreasing at the rate of approximately .0008 every 2,000 years.[135]

b. Earth's Elliptical Orbit and the Step Pyramid Complex

i. The dimensions of the Step Pyramid complex's *temenos* wall are: length: 1040 cubits and width: 530 cubits. These dimensions surprised the archaeologists, especially Lauer, who thought the complex's designers would have used dimensions that were somewhat simpler, such as 1000 cu. X 500 cu. or even 1040 cu. X 520 cu., both of which would have defined two exact squares. In analyzing this issue and commenting on Lauer's puzzlement in this regard, Livio Catullo Stecchini concluded that the designers had intended to create two near-squares in order to have their diagonals invoke the golden mean or *phi*. Stecchini, a renowned authority on ancient measures and ancient science, offers convincing proof of his opinion, but the designers may have had another purpose in mind altogether and

[134] Meeus, Jean; *Astronomical Algorithms* (William-Bell, Richmond, VA) Second Edition 1998; pp193-194

[135] Meeus, Jean; *Mathematical Astronomy Morsels III*; p 106.

the fact that the complex's dimensions incorporates or invokes the golden mean may be either a secondary consideration or an unintended coincidence.[136] The square/near-square issue, though, is relevant to what follows.

ii. The broad dimensions of the complex can be used as the major axis (1040 cu.) and minor axis (530 cu.) of an ellipse. The center point of this ellipse divides the major axis into two equal segments of 520 cu. each. The face of the Southern Tomb of the two royal princesses appears to mark the center line of the length of the complex. If this ellipse is then overlaid with a square: 530 cu. X 530 cu. and a near square of dimensions: 510 cu. X 530 cu. and the point of the intersection of the square with the major axis is taken as a focus of the ellipse, the resultant eccentricity of this ellipse is very significant to our discussions.

iii. Using the eclipse formula, CS/CP, and plugging in the values from the complex, 10/520, [137] we arrive at an eccentricity value of .01923. Next, deduct .0008 for each 2,000-year period that has elapsed since the complex's construction, which is estimated at approximately 4,600 years of age, and the adjusted value is .01923 - .002 = .01723. The present eccentricity of Earth's elliptical orbit is .01671, so the difference is .00052, which represents a deviation of some .03. (Earth's orbital eccentricity is a factor of apsidal precession and varies between .0034 to .058 over a period of hundreds of thousands of years.[138]) This is a very fine measurement for a stone complex that was built on the scale that it was. The foot of the north face of the 4-step pyramid appears to be the memorial for this value as it is at or near the location of the focal point of the ellipse described above.

Is there any evidence that the ancient Egyptians were aware of Earth's elliptical orbit? No, there's none. There is, though, an interesting drawing from the Egyptian Book of the Dead in a 18th dynasty tomb that shows the god, Sokar, who was identified with the necropolis of Saqqara, including the Step Pyramid complex. The god stands in an 'oval-shaped' figure. I will have more to say on this later.

The exact age of the Step Pyramid complex is not irrelevant to our discussions, but no further comment in this regard will be offered for the time being. However, I will state that there is no

---

[136] Stecchini, Livio Catullo; pp, 376-377

[137] The figures cited for the complex's length (1,040 cu.) and width (530 cu.) are 'nominal' dimensions, used for ease of recall and to reflect the fact that the complex's perimeter is directly related to a circle of 1,000 cu. The exact figures, though even these are also rounded, are shown in Appendix 1; they are: length: 1,040 cu. and width: 528 cu. Using these latter figures, the figure for the eclipse's eccentricity is 8/520 or .01538. This figure is also within the predicted range of values over the eccentricity's long-term secular cycle. The latter value, though, may reflect a historically earlier eccentricity value considered important to the Ancient Egyptians. We will have more to say on this issue later in this work.

[138] Wikipedia Contributors, "Orbital Eccentricity", Wikipedia, the Free Encyclopedia, https://en.wikipedia.org/wiki/Orbital_eccentricity (Accessed April 27, 2017)

reason to doubt or discredit the current estimates of the age for the complex; it is very much a monument of early dynastic ancient Egypt.

c. The Moon's Elliptical Orbit and the South Tomb

First off, let me say that there was nothing to suggest that the elliptical orbit of the Moon was memorialized in the complex, similar to the suggestions that were raised by the square/near-square issue for the Earth's elliptical orbit. I had to go looking for it. But the issue naturally arose once the Earth's elliptical orbit was found and the logical place for looking for it was on the *temenos* wall, since it provided the broad outline of Earth's orbit around the Sun. In other words, if the Earth was represented here, the Moon must be, too, and the logical place or places to go looking for it was along the north or south walls.

i. The width-wise centerline of the complex appears to be marked by the eastern face of the South Tomb's chapel, the face with the restored cobra frieze. This point divides the south wall into two segments of 265 cu. each. The distance from the centerline to the west face of the shaft of the South Tomb is 13.68 cu. If this point is taken as the focal point for an ellipse with its semi-major axis equal to one half of the length of the south wall (265 cu.) then the eccentricity of this ellipse is CS/CP or 13.875 cu./265 cu. = .0524, which is fairly close to the mean value for the Moon's orbital eccentricity: .055.[139]

ii. The principal structures along the north wall yield similar results. The distance between the centerline of the complex to the western face of the square inset area atop the north altar appears to be identical to the south wall's measurement, 13.68 cu.

The surprise in all of this, though, is why the measurement wasn't made from the centerline of the complex to the center of the south tomb's pit, which is what I had expected? This made the eccentricity measurement seem somewhat suspect, as taking it from the west wall of the pit seemed arbitrary, whereas using the distance between the center of the pit and the centerline of the complex seemed more logical. However, by accepting that the ancient Egyptians knew what they were doing when they incorporated these measurements in the way that they did, it lead to other very important and relevant discoveries, which are the subject of the next chapter.

[139] Meeus, Jean; *Mathematical Astronomy Morsels I*; (William-Bell; Richmond, VA) 1997; page 11. Meeus cites a mean value of 5.5, but gives instantaneous extremes for this value of .026-.077.

CHAPTER 31

BARYCENTERS MEMORIALIZED IN COMPLEX

a. Barycenter defined

The center of orbit of an orbiting body or satellite is not the center of the body being orbited, but is found at the center of mass of the two bodies—their barycenter or centroid. It is around this point that both bodies turn and their turns around it describe perfect circles. A useful analogy is to consider two unequal weights placed at opposite ends of a beam, which is then placed over a fulcrum such that the weights are in balance. The fulcrum, then, acts as the barycenter, or center of mass, of the two weights. Orbiting celestial bodies behave in much the same manner, as they turn around their barycenter. The barycenter is identical to the focus of their elliptical orbit.

The Earth-Moon barycenter is located inside of the Earth, some 1,000 miles beneath its surface. Unlike a true satellite, Earth's Moon does not orbit around Earth's equator, but follows a path that is inclined at approximately 5.1° to the ecliptic. This is because the Sun exerts substantial influence over the Moon. The solar system barycenter does not stay stationary at the center of the Sun. The solar system barycenter is subject to significant influence or perturbation from the largest planets: Jupiter, Saturn, Uranus, and Neptune. This perturbation can and often does move the solar system barycenter completely outside of the Sun, where it can spend a considerable amount of time—up to 60%—before returning to the interior of the solar orb. The perturbation caused by Earth-Moon on the solar system barycenter, in contrast, is so small as to be negligible. It is around the solar system barycenter, that the Sun-(Earth-Moon) circle one another.

b. The Earth-Moon barycenter and the South Tomb

i. The dimensions of the pit of the south tomb are: a square opening, 7 m. (13.37 cu.) on a side, with a depth of 28 cu. (53.48 cu.). If a circle is inscribed around the square of the opening, then the diagonal of the square is 10 m. (19.1 cu.) and defines the diameter of the circle, and the circumference of this circle is then 19.1 cu. X π = 60 cu. This is the key number in all sexagesimal counting systems, especially for reckoning time—60 seconds in a minute, 60 minutes in an hour. It is also a sacred number for many cultures, including the ancient Babylonians. And, according to Peter Tompkins, there were 60 Egyptian sexagesimes in the circumference of the Earth.[140] Does the number 60 in the circumference of this circle invoke the circumference of the Earth? It seems all-too-perfect, but that is exactly what it seems to do, in light of what follows.

ii. The diameter of this circle, 19.1 cu., is equal to the diagonal of the square of the pit opening. If the width of the square, 13.37 cu., is subtracted from 19.1, the result is 5.73 cu., which when divided by 2 gives the distance from the edge of the circle to the edge of the square, along a line perpendicular to the center of the square, 2.865 cu. (Recall that the western edge of the square is the point from which the eccentricity of the lunar orbit is determined.) Then, dividing this measure 2.865 by 19.1 yields .15 or 15%. Next, using a figure of 8,250 miles for the diameter of the Earth (25,920/ π) and multiplying it by 15%, a figure of 1,237 miles is arrived at. This is within the range of values for the location of the Earth-Moon barycenter, which are given as 10-15% of Earth's diameter, although it is at the upper end of the range. At present, the location of the barycenter is approximately 1,000 miles beneath the surface of the Earth. However, in order to accurately affix the location of the barycenter, other factors, such as the magnitude of the Earth's axial tilt, need to be taken into consideration.[141]

iii. If a circle is inscribed inside the square of the opening of the pit, its circumference is 13.37 cu. X π = 42 cu. The number 42 is significant to eclipses, as there generally are some 40 Saros series active at any one time. The ancient Egyptians likely selected this number because it allows for several important septenary calculations that are relevant to lunar eclipses. It is neither an average nor a mean, but an ideal. Recalling that the lunar orb is considered complete in Egyptian religion when it's parts add up to 63/64ths, the number 42 is related to 63 as 2/3. Also, 2/3rds of 42 = 28, which is approximately equal to the number of active Saros series that are producing lunar umbral eclipses, including partial and total. Again, this number is neither an average nor a mean, but an ideal, because of its relation to 63 and 42.

---

[140] Page Tompkins, Peter; *Secrets of the Great Pyramid* (Harper and Row,; New York, NY) 1971; pages 209 and 212. According to Tompkins, there are 3600 stadia of 600 Egyptian feet/stadia (2,160,000 Egyptian feet) in a sexagesime. In the circumference of the Earth, then, there are 60 X 2,160,000 E. feet = 129,600,000, which when divided by 6,000 E. feet/mile = 21,600 geographic miles. Both Tompkins and Stecchini use an Egyptian foot of .308 meters for geographic measurements, instead of the regular Egyptian foot of .300 meters, which was the length of the Egyptian foot in the Step Pyramid complex.

[141] A general presentation on Earth-Moon barycenter: http://qbx6.ltu.edu/s_schneider/astro/astroweek_2006.shtml "Who's 'Bary Center' and what does he have to do with the Solar System or the Earth-Moon system?" Also see: http://en.wikipedia.org/wiki/Center_of_mass, for good animation of the Earth-Moon barycenter under, "Barycenter in astrophysics and astronomny."

This number, 28, can be further divided into 14 active Saros series at each of the lunar nodes. The number 14 will come up again in our discussions.

iv. But why the use of ideal numbers, when the ancient Egyptians appear to have been very comfortable with and competent in using fractions of numbers in their measurements and calculations? There are two answers to this, both of which are important. First, lunar Saros series eclipses are based on the beat period of the synodic, draconic, and anomalistic months, which ideally are either integers or close to integers. This is one reason why average or mean numbers were not used. Second, these numbers are septenary versions of other ideal numbers that were considered to be related to the foundation of cosmic order, a concept that was one of the most fundamental principals of the ancient Egyptian religion which they called ma-at. I will develop this last point, which is very relevant, later.

c. The Solar system barycenter and the Step Pyramid

There are a number of supposed predecessor structures embedded in the Step Pyramid. They consist of an original square, *mastaba*, structure that was 63 m. (120.33 cu.) on a side. This was subsequently enlarged by extending it 3 m. (5.73 cu.) in all directions, and later an additional extension was built on the east side, which made the original *mastaba* an oblong. Then, a four-step pyramid was built atop this structure. Finally, a six-step pyramid was built atop this structure, which was extended considerably to the west and north before the addition of the final two steps.[142]

The great pit with its granite burial chamber does not lie under any of these structures or the modifications to them. Each of the major structures has a distinct center point that lies outside of the pit. This pit is similar in its dimensions to the pit under the south tomb.

If the pit under the south tomb can be identified with the Earth-Moon barycenter, it seems logical that the pit under the pyramid would be related to the solar system barycenter. If so, we would expect to find it centrally positioned under the pyramid, but as was previously noted it is not. Recall that the solar system barycenter moves around due to the perturbing influence of the four largest planets: Jupiter, Saturn, Uranus, and Neptune. In light of this, the fact that the location of the pit does not coincide with the center point of any of the structures above it makes sense, as it demonstrates that the solar system barycenter moves around due to the perturbing influence of these four planets. In the case of the Step Pyramid, presumably the center point of the original square *mastaba* is the center of the solar orb, and the center points of the subsequent structures built around and over it are the locations of the solar system barycenter at different times. The question that arises, though, is the pit merely symbolic of this fact or does it mark the actual location of the solar system barycenter at a particular moment in time? I will again defer on this point until later. But there is one further point on the solar system barycenter that bears remarking.

[142] Drioton and Lauer; pages 10 and 11. Also see Lehner, pages 84-90.

d. Solar System Barycenter and the *Temenos* Wall

Recall that the total number of *panneauxs* in the *temenos* wall is 4,383, which when divided by 365.25 equates to exactly 12 years. This number is directly related to the orbital period of Jupiter, Saturn, Uranus, and Neptune, all of which are either multiples of this number in integers or in integers and whole fractions. The orbital periods of these planets are: Jupiter-12 years, Saturn-30 years, Uranus-84 years, and Neptune-165 years. As multiples of 12, these numbers are: Jupiter-1, Saturn-2.5, Uranus-7, and Neptune-13.75. Although there is no historical record that the ancient astronomers were aware of the existence of Uranus and Neptune, they were well aware of Jupiter and Saturn. In any case, they could readily track the movements of any and all of these planets in their orbits around the Sun with the *temenos* wall. With this information, they could easily locate the exact position of the solar system barycenter at any given moment.

The perturbing influence of Jupiter is almost as great as that of the other three large planets combined. If Jupiter is in opposition to the other three, then the solar system barycenter is inside the solar orb, close to the center. If Jupiter is in conjunction with one or more of the others, but especially with Saturn, then the barycenter moves outside of the solar orb.[143]

[143] Meeus, Jean; *Mathematical Astronomy Morsels*, (William Bell; Richmond, VA.) 1997; pages 165-168.

CHAPTER 32

SIGNIFICANCE OF THE MEASUREMENTS OF THE GRANITE VAULTS

In many respects, it is the two great pits, the one under the south tomb and the other under the pyramid, which are at the very heart of the complex. And it is the enigmatic granite vaults in the pits, one at the bottom of each, that are the focal points of the pits themselves. The one under the pyramid is widely believed to have been the sarcophagus of Pharaoh Netjerikhet. Indeed, mummified parts of a male body found in this vault support this. However, the purpose of the vault at the bottom of the south tomb pit was a profound mystery when it was found and opened, and it remains so to this day. The vault was full of stone debris, but nothing else. There were no inscriptions, markings or embellishments on it of any kind—nothing that would shed light on its purpose or significance.

One of the purposes considered at the time of its discovery and later dismissed on closer analysis was that it was the container for the canopic jars typically used for the viscera of a dead king. Removing and placing the viscera in a separate vault was a customary royal burial practice for thousands of years in ancient Egypt. The problem with this interpretation, though, was that there was substantial evidence that the south tomb had been completed and sealed before the death of Netjerikhet. Alternatively, it was theorized that the king's placenta was buried in the vault, but if so it was a singular event as there is no other evidence of such a practice having been observed in ancient Egypt. Another alternative, and the one that is now generally accepted, was that a statue of the king was interred in the vault. However, this latter alternative also suffers from the fact that the king was apparently still alive when the south tomb was sealed. Why he would inter an image of himself before his death is not explained.[144]

So what of these two vaults, especially the one in the south tomb pit? Do they embody some further astronomical facts pertaining to lunar eclipses relevant to those found elsewhere in the complex? The logical assumption is that whatever these vaults may have once contained it would be their dimensions, orientation, etc. that would provide the evidence in support of this. This may indeed be the case, but it must be stated that the evidence available is

[144] Firth and Quibell; page 57

problematic. The detailed measurements and scaled drawings from the archaeologists on which almost all of our earlier analyses rested are insufficient at this point. The dimensions of the vaults are provided by the archaeologists and we will certainly consider them. However, the vaults are so small comparatively speaking that measurement errors and rounding factors, issues that were considerations earlier, are now critical matters. Using meters and cubits was appropriate when we were dealing with large objects and substantial distances, but not when measurements on the order of 1-3 m. (1.9-2.67 cu.) are involved. The scale is simply inappropriate. Rounding factors are also important, because they can dramatically affect the reliability of the measurements provided. Nevertheless, we will use what we have and proceed.

a. Step Pyramid Granite Vault

The granite vault in the pit under the Step Pyramid is the larger of the two, being over twice as big. It dimensions are as follows: height: 1.66 m. (3.17 cu.), width: 1.66 (3.17 cu.), length: 3 m. (5.73 cu.). What if any rounding was applied to these figures is unknown. The following quote from Firth and Quibell provides their impressions of this vault and its use as a burial chamber:

> "The comparatively small size of the interior of the sarcophagus chamber, 3 m. X 1.66 m. X 1.66 m., shows that it probably never contained an inner wooden coffin and it is unlikely that the pieces of wood composing it could have been introduced through the hole fitted with the granite plug, while the aperture is in fact so small that a body wrapped in linen could only be introduced with a certain amount of difficulty." [145]

Of course, the granite beams in the ceiling of the sarcophagus could have been put in position after the internment, but the burial would still have been cramped, given the small space available inside the chamber. Certainly, there would have been little or no room for anything else, other than a coffin. The simplicity of the sarcophagus seemingly stands in rather stark contrast with the elaborateness and grandiosity of the rest of the complex.

If we multiply the cubic dimensions (cubit of 524mm) of this vault, as given by Firth & Quibell, we obtain: 3.17 X 3.17 X 5.73 = 57.58 ≈ 57.6. However, if we assume that the ancient Egyptians intended for there to be an even number of cubits in the square: 3.17 X 3.17 = 10.05, then rounding would make the product 10, instead, which when multiplied by the length, 5.73, would make the volumetric measure in cubits of the vault, 57.3, instead of 57.6.

b. South Tomb Granite Vault

Turning now to the granite vault in the pit of the south tomb, it has dimensions: height: 1.40 m (2.67 cu.), width: 1.60 m. (3.06 cu.), and length: 1.60 m. (3.06 cu.). What if any rounding

---

[145] Ibid, page 3.

was applied to these figures is unknown.[146] If these dimensions are multiplied to produce the volumetric measure in cubits of the vault, the product is, 3.05 X 3.05 X 2.67 = 24.83. However, if we assume that the ancient Egyptians intended for there to be an even number of cubits in the square: 3.05 X 3.05, then the dimensions would be 3 X 3, instead, which is exactly equal to 9, and which when multiplied by the depth, 2.67, would make the volumetric measure in cubits of the vault equal to 24.03 ≈ 24.

c. Significance of the Combined Volumetric Measures of the Vaults

Summarizing the two preceding sections, there are at least three different measures for the volumetric measure in cubits of the two vaults:

i. Step Pyramid Granite Vault Volumetric Measures:

a. 3.17 X 3.17 X 5.73 = 57.58 ≈ 57.6 (Firth & Quibell's figures)
b. 3.17 X 3.17 ≈ 10, x 5.73 = 57.3
c. 10 X 5.7 = 57

ii. South Tomb Granite Vault Volumetric Measures:

a. 3.05 X 3.05 X 2.67 = 24.83 (Firth & Quibell's figures)
b. 3 X 3 X 2.67 = 24.03
c. 3 X 3 x 2.67 ≈ 24

Adding these measures together we obtain a combined volumetric measure in cubits as follows:

a) 57.6 + 24.83 = 82.43
b) 57.3 + 24.03 = 81.33
c) 57 + 24 = 81

The last figure, 81, is attractive on a number of levels. It is a whole number. It is the square of 9, which is one of the most sacred numbers of the ancient Egyptians (there are 9 gods in the Egyptian Ennead). It is composed of three cubes with sides of 3, which in turn yields a volume of: 3 X 3 X 3 = 27 for each cube. The combined volume for all three cubes is: 3 X 27 = 81.

However, of far more importance for our purposes is the fact that 57 is the number of eclipse years in an Exeligmos cycle (3 Saros cycles), which is the time interval for a Saros series to return to roughly the same geographic location as a predecessor eclipse of 57 eclipse years prior. Also, there are approximately 24 eclipse members in each of the 3 Exeligmos series that compose the typical Saros series. Thus, there are 57 eclipse years X 24 Exeligmos cycles

[146] Ibid; page 18.

= 1,368 eclipse years in the typical Saros series. Recalling that there are 18 tropical years (19 eclipse years) between each Saros series member eclipse and that there are approximately 72 eclipse members in a Saros series, there are then 1,296 tropical years in each typical Saros series.[147] If we use the Exeligmos cycle, instead, then there 54 tropical years between each Exeligmos cycle member in this typical Saros series and 24 member eclipses in each of the three Exeligmos series in this Saros series, for a total of 54 X 24 = 1,296 tropical years in each Saros series. Thus 57 and 24 are both of very great significance for our purposes. The figure 1,296 for the number of years in a typical Saros series is also very significant, but I will defer further discussion on this point for the moment.

It may also be noteworthy that the number of cubes with sides of 3 in the combined volumes of the two vaults is also 3. Could this be a further means for drawing attention to the Exeligmos cycle, the triple Saros cycle?

The numbers 57, 24, and 81 are very neat and very supportive of my argument. But are they reflective of the 'real' dimensions of these vaults, and not just dimensions that I derived by rounding certain figures? Not knowing what if any rounding factors were used by Firth & Quibell in recording their measurements, no definitive statement can be made in this regard and the issue must remain very much an open one.

I will make one final observation, though, before moving on. If the combined volume of the two vaults is assumed to be 81.33 (2nd volumetric measure in cubits) instead of 81, this number also has very great significance for my argument. It is the ratio of the Earth's mass to that of the Moon: 81.3/1, which is directly related to the location of the Earth-Moon barycenter.[148] However, it is recognized that this number was also derived by rounding and it is no more reliable than is 81. There will obviously have to be further detailed measurements of both vaults to settle the matter.

The association of these two numbers with the solar system and Earth-Moon barycenters, as represented by the pits under the pyramid and south tomb respectively, is significant. It demonstrates a belief that lunar eclipses originate from the barycenters, starting with the influence of the solar system barycenter and then arising directly in response to the influence of the Earth-Moon barycenter. The Exeligmos sub-cycle (57 eclipse years) of the Saros series begins with the solar system barycenter and then manifests at the geographic location determined by the Earth-Moon barycenter. The volumetric measure of the granite vault in the south tomb indicates that there will be 24 eclipse members in each Exeligmos sub-cycle.

---

[147] Espenak, Fred; *Eclipses and the Saros* and *Periodicity of Lunar Eclipses*; NASA Eclipse Web Site; http://eclipse.gsfc.nasa.gov/SESaros/SESaros.html; and, http://eclipse.gsfc.nasa.gov/LESaros/LEperiodicity.html; See in particular, Section 1.4 of the second reference, "Saros." Espenak finds that there are approximately 73 member eclipses and 1,300 years in the "typical" lunar Saros series.

[148] Meuus, Jean. *Mathematical Astronomy Morsels* 1997; page 170.

It is interesting to note that if one is standing in the maneuvering chamber above the granite vault in the south tomb, at the birth of a new Saros series and its Exeligmos sub-cycles that the exit is through a doorway to the west. This is traditionally the direction of death in the ancient Egyptian religion, but it is also the direction that almost all of the heavenly bodies proceed towards as they conclude their daily transit across the heavens, and begin their long journey to a new beginning. In this sense, the west is also the direction of renewal and re-birth. Beyond this doorway, a long flight of stairs rises to the west, exiting the *mastaba* of the south tomb at the place where the eclipse year begins on the *temenos* wall. The measurements and symbolism are almost all too-perfect.

CHAPTER 33

FOR LOVE OF THE WHOLE WORLD

Is there anything in the ancient Egyptian record that would support any of this, especially knowledge of the barycenters and their relationship to the Saros and Exeligmos cycles? No, there is nothing explicit in this regard, but there is an interesting passage from the Middle Kingdom that may be relevant. It concerns a discussion between Atum, the great god, and Thoth, the messenger and scribe of the gods, regarding instructions for Geb, the earth god, on dealing with the serpents that are in him.

This God (i.e. Atum) called to Thoth, saying, 'Summon Geb to me, saying, "Come, hurry!" So when Geb had come to him, he said:

> 'Take care of the serpents which are in you. Behold, they showed respect for me while I was down there. But now you have learned their [real] nature. Proceed to the place where Father Nun is, tell him to keep guard over the serpents, whether in the earth or in the water. Also, you must write it down that it is your task to go wherever your serpents are and say: "See that you do no damage!" They must know that I am still here (in the world) and that I have put a seal upon them. Now their lot is to be in the world for ever. But beware of the magic spells which their mouths know, for Hike is himself therein. But knowledge is in you. It will not come about that I, in my greatness, will have to keep guard over them as I once did, but I will hand them over to your son Osiris so that he can watch over their children and the hearts of their fathers be made to forget. Thus, advantage can come from them, out of what they perform for love of the whole world, through the magical power that is in them.'[149]

As is the situation with many of the texts that survive from ancient Egypt, there is ambiguity in it that confuses its exact meaning. On one hand, it could just be referring to snakes, but if so, a number of passages from it are curious. Why is Geb enjoined to "...**write it down**...?" What do these sentences mean: "But beware of the magic spells which their mouths know, for

[149] Clark, R.T. Rundle; *Myth and Symbol in Ancient Egypt*; pages 243-244.

Hike **[Divine Word[150]]** is himself therein. But **knowledge is in you**."? And what of the final observation from the text regarding these dangerous serpents: "Thus advantage can come from them, out of what they perform for **love of the whole world**, through the magical power that is in them."? ([Item] and bolding added for explanation and emphasis.)

In the alternative, it seems that the text makes perfect sense if it speaks to the 'cobra' imagery associated with the Saros series by the ancient Egyptians. Saros series are born, manifest themselves over some thirteen centuries, and then die, only to be reborn some thirty years later. Also, they are indeed in the earth (Earth-Moon barycenter) and what they do is a manifestation of the order that exists between heaven and earth.

[150] Ibid; page 249.

CHAPTER 34

ORIGINS OF THE SAROS CYCLE AT THE SOUTH TOMB

The granite vault in the south tomb is composed of 14 granite blocks: 4 in the ceiling and 2 in each of the other faces. As previously noted, there are no inscriptions, markings or embellishments anywhere on it, neither inside nor outside. However, there is a curious copper wash on one half of one of the blocks. It is found on the inside face of the lower block on the south wall of the vault, on the southeast side. While traces of this same copper wash can be found on all four side walls, it is only on this one face that it is uniform. Firth and Quibell remarked on it as follows:

"The surface of the granite on the east side of the bench below the plug hole is covered with the green stain of copper, covered so uniformly that I at first thought that a sheet of metal must have been in contact with it; this is most improbable, an elaborate final dressing with copper tools equally unlikely. Traces of this colour are to be seen on all four sides of the chamber. Probably the walls were covered with a paint containing a salt of copper. This seems at first surprising, for we never find a surface of green wash in wall reliefs; this is however due to the coarse grain of the green powder used; if ground fine it would lose its colour; ground coarse it would not adhere well."[151]

In many respects, this curious "green stain" may be one of the most important features of the entire complex. It covers roughly 1/28th of the interior of the vault, a number that is very important for Saros series lunar eclipses, as it represents one of the some 28 series that are producing partial and total umbral eclipses, 14 at each node. (These numbers will be explained in more detail and justified later.) Southeast is the direction that a new ascending node Saros series arises from after the death of an earlier series at this node. (Recall that the various members of a Saros series at the ascending node move progressively toward the northeast after arising from the Antarctic region and moving toward the Arctic region.) Once the series dies in the Arctic, another will begin some 31 years later <u>on the same longitude</u> in the Antarctic. Thus,

[151] Firth and Quibell; pages 57-58.

from the perspective of the dead series, it is 'reborn' south of it and arises from the southeast, where its successor series will begin its long journey from the Antarctic to the Arctic.

a. Possible Significance of the Copper Wash in the South Tomb Vault

The ancient Egyptians were very fond of using similarly sounding words to substitute objects, creatures, events, or ideas for one another. The Egyptians believed that there was a relationship between such words and for them this relationship, which could never be fully defined nor completely understood, provided a suggestion or hint of the Divine Word at the beginning of time, when God first called forth the names of all things in creation. It was not simply an exercise in rhyming for them, but a means for subtly enabling consciousness or awareness to move ever closer and deeper toward an understanding of the original, underlying source for such words, the Divine Word.[152]

An example relevant to our line of inquiry is found in the association of Thoth, the Egyptian god of measurement and time who is identified with the Moon and lunar cycles, with the ibis. Thoth is often depicted in human form with the head of an ibis, which is often colored blue. The transliteration of Thoth's name is *Teḥuti*, which means "the measurer." One of the Egyptian names for the ibis is, *Tekh*, which is quite similar in sound to their name for the Moon when used to measure time, *Teḥu*. This is probably the reason that Thoth is often depicted as an ibis.[153] The fact that the Thoth's ibis head is often blue is also relevant to our inquiry. It should also be noted in passing that Thoth is the god who called forth the names for the creatures, things and ideas conceived of by Atum, which gave rise to creation.

It is probably no coincidence that a vast catacomb of ibis mummies is found nearby, which by some estimates may contain over a million ibises, dating from the earliest to the latest eras of the ancient Egyptian civilization. A catacomb of baboon mummies is also found nearby. Both the ibis and the baboon were identified with Thoth.[154]

b. The Color Green and the Eye of Horus

The ancient Egyptian word for green is *w3ḏ* or *uatch* [155] —both are pronounced and sound the same—which can also mean youthful or fresh. This same sounding word was also used for any

[152] Clark, R. T. Rundle; Pages 266-268.
[153] Budge, E. A. Wallis; *The Book of the Dead*; (Copyright, University Books; Published by Gramercy, a trademark of Random House, New York, NY) 1999; pages 183-184, entry under Teḥuti or Thoth.
[154] Lauer, Jean-Philippe; *Saqqara, The Royal Cemetery of Memphis, Excavations and Discoveries Since 1850*; pages 217-224.
[155] Gardiner, Sir Alan; Egyptian Grammar; (Griffith Institute, United Kingdom, reprint 2005); p 480 item M 13; see also: Budge, E. A. Wallis; *An Egyptian Hieroglyphic Dictionary*, Volume I; Dover Edition; Mineola, N.Y.) 1978; pages 150; (Note: Gardiner uses the spelling, *w3ḏ, while* Budge uses the spelling, uatch)

green stone in general. A similarly sounding word, *W3dyt* or *Uatchit*[156]—both are pronounced and sound the same—was the name of the Egyptian cobra goddess, who graced the uraei of the crown of every king of ancient Egypt from the very beginning. Another similarly sounding word, *uatch*, was the name for green eye shadow, a makeup which contained sulphate of copper.[157] However, the word for copper, *ḥemt*, is entirely different, both in form and sound.[158] Similarly, the word *khesbeṭ*, which was used for both the color blue as well as for the blue stone, lapis lazuli, is different in these same respects.[159] It is not known whether or not the ancient Egyptians distinguished between the colors blue and blue-green.

Wedjat or *Utcha-t*[160]—both are pronounced and sound the same—was the name of the divine eye or eye of Horus, which when depicted as the left eye symbolized the Moon. (The same hieroglyph when depicted as the right eye was the Sun.) Both were called *Wedjat*.[161] When this word is used for the Moon, it is the same eye that we discussed earlier, which could be taken to pieces and re-assembled to produce the fraction 63/64. Interestingly, the word *Khesbetch* means, the blue-eyed god or blue god, who may be the god Horus.[162] (I will use the spelling *wedjat* when referring to the *wedjat* eye from here forward.)

The substitution of the eye for the cobra, especially the rearing cobra, and vice-versa is well attested.[163] But what of the color green? There is nothing in the record that I know of that indicates that the color green was ever used as a substitute for either the eye or the cobra goddess, or for the Moon. It clearly could have been used in this manner and such a substitution would seem to have been entirely permissible, given the linguistic practices of the ancient Egyptians. However, as I said, there is no evidence to support that this was ever done. Then again, there is that blue color which we noted earlier was used for the ibis head of Thoth. Could this be significant? There are a number of speculative avenues open to us in this direction, but I will not pursue the issue further. It would only produce opinions and nothing more. However, there is one further feature of eh Step Pyramid complex related to this issue that bear's remarking.

Beneath both the Step Pyramid and South Tomb there are a number of beautiful bas-reliefs of Pharaoh Netjerikhet, located in or off of the passageways surrounding the granite vaults in the pits. The reliefs are grouped together in threes, side-by-side, each in its own alcove or closet. The reliefs are found on the back wall of each alcove, while the sides and ceilings

---

[156] Ibid: Gardiner p 476 item I 13; and Wallis p 151 b; (Note: Gardiner uses the spelling *W3dyt, while* Budge uses the spelling, *Uatchit*)

[157] Budge, E. A. Wallis; *An Egyptian Hieroglyphic Dictionary*, Volume I, p 150

[158] Ibid, page 485.

[159] Ibid, page 564.

[160] Gardiner, Sir Alan; *Egyptian Grammar*; p 476, item D 10; and Budge, E.A. Wallis; *An Egyptian Hieroglyphic Dictionary*, Volume 1, p 194; (Note: Gardiner uses the spelling, *Wedjat, while* Budge uses the spelling, Uatcha-t)

[161] Ibid

[162] Budge, E. A. Wallis; *An Egyptian Hieroglyphic Dictionary*, Volume 1, page 564

[163] Clark, R. T. Rundle; pages 218-230.

of each alcove are decorated with blue-green glazed faience tiles. These blue-green tiles are also used for decoration in several other areas in these passageways. While each bas-relief carving is an outstanding work of art, the exact meaning or significance of them is unknown. Similarly, the significance of the blue-green tiles is unknown, and their use in this manner was unprecedented and apparently a singular event; there is no known subsequent use of or parallel to them. The possible connection of the color blue-green of these tiles to Thoth is readily apparent, as it is to the Eye of Horus. Were the two meanings deliberately conflated with the blue-green color of these tiles? The possible meaning and significance of the appearance of the bas-reliefs in groups of three, though, will be discussed in Part V. However, this triple grouping may be related to the dimensions of the granite vaults and the number 3, which in turn may be related to the Exeligmos cycle—the triple Saros cycle—as discussed earlier.

CHAPTER 35

THE INEX CYCLE AND THE *HEB-SED* FESTIVAL

a. *Heb-Sed* Festival

As I detailed earlier, in Chapter 2, the *Heb-Sed* or *Sed*-festival is one of the most important ceremonies associated with kingship in ancient Egypt, yet it is one of the least understood, despite numerous references to it throughout ancient Egypt's long history. The Ptolemaic Greeks referred to it as the thirty-year jubilee, suggesting that it marked the 30th anniversary of a king's reign, but in earlier dynasties there are instances of kings celebrating a *Sed*-festival early in their reign, while others, who are known to have had long reigns, did not celebrate it at all. Still other kings are known to have celebrated the festival several times during their reign. The historical record clearly does not support the possibility that the festival marked a king's 30th anniversary on the throne. But if it didn't, then what event did it mark? Could it have celebrated the king's vigor and ability to reign, as some Egyptologists have suggested, which would mean that it was a matter entirely within the prerogative of the king or his court to call for a *Sed*-festival, at their discretion? This hardly seems logical, given the great age of the festival and the ceremony attending it, and yet the sporadic and seemingly unpredictable nature of the occasion has baffled Egyptologists for generations.

One of the most important aspects of the Step Pyramid complex is the *Heb-Sed* court, which is located in its southeast corner, as described earlier in Chapter 2. The Court takes its name from the presence of a low stone dais or platform, near the southern end of the court. This platform has two stairways that provide access to its platform. One has five steps and the other has four. Both are on the east flank of the platform, one at the northern and the other at the southern end. This structure is the definitive symbol of the *Sed*-festival, and its image was used in ancient Egyptian hieroglyphs whenever the festival was written about.[164]

[164] Wilkinson, Toby A. H.; *Early Dynastic Egypt*; the *Heb-Sed* is described on pages 212-215

b. The Inex Cycle

The Inex or In-exeligmos cycle[165] is one of the principal lunar eclipse cycles used for long-term record keeping and forecasting for future eclipses. It is seldom used by itself, because it does not have the unerring track record of the Saros, but the two cycles are often used together with quite good results. Detailed tables of Saros and Inex eclipses have been drawn up that go back thousands of years, which allows for dating historical references to eclipses. These tables also allow for very reliable forecasting of future eclipses that was unrivaled until the advent of modern computers. The various member eclipses of the two cycles have been plotted out in a stunning panorama, with the Saros series in columns and the Inex series in rows and the columns separated by the Inex period. I will have more to say about this panorama in a later chapter.

The period of the Inex cycle is as follows:

Days:	10,571.95
Synodic months:	358
Draconic months:	388.5
Anomalistic months:	383.674
Eclipse years:	30.5

The reason that the Inex is used in conjunction with the Saros cycle is because very often a Saros series member eclipse will be followed one Inex period later by a similar eclipse on the same meridian, but at the opposite latitude in the other hemisphere. This is what makes the two periods so useful, when used together. This occurs because the draconic and anomalistic month periods of the Inex are not whole integers, as is the case with the Saros. The Inex is also useful, because when an old Saros series dies off at one pole, an Inex eclipse will often occur at the opposite pole, one Inex period later, and begin a new Saros series along the same meridian. This is not an unerring phenomenon, because Inex eclipses sometimes fail to appear after an Inex period has elapsed, for reasons that are beyond the scope of this work. However, it is the birth of a new Saros series, one Inex period after the death of the old Saros series that is relevant to the immediate topic. In my opinion it is this event, the birth of a new Saros series, which lies at the heart of the *Heb-Sed* and provides the festival with its meaning and purpose.

c. The Inex Cycle and the Step Pyramid Complex

It is tempting to find in the number of eclipse years in an Inex period, 30.5, the source for the reason that Greeks referred to the *Heb-Sed* as the, 'thirty-year' festival, but this number is not memorialized anywhere in the Step Pyramid complex, to my knowledge. Rather, it is

[165] A good succinct summary of eclipses and the principals behind them can be found at the Wikipedia article found at, http://en.wikipedia.org/wiki/Eclipse_cycle. This article is very well sourced. Much more information in far greater detail can be found at NASA's web site, http://eclipse.gsfc.nasa.gov

the number of synodic months in the Inex period, 358, that is the source, as a period of 358 synodic months is only two months shy of thirty lunar years of 360 synodic months. This number, 358, is memorialized several times in the complex, but its presence is not immediately obvious, although in several instances, it presents itself in somewhat dramatic fashion and in very appropriate locations.

Recall from an earlier chapter, where it was demonstrated that the number of shrines along the east and west sides of the *Heb-Sed* court, 12 and 13 respectively, when multiplied by the difference between the length of the tropical year, 365.25 days, and the eclipse year, 346.62 days, or 18.63 days, provide the lengths for the Saros period in synodic and draconic months.

$12 \text{ X } 18.63 = 223.56 \approx 223$

$13 \text{ X } 18.63 = 242.19 \approx 242$

The measure of 18.63 days is, as was noted earlier, equal to the length of the South Tomb *mastaba* along the south *temenos* wall, as measured in *panneauxs* ÷ 12 (223.5 *panneauxs* ÷12 *panneauxs*/day = 18.63 days). This factor can also be used to find the number of synodic months in the Metonic cycle, by dividing it into the total number of *panneauxs* in the *temenos* wall (4,383 ÷ 18.63 = 235.26 ≈ 235). (I am not introducing the Metonic cycle here; merely using the measure of its period to demonstrate an important point.)

<u>The Duration of the Inex Series</u>: The <u>total</u> number of *redans* along the temenos wall is 226. This is a number that is related to the duration of a typical Inex series, which is approximately 225 centuries.[166]

My purpose in delving into this material is to remind the reader of how the various counts and measures found in the Step Pyramid complex often relate to one another, and can be used in simple mathematical calculations to quickly determine other useful and important numbers that are connected to lunar cycles and eclipses. I believe this was all done by design, when the Step Pyramid was built. In this respect, the number of *panneauxs* (4,383) in the *temenos* wall will prove very useful to our immediate purposes in finding the number of synodic months in the Inex cycle.

d. <u>The Granite Vault of the South Tomb and the South Grand Court, and the Inex Cycle</u>

Earlier, it was noted that the dimensions of the small granite vault buried at the bottom of the shaft of the south tomb had the following dimensions:

---

166 NASA website, "Periodicity of Lunar Eclipses", page 7. https://eclipse.gsfc.nasa.gov/LEsaros/LEperiodicity.html

Width: 1.60 m. (3.056 royal cubits)
Length: 1.60 m. (3.056 royal cubits)
Height: 1.40 m (2.67 royal cubits)

We will use the measures as given in royal cubits for what follows. The length and width are clearly equal, forming a square that is 3.06 royal cubits on a side and having a perimeter of 12.24 royal cubits. If this number is divided into the total number of *panneauxs* in the *temenos* wall, the result is: 4,383 ÷ 12.24 = 358.08, which is almost exactly equal to the number of synodic months in the Inex cycle, 358. That this number derives from the south tomb's granite vault, I believe, is quite significant. As was detailed in Chapter 9, this vault arguably represents the Earth-Moon barycenter, the source of the Saros cycle.

The number 12.24 can also be derived from the columns in the entryway colonnade. There are 36 columns with 17 ribs each in the colonnade, for a combined total of 612 ribs in all of these columns (ignoring, for the moment, that there are actually19 ribs in the four southern columns of the colonnade). If this number is doubled and divided by 100, the result is 12.24, which when divided into 4,383 yields 358. In another example, if we take half of the number of ribs in all of the columns with 19 ribs in the entryway colonnade and in the adjacent transept, 12 columns total, the result is: (19 X 12) ÷ 2 = 114. If this number is multiplied by π, the result is 358.14. While these two examples both generate the Inex period in synodic months, they are less convincing in the processes used to do so, than was the case with the granite vault.

Another ritual associated with the *Sed*-festival was called, 'encompassing the field', which involved the king running around the walls. Egyptologists have always assumed that 'the walls' referred to were the walls of Memphis, the nearby capital of ancient Egypt during the old kingdom. This is quite plausible. But, if the granite vault in the south tomb, as the symbolic Earth-Moon barycenter, was the source for the Saros cycle, then couldn't this ritual just as plausibly have referred to the running of the walls around the Step Pyramid complex? A run around the walls of the complex would seem to quite admirably mimic the complete circle that the Inex cycle makes in returning to the same meridian as the Saros series that died 30 lunar years earlier. And the same granite vault would also seem to be the most appropriate spot for the beginning of a new Saros series after the completion of the Inex, as symbolized by the king's run around the walls.

Significantly, the number 358 is found in the measure of the width of the South Grand Court, which Lauer found to be 107.4 meters or 205 royal cubits. The cubit measure converts to 358.76 Egyptian feet (205 · 524mm / 299.42mm = 358.76), which is very close to the number of Synodic months in an Inex cycle. This appears to be very relevant considering the foregoing discussion of the 'encompassing the field' or running of the walls ritual. The court is clearly surrounded by walls! It may be further significant that the length of the South Court—335 cubits, measured from the interior surface of the south *temenos* wall to the foot of the south side of the Step Pyramid— converts to 586 Egyptian feet (335 · 524mm / 299.42mm = 586 Egyptian feet, which is approximately 584, the Venus inferior conjunction period, but interestingly and perhaps more significantly a measure that is also approximately the circumference of the Earth's orbit about the Sun: $\approx 584 \cdot 10^6$ international miles. Also, the length of the Court divided by its width (335 cubits / 205 cubits) is 1.634, a close approximation of *phi*, 1.618.

CHAPTER 36

RECORDS OF *HEB-SED* CELEBRATIONS AND ECLIPSES COMPARED

New Saros series are not born with regular intervals. In some eras, new ones are born with relative frequency, while in other instances there are few by comparison. For example, if we take the 14 Saros lunar series that were born between the years 2,646 BCE and 2,230 BCE, an era that begins with the time frame for the Step Pyramid complex, we find that there were fourteen series born during this time period.[167] (The dates do not run sequentially, because a new series is not numbered until it peaks, which occurs when the axis of its umbra shadow passes closest to the center of the Earth.) The chronological interval between new series varies considerably in this set, from one year (series #5: 2,455 to series #10: 2,454) to eighty-three years (series #10: 2,454 to series #11: 2,371). A review of the series listed in the "Five Millennium Canon of Lunar Eclipses" will show that this variation is quite typical.

1 – 2570 Mar 14
2 – 2523 Mar 03
3 – 2567 Dec 30
4 – 2646 Oct 06
5 – 2455 Dec 22
6 – 2624 Aug 04
7 – 2595 Jul 16
8 – 2494 Aug 08
9 – 2501 Jun 26
10 – 2454 Jun 17
11 – 2371 Jun 29
12 – 2360 May 28
13 – 2313 May 20
14 – 2230 Jun 01

[167] "Five Millennium Canon of Lunar Eclipses: -1999 to +3000 (2000 BCE to 3000 CE)"; Table 5-3

But, how do these dates compare with the historical record of *Sed*-festivals? While there are references to *Sed*-festivals celebrated during the years covered by this period of time, 2,646-2,230, in contemporary ancient Egyptian sources, firmly dating them has been and continues to be a very serious problem. Although the magnitude of this problem improves substantially in later times, especially in the New Kingdom dynasties, it is still significant. In fact, the entire chronology of ancient Egyptian history is at best tentative and in many cases subject to assumptions. It relies primarily on summing up the number of years assigned to the reigns of individual kings, which is plagued by the facts that: 1) kings sometimes are omitted from the available lists of kings, 2) their length of reign is not always known with certainty, and 3) younger kings often were co-regents with older kings. Thus, there is no firm chronology of ancient Egypt with which to compare the list of dates of the first eclipse of the Saros lunar series from the Canon. In fact, the primary reason the Canon was drawn up in the first place was to assist archaeologists in establishing a firm chronology for ancient civilizations, including ancient Egypt.[168]

The chronology of the reign of Hatchepsut, female Pharaoh of the Eighteenth Dynasty, embodies all of the weaknesses described above. A brief outline of her life follows.

1) Date of birth is unknown
2) Married to Thutmoses II and may have been his co-regent later in his reign
3) On the death of Thutmoses II (date unknown), adopted a putative son of his, Thutmoses III, who became heir apparent, while she acted as his regent
4) At some subsequent but uncertain date, she ascended the throne of Egypt and reigned as king; actively attempted to obliterate any memory of the reign of Thutmoses II and have herself declared the direct successor of her father, Thutmoses I
5) At some subsequent but uncertain date, she either died or was deposed and was succeeded on the throne by Thutmoses III, who either initiated or condoned an effort to eradicate the memory of Hatchepsut's reign and have himself listed as the direct successor of Thutmoses II

So, how long did Hatchepsut reign and during what years? Most Egyptologists are fairly certain that her reign lasted some 22 years. However, the beginning and ending dates for her reign vary considerably (all dates are BC): 1504 – 1482, 1490 – 1468, 1479 – 1457, or 1473 – 1458.[169] Egyptologists are also in agreement that she celebrated a Sed-festival in year 15 of her reign.[170] There were two Saros series born during the range of years for her reign 1504 – 1458 and one that died (the reason for including this series will become obvious in a moment):

[168] Ibid; page vi

[169] Tyldesley, Joyce; *Hatchepsut; The Female Pharaoh*; Viking Publishing, Penguin Group (New York, N.Y.) 1996; page 12.

[170] Ibid; page 110

Series born:
#37 – 1492 Apr 03
#43 – 1463 Sept 07

Series died:
-3 – 1478 May 28

There are several attractive possibilities for synchronizing one or another of these dates with those from her reign, but the dangers in doing so are obvious.

There are other issues, as well. The nature of the *Sed*-festival itself is imperfectly understood. In some contemporary references it is referred to as the, "end festival" or "tail festival"[171], obviously implying that it marked the end of something, rather than a beginning. We could speculate that the end of a Saros series, something that lasts for some thirteen centuries, was also important enough to warrant some sort of recognition, but this only compounds our difficulties in assessing whether or not there is some firm prove that the Inex cycle was associated with the *Sed*-festival. Admittedly, there is none; it is only an attractive theory, and one that should be considered within the context of the underlying purpose of the Step Pyramid complex and its strong connections to lunar cycles, and especially to lunar eclipses and the Saros cycle. In other words, it fits.

There are several other questions that are relevant to this issue, if the *Sed*-festival is based on the Inex cycle. Does it mark the first eclipse of a new series or the first visible eclipse in the series, i.e. an eclipse in the umbra? To what extent if any did 'royal prerogative' influence the issues of when, where and how often *Sed* was celebrated? (This could be a serious issue, especially in later periods of ancient Egypt's civilization, if precise knowledge of the underlying astronomical phenomenon weakened over time, yet assumptions about its importance survived. This seems to be the case with the reign of Ramses II in particular, who celebrated a Sed-festival once every three years, in the later part of his long reign.) And finally, did the *Sed*-festival mark all new Saros series or just those at the ascending node?

---

[171] Petrie, W. M. Flinders; *Researches in Sinai* (Originally published by John Murray; 1906; Elibron Classics) Chapter XII, "The Revision of Chronology"; pages 163 – 185.

CHAPTER 37

THE INEX CYCLE AND THE VULTURE

The cobra on the king's crown or headdress, which was positioned immediately above the brow of the king, was one of a pair of royal symbols called the uraeus. The other symbol was the vulture. If the cobra is the symbol for the Saros cycle, as I have argued, then could the vulture be the symbol for the Inex? I believe that this is indeed the case, and that it takes the senior position to the right of the cobra for a reason rooted in the nature of the Inex and its connection to the continuity of the Saros. Once a Saros series dies at the Earthly pole opposite to the pole of its birth, a new Saros series will arise one Inex cycle later, at the same pole where its predecessor was born. The connections between this cycle and the ancient Egyptian religious belief in a cycle of life, death, and resurrection are immediately obvious. And the aptness in the use of the symbol of a vulture, bearing the body of the deceased Saros series to the pole of its birth, where a new one arises, is a compelling one for such a belief. The vulture, then, represents not just death, but the continuity of life, and the potency of such a symbol when used in royal iconography, such as the uraeus, is immediately obvious.

One of the earliest depictions of an Egyptian king is found on the head of the ceremonial mace of the first king of Egypt's first dynasty, Narmer. The king is shown sitting on a throne in a pavilion, atop a stairway, in the tight-fitting robe that is usually associated with the *Sed*-festival. Immediately above the pavilion a vulture is shown.[172] Does this scene depict the *Sed*-festival or merely one part of it, i.e. the death of a Saros series? It is tempting to conclude that it does indeed represent the death of a Saros series, but as stated earlier, it is impossible to say so with certainty, without definitive knowledge of exactly how and when the *Sed*-festival was celebrated.

[172] Wilkinson, Toby A. H.; *Early Dynastic Egypt*; page 193

CHAPTER 38

POSSIBLE DATES MEMORIALIZED IN THE STEP PYRAMID COMPLEX

Earlier in this work it was noted that because of certain numerical figures memorialized in the Step Pyramid complex that the structure could have been designed to memorialize a specific date. The figure for the eccentricity of Earth's elliptical orbit from Chapter 7, 'Earth's Elliptical Orbit', is one example of such a date, and the location of the pyramid within the complex and its possible relationship to the gamma factor of a particular member of a Saros series from Chapter 6, 'Gamma and the Flare of the Cobra's Hood', is another. While the latter example raises interesting possibilities in this regard, it is the former that is by far the more significant. The eccentricity of Earth's elliptical orbit changes very gradually over time, widely believed to be approximately 100,000 years, and if the Step Pyramid memorializes a specific value for it, which I believe it does, then that value could be equated to a certain date, which I calculated in Chapter 7 to be approximately 2,600 BCE. This date agrees nicely with the currently accepted chronology for the date of construction of the Step Pyramid complex.

There are several other dates that are possibly memorialized in the complex, too, and it is these dates that are of primary interest to me at this point and the purpose for including this chapter. All of these dates are associated with that series of long-term Earth motion cycles that are grouped together under the term Milankovitch cycles, in honor of Milutin Milankovitch, the Serbian astrophysicist, who theorized that these cycles affected Earth's long term climate in dramatic ways.

a. Milankovitch Cycles

Changes in Earth's orbital eccentricity is one of three long term cyclical factors that Milutin Milankovitch[173], a Serbian astrophysicist, identified as causing large scale changes in Earth's climate, including glaciations. The other two long term cyclical factors Milankovitch identified, include: 1) the change in obliquity or tilt to the Earth's orbital axis, and 2) the precession of the

[173] Milutin Milankovitch (1879-1958); NASA Web site, Earth Observatory; http://earthobservatory.nasa.gov/Features/Milankovitch/milankovitch_2.php

line of apsides of Earth's elliptical orbit. Together these three factors are commonly referred to as the Milankovitch Orbital Cycles. The time periods involved with each of these cycles are as follows:

i. Orbital eccentricity: Earth's orbital eccentricity varies over a period of some 90-100,000 years, from maximum elliptical to near circular. At present, the eccentricity of Earth's orbit is decreasing and it is gradually becoming more circular.

ii. Obliquity of the orbital axis: The obliquity of Earth's orbital axis, or its tilt, varies over a period of some 40,000 years, from 22.1° to 24.5° from the vertical. At present it is gradually becoming more vertical and is now at approximately 23.7°.

iii. Precession of the line of apsides:[174] (also called perihelion or apsidal precession) The line of apsides of Earth's elliptical orbit, the line connecting its aphelion or Earth's furthest distance from the Sun and its perihelion or Earth's closest approach to the Sun, gradually shift over a period of some 20-21,000 years. At present, perihelion occurs near the winter solstice or January 4th in the northern hemisphere, i.e. the northern hemisphere is closest to the Sun. In approximately 10,500 years, the line of apsides will precess, with the line rotating some 180°, and aphelion will occur near or at the winter solstice in the northern hemisphere, i.e. the northern hemisphere is furthest from the Sun, while perihelion will occur near or at the summer solstice in the southern hemisphere.

The Milancovitch cycles figure very prominently in the gradual warming and cooling of the Earth, and in the increase and decrease in the corresponding degree of glaciation present. Though interesting, this aspect of the Milankovitch orbital cycles are beyond the scope of this book. As noted earlier, our interest in them lies in their relevance to the possibility that a specific date was memorialized in the Step Pyramid complex. All three of the Milankovitch cycles figure in this issue.

b. Milankovitch Orbital Cycles and the Step Pyramid complex

i. Orbital eccentricity: As noted above, the Step Pyramid complex has evidence of the value of the eccentricity of Earth's elliptical orbit at the time of its construction, as was discussed in Chapter 7. This eccentricity value indicates that the Step Pyramid was constructed some 2,600 years BCE or 4,600 years ago.

ii. Obliquity of the orbital axis: The geographic location of the Step Pyramid complex figures prominently in this regard; it is located at: latitude: 29°—52'—16" North; longitude: 31°—12'—59" East. This location would mark the upper reach, or

---

174 Miltuin Milankovitch; Orbital Variations; NASA Web Site, Earth Observatory: http://earthobservatory.nasa.gov/Features/Milankovitch/milankovitch_2.php

major standstill[175] of the Moon's orbital obliquity of 5.1° under the following circumstances:

1) the obliquity of Earth's orbital axis is at the upper limit of its long term cyclical range, 24.5°, during the summer solstice in the northern hemisphere

2. the Moon is at the upper limit of its orbital obliquity, 5.1° to the ecliptic—the plane of Earth's orbit about the Sun—or major standstill.

Given these circumstances, the Moon will appear directly over the Step Pyramid. The proof is as follows (all locations are on the celestial sphere):

Ecliptic's location at summer solstice in the northern hemisphere:	24.5°
Moon at the upper end of its orbital obliquity:	5.1°
Moon's radius in arc minutes (15'):	0.25°
Moon's latitude (declination) on the celestial sphere:	29.85°

This equates to 29°—51' on the celestial sphere, which compares with the Step Pyramid's latitude: 29°—52'. Under these circumstances, the Moon will be directly overhead at the Step Pyramid's location.

If the current obliquity of Earth's orbital axis is on the order of 23.7° and gradually increasing toward the vertical, then the time elapsed since it was at its extreme of 24.5° is calculated as follows (this is a simplistic calculation and assumes that changes to Earth's orbital axis obliquity occur at a regular, predictable rate:

Givens:
Obliquity range: 2.4° (24.5° - 22.1°)
Obliquity cycle period: 40,000 years
Obliquity range over one complete cycle period: 4.8°
Change to obliquity since last limit reached (24.5°): 1° (24.5° - 23.7°)

Given these values, it takes approximately: 40,000 years ÷ 4.8° = 8,333.33 years for a 1° change to Earth's orbital axis obliquity to occur. For a 0.8° change, the figure is, approximately 6,700 years. This means that it has taken some 6,700 years for the obliquity of Earth's orbital axis to reach its present measure of 23.7°, from the time that it last reached its greatest angle of 24.5°. The Moon, on the other hand, reaches the highest point in latitude on the celestial sphere once every 18.6 years, which is its precession rate.

[175] Orbit of the Moon: http://en.wikipedia.org/wiki/Orbit_of_the_Moon#Elongation

This would indicate that the date memorialized in the Step Pyramid complex is approximately 4,700 BCE, which is the estimated date when the obliquity of Earth's orbital axis was last at 24.5°, the angle required for the Moon to appear directly overhead at the complex's location.

c. Precession of the Line of Apsides:

For the sake of argument in this regard, it is assumed that the *temenos* walls of the Step Pyramid complex do, indeed, define Earth's elliptical orbit. Given this, the center of the complex represents the center of Earth's elliptical orbit. In Chapter 7, 'Earth's Elliptical Orbit', I assumed that the north end of the square formed by the northern half of the complex marked the aphelion of Earth's orbit, and that the northern foot of the mastaba embedded in the Step Pyramid marked the location of the Sun, or the first foci of Earth's elliptical orbit. Several facts support affixing the focus of Earth's elliptical orbit about the Sun in this manner. The first is the use of the center points of the several structures embedded in to symbolize the movement of the solar system's barycenter, as was discussed in an earlier chapter, which would tend to suggest that the Earth's orbital focus about the Sun was somewhere in this vicinity. And the second is the fact that the southern half of the complex has many more structures than does the northern half, which is distinguished by a large, empty desert field to the north of the Step Pyramid temple. In my opinion, this emptiness symbolizes the second focus of Earth's elliptical orbit, the one that is empty.

The lengthwise dimension of the complex, then, represents the line of apsides of Earth's elliptical orbit, with perihelion on the southern wall and aphelion on the northern wall. This would be an interesting piece of evidence if we could determine the times of the year when Earth reached its line of apsides, i.e. perihelion and aphelion. If we knew this, we could establish a firm date for these events and provide a date certain for the complex. In fact, this can be done. The previous discussion on the obliquity of Earth's orbital axis provides the missing piece of the argument.

There are two ways of viewing the evidence of a memorialized date that may have been suggested by the geographic location of the Step Pyramid complex. One, it suggests a memorialized date, as discussed. And two, it simply suggests that the complex's location points to or emphasizes the summer solstice. If it is the latter, then the complex, and by implication Earth's line of apsides, are oriented toward the summer solstice in the northern hemisphere, with aphelion occurring at the summer solstice, while winter solstice would occur at perihelion. The prospective dates associated with these circumstances can be determined with a great deal of certitude, as the line of apsides precesses at a fairly regular and predictable rate, over some 21,000 years.

As earlier noted, at present Earth reaches perihelion around January 4th and reaches aphelion six months later, around July 4th. This is interesting. According to the evidence of the line of apsides from the complex, either the complex memorializes a current date, ca. 2010 CE, or one that occurred some 21,000 years ago. The questions that arise, though, are, which of these two dates did the complex's designers intend to memorialize, and what does it signify, if anything?

d. Summary and Conclusions

There are at least four possible dates memorialized in the Step Pyramid complex. They are:\

i. 2,600 BCE, based on the eccentricity value for Earth's elliptical orbit at the time of the complex's construction;
ii. 4,700 BCE, based on the Moon's appearance directly overhead at the complex's location with Earth's orbital axis obliquity at 24.5°;
iii. 19,000 BCE, based on the orientation of the line of apsides in the complex and its relationship to the summer solstice in the northern hemisphere;
iv. the present era, ca. 2010 CE, based on the orientation of the line of apsides in the complex and its relationship to the summer solstice in the northern hemisphere

The first date is the least controversial, as it is in full agreement with currently accepted chronology and the date of construction it assigns for the complex. This date can probably be accepted with little argument or controversy. The second is rooted deeply in the past, beyond currently accepted chronology. This is a problem that can be resolved by interpreting the data on which it is based as signifying not a date, but an astronomical event—summer solstice in the northern hemisphere—as discussed earlier. This is a convenient resolution to the problem, but it is not one without its weaknesses.

The third and fourth dates are both highly problematic. The third, 19,000 BCE, is mired in remotest antiquity, beyond any currently accepted chronology, and its possible significance can only be categorized as speculative in the extreme, as there is no historical context for it. But it is the fourth date, the one pointing to our current era—ca. 2010 CE—that is most problematic. Clearly it has no historical context and this date or rather 'dating' can only be categorized as prophetic or forecasting in nature, but to what purpose? In some popular circles it would most likely be interpreted as pointing to the same doomsday, end-of-the-world scenario that the Mayan Long Count calendar appears to suggest for December 21st, 2012.[176] There is certainly a symmetry in the dates or dating, between the Step Pyramid complex and the Mayan Long Count calendar, which is quite interesting in and of itself, because it suggest a possible common intellectual lineage between the two civilizations. But does this dating signify the end of the world? I don't know, but the pattern of the fourteen *faux portes* on the *temenos* wall may be suggestive in this regard.

[176] Wikipedia offers a good, well-sourced summary of the Mayan Long Count calendar and the possible interpretations of the date, December 21st, 2012. http://en.wikipedia.org/wiki/Mesoamerican_Long_Count_calendar

Recall that there are fourteen *faux portes* or gates on the *temenos* wall and that there are 24 *panneauxs* in each of them. I believe that these fourteen *faux portes* represent fourteen Exeligmos cycles of 24 lunar eclipse members each and, since there are three Exeligmos cycles in each Saros series, these *faux portes* also represent fourteen Saros series. If this is so, then their arrangement on the *temenos* wall creates an interesting pattern.

There are three *faux portes* each on the north and south walls of the *temenos* wall, while there are four each on the east and west walls. The three on the south wall are somewhat irregular in their arrangement. The four found on the west wall and the three found on the north wall, on the other hand, are very regular in their placement and in their separation from one another. However, it is the four on the east wall that are most irregular. Three of them are found in very close proximity to one another, while the fourth is separated by a wide distance from these three, which makes the arrangement of these four *faux portes* on the east wall very irregular as a whole.

If the *temenos* wall is a reflection of Earth's elliptical orbit about the Sun and the fourteen *faux portes* are fourteen Saros series, then the irregularity of the arrangement of *faux portes* on the east wall, the part of Earth's elliptical orbit that is in the process of turning toward the northern hemisphere at the Summer Solstice, as the line of apsides precesses, may suggest that the occurrence of the Saros series on this part of the ellipse may be about to change in some manner. The close placement of the first three *faux portes* on this portion of the *temenos* wall may suggest that the Saros series will occur more frequently at first and then stop almost entirely thereafter.

However, if the total number of *redans* along the *temenos* wall, 226, represent the duration of the typical Inex series, which is 225 centuries, then the pattern of the *faux portes* may simply reflect the fact that the Inex cycle is erratic at its beginning, then settles down to a long stable pattern, and then towards the end of its life becomes erratic again. If this is the case, then the location of the *faux portes* and their pattern may signify no particular date, either from the past or the future.

Beyond this admittedly, very speculative assessment, though, I can offer no further comment. Does any of this portend disaster? I can't say, but a change in the Saros series could very well be accompanied by catastrophic events, given their close connection to Earth's astronomical movements.

CHAPTER 39

PHI, THE GOLDEN MEAN

a. Five-pointed Star Blocks

When the granite vaults under the Step Pyramid and the South Tomb were investigated by the archaeologists Firth and Quibell, they found a number of curiously decorated limestone blocks lying loosely close by. Most of these blocks were decorated with two, five-pointed stars on one side, while their other sides were blank. However, a number of similar blocks were found in the same areas with two stars on two opposite sides. The stars are approximately 0.3m (≈ 1 Egyptian foot of 299.42mm) in diameter. The limestone blocks that these stars are carved on are each approximately: 0.52- 0.54 m X 0.26 m (≈ 1 cubit X 1 cubit X .5 cubit), with a space of 0.38 m (≈ 1.27 Egyptian feet) between the stars.

At first, the archaeologists thought that these blocks were from an earlier structure that was pulled down and its stones recycled into the Step Pyramid complex. But, the stones were found nowhere else in the complex and the archaeologists were left baffled by their presence. A very interesting finding made by the archaeologists, from our standpoint, is the fact that, "...a count showed that those in the south tomb with stars on one side and on two opposite sides were about equal in number (13 and 12)...."[177] The ramifications of these two numbers (13 and 12) in terms of the lunar eclipse cycles that are memorialized elsewhere in the complex are readily apparent. We have met their counterparts earlier, in the shrines surrounding the *Heb-Sed* Court, where 13 shrines were located on the west side of the court while 12 were on the east side. These star blocks also have very great significance involving the planet Venus and its transit cycle, which will be discussed in detail later on in this work. For the moment, though, we will stay focused on the numbers 12 and 13.

Recall from earlier in this work where it was demonstrated that: 13 X 18.63 = 242.19, and 12 X 18.63 = 223.56, which are, respectively, the approximate number of draconic and synodic month periods of the Saros cycle—242 and 223. (The number, 18.63, is the difference between the tropical year and the eclipse year.) This would probably be no more than interesting corroboration for an earlier finding in this work and a passing point of interest, but for the fact

[177] Firth and Quibell; *Excavations at Saqqara; The Step Pyramid*; page 46.

of the five-pointed star's connection to *phi*, the golden mean, which is symbolized by *phi*, and *phi*'s connection to the Saros cycle. *Phi* is related to the Saros cycle in a most profound way.

b. *Phi* and the Saros Cycle

Each line in the five-pointed star intersects another at exactly the ratio of *phi*, 1.618. Each point of the star also incorporates a 'golden triangle' that is composed of the angles, 36°, 72°, and 72°, from which similar triangles can be generated, ad infinitum. The golden triangle also is related to *phi*.[178] I will have more to say about the golden triangle and its relation to ancient Egypt in a moment.

It is the circle that connects to the Saros cycle. In many respects, the circle is a particularly apt symbol for God in that it has no beginning and no end, and its shape is perfect. In ancient Egypt, the High God declares, "I was he who came into existence as a circle, he who was the dweller in his egg."[179] In the mind of the ancient Egyptians, the primacy of the circle in creation and its connection to God seem apparent. It was the circle that preceded all else. It is also one of the most fundamental of shapes.

If the golden mean conjugate, 1/*phi*, is multiplied by the number of degrees in a circle, the result is, .618 X 360 = 222.48, which is approximately equal to the number of draconic months in the Saros cycle, 223. (Recall that the length of the *mastaba* atop the south *temenos* wall, as measured by the corresponding number of *panneauxs* of the wall at this location, is 223.5, which demonstrates that this structure is also related to *phi*.) The presence of the star blocks around the two granite vaults in the Step Pyramid complex thus seems to imply that the Saros cycle arises from the Sun-Earth barycenter and Earth-Moon barycenter, which themselves define circles, through the mechanism of *phi* Recall that the Sun and Earth revolve around the Sun-Earth barycenter and that the Earth and Moon revolve around the Earth-Moon barycenter, both in perfect circles, and that the two granite vaults apparently memorialize these facts. In the sense that the Inex cycle also completes a circle and begins a new Saros series after the death of an old one, the birth of the new one is also related to *phi*. This apparent relationship of *phi* to the Saros cycle and its rebirth would seem to somewhat justify the almost mystical beliefs held by many people throughout history that *phi* is intimately connected to the generative force behind life itself.[180]

All of this, though, begs a perennial question that has vexed scholars of ancient history for generations. Did the ancient Egyptians know about *phi* and if so when? Most scholars believe that such knowledge originated with the Greeks, but there are a few who believe otherwise. It is not my intent to delve deeply into this debate, but I would like to offer a brief opinion in this regard.

---

[178] Phi and Fibonacci in Golden Triangles: http://www.goldennumber.net/triangles.htm

[179] Clark, R.T. Rundle; page 74; quote is from Coffin Text 714.

[180] Tompkins, Peter; *Secrets of the Great Pyramid*; page 191

c. Memphis (Mem·phi)

The Greek name for the ancient Egyptian capital city is Memphis, which obviously seems to incorporate the word *phi*. Egyptologists believe that this pronunciation was the result of the Greeks inaccurate translation of the Egyptian name for the city, which was *Menofer*, a name that derives from the nearby pyramid town of the sixth dynasty pharaoh, Pepi I and that is translated as "Enduring and Beautiful is the Pyramid [of Pepi]".[181] The glyphs for *Menofer* are shown below:

It should be readily apparent that, phonetically, there is little similarity between the words, Memphis and *Menofer*, and that either the scholars are correct about the phonetic misunderstandings of the ancient Greeks when it came to the ancient Egyptian language or that something else entirely is at play to explain this discrepancy. In my opinion, something else is at play and it involves *phi.*

The glyphs shown below are those from the *Menofer* name that have the phonetic value of *'nofer'*, meaning beautiful, and other similar meanings:

There is no dispute with this. The phonogram of the heart-lung symbol is used for the 'n' sound and is followed by the glyphs for the 'f' and 'r' sounds respectively. The first glyph is generally understood to be a heart with lungs, but it also appears to be a lute or similar musical instrument with a sound box, throat, and tuning pegs. Among the arts, music can easily be symbolized by the instruments used to play it. A very similar instrument is often depicted in ancient Egyptian scenes depicting beautiful female musicians.[182]While this instrument is not attested until the Middle Kingdom, there is no reason to belief that it could not have originated in earlier times. From the available evidence, it seems to have been an instrument played mainly by women. If this association is correct, then it may symbolize not only music, but more specifically, music played by beautiful women, which would strongly suggest that the lute glyph is used as an ideogram or picture word for the concept of beauty and harmony, instead of a phonogram or sound sign. In light of this, if the lute stands for beauty, then the two remaining signs become sound complements for 'f' and 'r', meant to visually 'balance' the written glyph for '*nofer*.'

---

[181] Gardiner, Sir Alan; *Egypt of the Pharaohs*; page 91

[182] Wikipedia Contributors, "Music of Ancient Egypt", Wikipedia the Free Encyclopedia; https://en.wikipedia.org/wiki/Music_of_Egypt ; note the several illustrations included.

However, if the 'f' and 'r' glyphs are <u>not</u> sound compliments, but glyphs with associated but independent meanings, instead, then an entirely different meaning emerges for '*nofer*.' If 'f' is pronounced *phi* and is also understood to mean *phi*, the golden mean; and the 'r' sign is phonetic, as is the standard meaning attached to it, then '*nofer*' could literally mean: beautiful is phi-r (pyre, or fire?). *Menofer,* then, can be understood to be: **"Enduring and Beautiful is the *Phi-r* (pyre) at the Place of the Pyramid."** That this meaning is likely correct is supported by the numerous and extremely important manifestations of *phi* at the Step Pyramid complex, as well as at site of the Great Pyramid that are discussed in detail in several chapters in Part IV. But why is fire associated with this meaning, which it clearly is? Possibly because *phi* is a static concept, but when *phi* generates itself endlessly, then it becomes an ongoing, generative force, similar to fire. The five-pointed star symbolizes such an ongoing, generative force. There is much more on this point in the following section.

Given all of the foregoing, it is relatively easy to envision a simpler rendering of the name *Menofer* as: Memphis (Mm-*Phi*), which would translate as, 'Among *Phi*' or 'In the Midst of *Phi*', hence, 'City in the Midst of *Phi*.' Memphis, then, would be an accurate rendering in Greek of the Egyptian abbreviated name for the City.

d. <u>Pyramid; Origin and Meaning of the Word</u>

In *Menofer*, the penultimate glyph is:

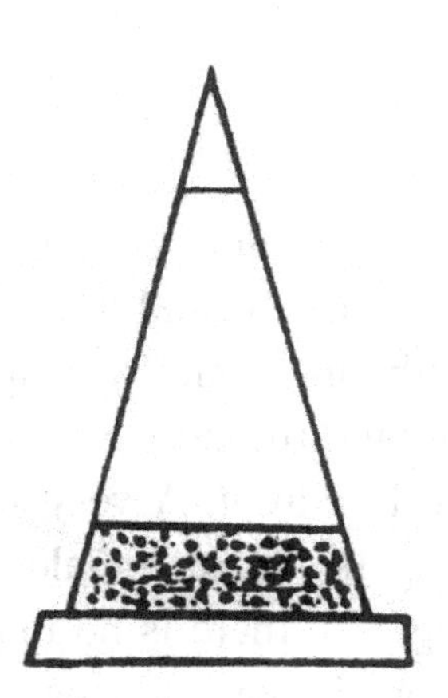

This glyph[183] has several parts, all of which are important. The base is usually understood as a glyph for *Mayet*, or order. (Note the slight angle on the leading edge of the pedestal.) The component immediately above is the usual glyph for granite and is usually understood to mean permanent or enduring. And the component immediately above this is the key to understanding the glyph as a whole. The very noticeable height of the triangle in relationship

[183] Lehner, Mark; *The Complete Pyramids; Solving the Ancient Mysteries*; Thames and Hudson, (New York, 1997) page 34 and Half Title Page. (Re-printed with permission of Thames and Hudson.) The drawing is from the tomb of Ptahhotep I and is titled, "mr pyramid.")

to its narrow base creates very sharp slope angles, which are strikingly unlike those of any known ancient Egyptian pyramid. This was not a matter of artistic license or an error on the part of the artist in rendering an accurate image of a pyramid. Rather it was done by specific design.

In Egyptology, this glyph is broadly recognized as a 'determinative', or indicator as to the exact meaning intended by a writer using hieroglyphs to express himself. In this case, it was commonly used as a determinative that was associated with the word, mr, which is the ancient Egyptian word for 'pyramid' or 'tomb.'[184] In practice, this glyph is usually referred to as the "mr pyramid." However, in my opinion, there is much more to this glyph and that it actually embodies a representation of *phi*, or more specifically a golden triangle that was intended to stand for the five-pointed star, both of which were geometric representations or proxies of *phi.* If I am right, then, *Menofer* would not mean, 'Established and Beautiful is the Pyramid (of Pepi')[185], but something closer to, 'Established (Enduring?) [is the] Perfection (Beauty?) of the Place of the Pyramid.'

The word, 'pyramid' is widely accepted as being a Greek word, deriving broadly from the Greek word 'pyre', for fire. However, the word may be composed of two words: one referring to the glyph's triangle (pyre or fire), and the second referring to the glyph's base (a-mid, or i-ma-at or i-mayet); thus, pyr-i-ma-at or 'fire of mayet.' Interestingly, the Egyptian hieroglyph for fire is Gardiner's sign Q-7, which is used as a determinative for a number of words related to fire or heat, and is composed of a sharp-angled, isosceles triangle atop a base that is composed of three smaller triangles:

The apparent similarity of the shape of the isosceles triangle of this glyph to the triangle depicted in the mr pyramid glyph strongly suggests that the two are the same, and that the glyph for the mr Pyramid is also connected to the Egyptian determinative for fire. Given this, then, it would be reasonable to assume that the word 'pyramid', with its connections to fire, is a borrow word, that is Egyptian in its origin. But what of *phi*?

Measuring the angle at the apex of the foregoing drawing of the mr pyramid shows it to be approximately 30°. I would assume, though, that the angle was intended to be on the order of 36°, an angle that would create an isosceles triangle with angles 36°—72°—72°. <u>However, no such assumption is necessary, as the 36°—72°—72°. triangle is also embodied in the glyph.</u> Recall the slight angle at the leading edge of the pedestal of the glyph. If a right triangle is

[184] Budge, E.A. Wallis; *An Egyptian Hieroglyph Dictionary*; Dover edition published 1978 (Mineola, NY) page 314b

[185] Gardiner, Sir Alan; Egyptian Grammar (Griffith Institute, Oxford; third ed. revised, 2005), p 183, and pp 568-569

drawn from the middle of the base to the apex and then from the apex to the base, along the angle at the face of the leading edge of the pedestal, then the angle at the apex is 18°. If this triangle is replicated on the opposite side, an isosceles triangle is created with angles equal to 36°—72°—72°. This is a golden triangle, which incorporates the phi relationship between the side and base cords.

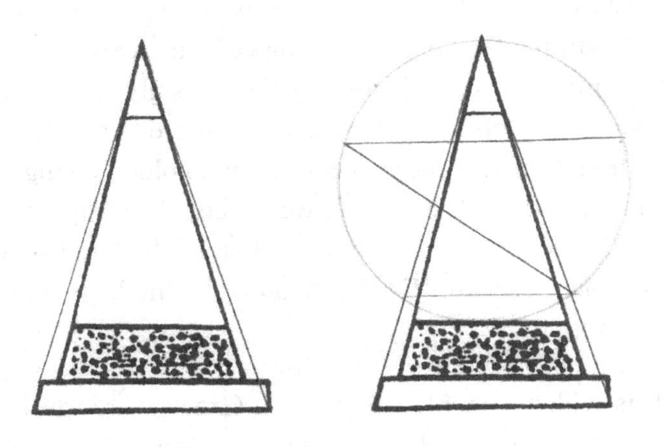

How important is this? Very. The 36°—72°—72° triangle, is central to many trigonometric functions, which is why it received a great deal of attention in Euclid's *Elements.*[186]It may also be assumed that the 'mr' pyramid, as depicted in the foregoing drawing, is a proxy for the five-pointed star, the straight lines of which cut each other to create phi, endlessly replicating *phi.* The five-pointed star is comprised of these triangles (see foregoing drawing). When drawn within a pentagram, the five-pointed star can be further rendered into an endless series of 36°—54°—90° right triangles, which is a close approximation to Pythagoras's 3—4—5 triangle. (**Note that the measures of the all the angles created by the five-pointed star in a pentagram—18°, 36°, 54°, 72°, 144°—are numbers from the Pattern.)**

In the following figure, right triangles are created by drawing a diagonal across the stippled portion of each pyramid. On pyramid "1", a 30°—90°—60° triangle is created at points ABC. On pyramid "2", a 36°—90°—54° triangle is created at points DEF. These two triangles are astronomically significant at the location of the Great Pyramid, a topic which will be discussed in depth in several chapters in Part IV.

[186] Tompkins, Peter; *Secrets of the Great Pyramid,* (Harper & Rowe Publishers, Inc. New York, NY; 1971); Appendix, *Notes on the Relation of Ancient Measures to the Great Pyramid,* by Livio Catullo Stecchini, Pages 377-378

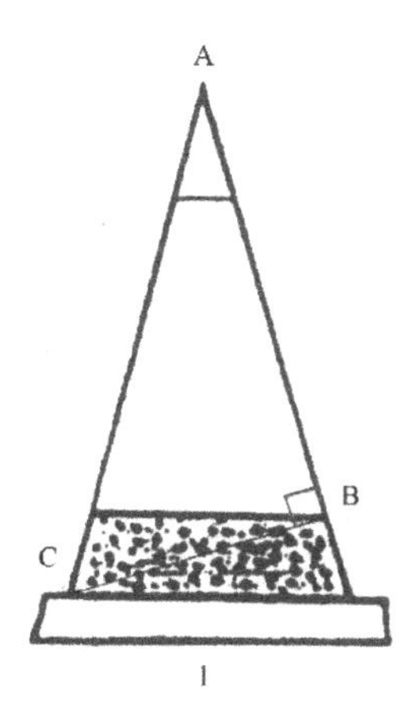

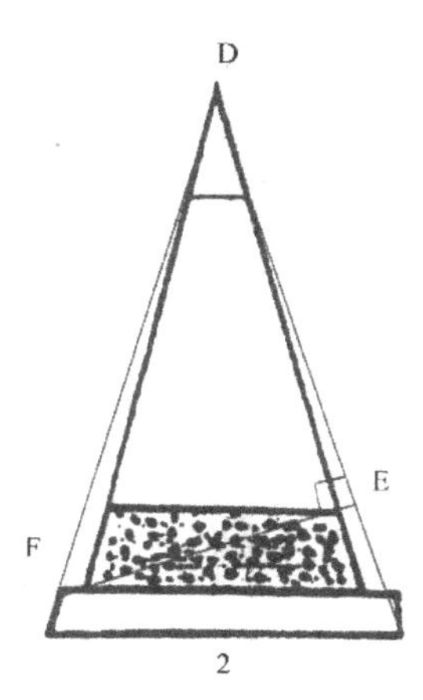

Livio Stecchini believed that the 36°—54°—90° right triangle can be referred to as a "mr triangle", and that a name of Egypt, 'To-mr', or 'Tomera', 'Land of mr',[187] is based on this triangle. The presence of the 36° triangle in the 'mr' pyramid would seem to support him.[188] However, there is much more to be said about this issue, but I will defer for the moment.

e. The 1:2 rectangle and phi

There are three stelae depicting *Netjerikhet* in a chamber under the South Tomb. One of these stelae provides almost perfect proof of *phi*'s connection to the generative principal. *Netjerikhet* is shown running, wearing only a belt. From the front of the belt, at the king's lower midriff area, a steep-sloped triangular image is faintly depicted, which reaches all the way to the ground. The triangular image is not part of the king's garb, as it appears to be almost transparent, more like a ray of light. This triangle is also almost identical in shape to the one from the glyph shown above.[189]

---

[187] Gardiner, Sir Alan; Egyptian Grammar (Griffith Institute, Ashmolean Institute, Oxford; University Press, London; 3rd Edition, 2005) page 569, upper right, entry for *mri; Ta-meri,* a name of Egypt.

[188] Tompkins, Peter; *Secrets of the Great Pyramid,* (Harper & Rowe Publishers, Inc. New York, NY; 1971); Appendix, *Notes on the Relation of Ancient Measures to the Great Pyramid,* by Livio Catullo Stecchini, pages 291-292; see also, pages 377-378

[189] Firth, Cecil M. and Quibel, J. E. *Excavations at Saqqara; The Step Pyramid,* Plate 42. (Reprinted with permission of Martino Fine Books)

But all of this is yet still more involved. The angle at the apex of these pyramids, including the depiction of the triangle on the stela of *Netjerikhet* found in the Step Pyramid complex, is invariably on the order of some 26° 34’. A right triangle with this angle can readily produce the *phi* relationship. Given a right triangle, ABC, with angle 26° 34’ is drawn (see following drawing), with side opposite in proportion to side adjacent as 1:2. Then, an arc is drawn, centered at point C with radius BC, and swung upwards from point B, such that it cuts the hypotenuse AC at point E. A second arc is then drawn, centered at point A with radius AE, and swung downward from point E to where it intersects the side adjacent at point D. Point D at the second arc’s intersection on the side adjacent defines the *phi* ratio, with the ratios AB:AD and AD:DB both being equal to *phi*, 1.618.

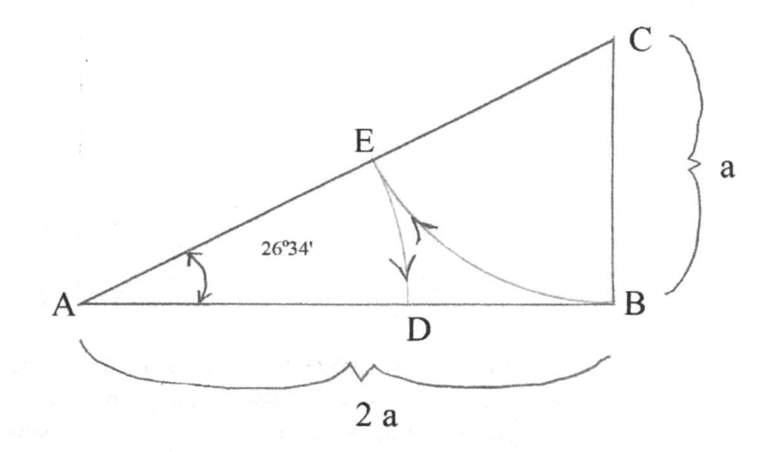

This triangle was used to establish the *phi* proportion in a great deal of Egyptian art and architecture, which is why the 26° 34' angle is found so often in ancient Egypt.[190] In my opinion, it also stood as a proxy for the golden triangle as well as for the five-pointed star, whenever the angle 26° 34' was depicted at the apex of any triangular object that was rendered in the shape of an isosceles triangle. Why use a drafting device as a twice-removed proxy for *phi* and the five-pointed star? Probably for the same reason that the name of Memphis was never written down: these matters were sacred and as such were not to be spoken of or written about directly; only hinted at. One distinctive difference, though, is that the 26°34'-triangle only produces *phi* once, unlike the golden triangle and the five-pointed star, which can endlessly replicate it. This may be an important consideration in depicting *phi* in this manner, because of *phi*'s relationship to creation and reproduction, which may have been the inviolable sacred principal.

Whether the word *phi* phonetically derives from an ancient Egyptian word is a far more difficult issue to address, and it is an issue that is beyond the scope of this work. However, on the basis of the *Netjerikhet's* stela under the South Tomb and the star blocks that were found around the granite vaults in the Step Pyramid complex and their significance, I would conclude that it appears the ancient Egyptians were very familiar with *phi* and its significance since time immemorial. So much so, that it may have played a very deep and profound role in their religion. An attractive line of inquiry, no doubt, but let us return to the subject at hand, the Moon. The circle has more to reveal to us about the Moon through the ancient Egyptian lunar ratio, 63/64, which we met earlier.

There is one more aspect of the stela of *Netjerikhet*, which we focused on earlier, that is of interest and relevant to our immediate discussion. The two glyphs that appear in the upper right hand corner of the stela, the one a grouping of three items bound together, usually described as animal skins, which generally means, birth; and the second a human face that is generally used as a preposition meaning, upon.[191] Interestingly, both appear immediately above a rendering of a corner of a wall that looks very similar to the *temenos* wall surrounding the Step Pyramid complex, and which may in fact be a depiction of the southwest corner of this wall, since the rendering of pharaoh on the Stela faces to the southwest. Could all of this, then, have the meaning that a 'birth or beginning' takes place at this location 'upon' the *temenos* wall? This is an interesting possibility and one that would fit nicely with the beginning of a new Saros series. Indeed, this is Firth's interpretation of these glyphs, although he does not make their connection with the southwest corner of the Step Pyramid complex. Instead, he explains "...that the king was supposed to renew his life at the south-west angle of <u>some building</u>."[192] (My emphasis added.)

---

[190] Temple, Robert; *The Sphinx Mystery; the Forgotten Origins of the Sanctuary of Anubis* (Inner Traditions, Rochester, Vermont) 2009; page 381

[191] Budge, E. W. Wallis; *An Egyptian Hieroglyphic Dictionary*, Volume I; pages 321 and 492

[192] Firth and Quibell; page 60.

CHAPTER 40

NETJERIKHET AND THE TWO ROYAL DAUGHTERS

During their extensive examination of the Step Pyramid complex, Firth and Quibell found a number of, "...cylindro-conical shaped limestone stelae (with a cup-shaped hollow at the top) which bore an inscription indicating that the locality belonged to *Neterkhet* (sic) and the two "royal daughters", *Intkaes* and *Hetep-hernebti*." These stones were recovered from the fill of the *temenos* walls, but Firth and Quibell were of the opinion that they had formerly marked the boundary of the complex, and had been taken down and re-used for fill after the complex was finished.[193] The two Egyptologists were also of the opinion that the complex may have contained the burials of all three of the named individuals. However, it is not the burials of the three that concern us here, but their names; or more to the point, their meaning.

Most Egyptologists are in agreement that the first glyph in *Netjerikhet's* name, *neter*, means god. The second glyph, though, is more controversial. This glyph, which is believed to represent an animal skin with a tail, is transliterated as, *khet*, a word that can have a number of different meanings. Without going into all of the arguments offered in favor of one interpretation or another, most Egyptologists have concluded that the word has a meaning that implies, 'body' or 'corporation.' Thus, the meaning attached to *Netjerikhet's* name by most Egyptologists is, 'Most Divine of the Corporation.'[194] Others interpret it to mean, 'Body of the Gods' or 'Divine of Body.'

As noted earlier, though, *khet* does have other meanings, but before moving on, I would observe that its animal skin glyph is a leopard skin, a garb reserved for the 'Keeper of the Secrets of Heaven', who was the head of the Heliopolitan priesthood and ancient Egypt's chief astronomer. Presumably, Imhotep, the man to whom the design and construction of the Step Pyramid complex is usually attributed, also wore the 'coat of stars'. If this was, indeed,

[193] Firth and Quibell; page 1.

[194] Wilkinson; page 202

the ancient Egyptians intent, then the glyph would have a far different meaning, as would *Netjerikhet's* name. In this respect, another meaning that *khet* holds is, 'a measure of land'.[195]

As measure, *khet* is found in a couple of very interesting Egyptian words. The first is *meru-khet*, which means 'measurer', or the 'Moon'. The second is *mer-khet*, which means, 'measurer of time' or 'clock'.[196] If these words and their meanings are the appropriate context for the meaning to be attached to *Netjerikhet's* name, then his name would have a proximate meaning of, 'God's Measurer'. Interestingly, 'God's Measurer' could also be used as an appropriate epithet either for the Moon or for the Egyptian god of measure, Thoth, who was closely associated with lunar cycles and measures. Also, according to the Turin list of kings, Thoth was one of the legendary first kings of Egypt.[197] I will have more to say about all of this in a moment, but first we need to talk about the two royal daughters.

Firth and Quibell translated the names of the two daughters as, *Intkaes*: 'She Who Brings Her Ka', and *Hetep-hernebty*: The Faces of the Two Goddess are at Peace'.[198] Nowhere in the complex are the two daughters referred to as the daughters of *Netjerikhet*; they are simply called daughters of the king. The presence of the word '*nebty*' in the name of the second daughter is interesting. *Nebty* is usually translated as the 'Two Ladies' and was part of the royal titulary of Egyptian kings from the very earliest times.[199] The king's 'Two Ladies' name was usually preceded by pictorial hieroglyphs of a vulture and a cobra, both in baskets. An alternative rendering for *nebty*, depicts two cobras in baskets, while still another has the two crowns of Egypt in baskets.[200]

[201]

Are the 'Faces of the Two Goddesses' fragment incorporated in *Hetep-hernebty's* name a reference to the two lunar nodes, where the lunar eclipses of the Saros series either descend or ascend? Are they synonymous with the 'Two Ladies'—the vulture and the cobra—which give rise to the *nebty* name of Egyptian kings? It is tempting to answer in the affirmative to both questions, but other than the fact that *nebty* is present in *Hetep-hernebty's* name, there is nothing else to support such a conclusion. Besides, if we were to conclude that the 'Two Goddesses' are the 'Two Ladies' of the king's *nebty* name, then what cosmic role would we assign to the other royal daughter, *Intkaes,* 'She Who Brings Her Ka'?

---

[195] Budge, E. A. Wallis; *An Egyptian Hieroglyph Dictionary*; page 567a

[196] Ibid, page 315b

[197] List of Pharaohs; http://en.wikipedia.org/wiki/List_of_pharaohs

[198] Firth and Quibell; page 15

[199] Wilkinson; pages 203-204

[200] Budge, E. A. Wallis; *An Egyptian Hieroglyph Dictionary*; page 357

[201] Wkipedia Contributors, "Two Ladies", Wikipedia the Free Encyclopedia: https://en.wikipedia.org/wiki/Two_Ladies. See graphic under Nebty Name. (Accessed May 12, 2017)

It is purely speculative, but could the ancient Egyptians have given the king his *nebty* name, based on the Saros series immediately preceding the king's birth, and could they have believed that his *ka*, or spirit came from this same series? Alternatively, could the king's *nebty* name be based on both the Inex series of his birth, as symbolized by the vulture, as well as the Saros series, as symbolized by the cobra? Such astrological calculations were not unknown to the ancient Egyptians, so some form of this practice may have been entirely possible. And associating the king with such manifestly divine phenomena as a Saros series seems entirely logical, especially in light of the fact that the Saros series is reborn after dying, having lived for over a thousand years. And yet, although such theories might be very appealing, they are all purely speculative.

Which brings us back to 'God's Measurer'; if this is, indeed, an epithet for Thoth, could the entire Step Pyramid complex be his monument and not the Pharaoh Netjerikhet's? In my opinion, this is a very distinct possibility. The sophistication of the complex, with its extensive focus on the Moon and lunar cycles—all prerogatives of Thoth—would seem to be embellishments more appropriate in a memorial to a god rather than in a funerary monument to a king, especially in light of the fact that the ancient Egyptians regarded Thoth as a lunar deity. There are other pieces of evidence at Saqqara that point in this same direction, which we will examine later, but for now we will conclude and move on. However, before we do there is one final issue that must be addressed in this regard and that is the identification of the Pharaoh Zoser with the name Netjerikhet.

If the name Netjerikhet has some other meaning than that currently attributed to it by Egyptologists, this does not mean that the name could not have been held by the Pharaoh Zoser. Indeed, there is solid evidence and every reason to believe that Netjerikhet was one of Zoser's names. The use of a god's epithet or a theological concept in one or more of Pharaoh's names and titles, including his Horus name, was a common practice, even in Ancient Egypt's earliest days.[202] Thus, there is no conflict with Zoser holding the name, Netjerikhet, while the same name also pertains to the name or epithet of a god such as Thoth. With this practice in mind, in my opinion there is a distinct possibility that Zoser may have taken the name, Netjerikhet, from the Step Pyramid complex rather than having imposed his name upon it. In other words, the complex was already in existence long before Zoser's reign and he took his Horus name, Netjerikhet, from it. This possibility, though, has very significant historical ramifications, which we will defer further discussion on for the time being.

This is not the final word on the name, Netjerikhet. There is far more to this name than can be presented here, which will be deferred until Part V.

[202] Wilkinson, Toby A. H.; *Early Dynastic Egypt*; p 202

CHAPTER 41

MAYET; DIVINE ORDER OF THE UNIVERSE

One of the deep mysteries of human knowledge is the origin of our systems for measuring time and distance. I have already investigated this issue at great length in Part I of this book, and have advanced the theory that there is a natural Pattern that underlies all manifested order and its measure. I have also advanced the theory that these concepts were taught to humanity at the dawn of civilization, with the help of certain ancient structures that embodied much, if not all, of this knowledge. I will refer to this order by its ancient Egyptian name, *Ma'at* or *Mayet*, of which the measures of time and distance of the Pattern are but a part of the much broader meaning that *Mayet* held for the ancient Egyptians that included moral order. In my opinion, the ancient Egyptians were aware of the Pattern and well understood its fundamental significance to cosmic order and *Mayet*.

A distinguishing feature of *Mayet* for the ancient Egyptians was that it originated with God and was a reflection of His universal plan. The ancient Egyptians believed that this plan underlay everything and was the foundation of all existence. This believe in *Mayet* was one of the central tenets of the ancient Egyptian religion, from the earliest of times to its last days. When the ancient Egyptians wished to invoke authority in support of a practice or belief of theirs, they would aver that it was from *Tep Zepi*, the first time.[203] This is not simply a harking back to the distant past of their civilization, but to the underlying order of the universe that was established at the beginning of time. Ptah-hotep, a pre-eminent sage of the Old Kingdom, declared:

> "Great is *Mayet*, lasting and penetrating, it has not been disturbed since the time of him who made it."[204]

In Egyptian cosmological terms, the architect of the universe was the god, Ptah, who was depicted as a bearded male standing or sitting upon a glyph signifying *Mayet*, or divine order. Ptah's plans were carried out by the lesser gods, including Atum-Re, who created the universe according to the plans. It was the divine scribe, Thoth, the god of measurement, calendars

---

[203] Clark, R. T. Rundle; *Myth and Symbol in Ancient Egypt*; page 263

[204] Ibid, page 64

and cycles of time, who articulated the will of the lesser gods in carrying out creation. And it was *Mayet*, the companion of Thoth, who was the underlying order of the universe, the divine plan.[205]

While *Mayet* was perfect, the created universe was not; it closely resembled the divine plans, but because it was constructed by the lesser gods, it was less than perfect. Had Ptah himself carried out the plans, then the created universe would have been perfect, too. The ancient Egyptians also believed that the *Tep Zepi* was a golden age, when the created universe was perfect for a brief period of time, but that was before, 'rage, or clamor, or strife, or uproar had come about' and corruption entered it.[206]

Ptah, the lord of eternity, sits atop the universe, on the great seat beyond time. I believe the ancient Egyptians identified the north ecliptic pole with the god Ptah, and that they believed that it was where the throne of God was located.

> "The Great God lives, fixed in the middle of the sky upon his supports, the guide-ropes are adjusted for the great hidden one, the dweller in the city."
>
> "I am that Creator who sits in the supreme place in the sky. Every god who does not now come down beside me, I have put him off until later."[207]

The north ecliptic pole, that place of darkness that forever obscures the "hidden one", is a place that is literally and figuratively beyond all time and measure. As Plato described it in Timaeus—a tale told by a Greek astronomer as related by the priests of Egypt:

> "Wherefore [God] resolved to have a moving image of eternity, and when he set in order the heaven, he made this image eternal but moving according to number, while eternity itself rests in unity; and this image we call time. For there were no days or nights and months and years before heaven was created, but when he constructed the heaven he created them also. They are all parts of time, and the past and future are created species of time, which we unconsciously but wrongly transfer to the eternal essence; for we say that he "was," he "is," he "will be," but the truth is that "is" alone is properly attributed to him, and that "was" and "will be" are only bespoken of becoming in time, for they are motions, but that which is immovably the same cannot be, older or younger by time, nor ever did or has become, or hereafter will be, older or younger, nor is subject at all to any of those states which affect moving and

[205] West, John Anthony, The Travelers Key to Ancient Egypt; Quest books, 1995; pages 67-69 and 79
[206] Clartk, R.T. Rundle; page 264
[207] Buck, A.D.; *The Egyptian Coffin Texts*, II, Chicago, 1939; pages: 239b, 258, 260, 265, and 267.

> sensible things and of which generation is the cause. These are the forms of time, which imitates eternity and revolves according to the law of number."[208]

And:

> "Time, then, and the heaven came into being at the same instant in order that, having been created together, if ever there was to be a dissolution of them, they might be dissolved together. It was framed after the pattern of the eternal nature that it might resemble this as far as possible; for the pattern exists from eternity, and the created heaven has been, and is, and will be, in time. Such was the mind and insight of God in the creation of time." (Author's underscoring)[209]

Modern science properly is focused on accuracy in measurement, and on averages and means for devising standards for the measurement of recurring phenomenon. These are essential practices of modern scientific inquiry, which have permitted tremendous advances across all fields of human intellectual endeavor. As we have already seen from our earlier analyses of the Step Pyramid complex, the ancient Egyptians were capable of very accurate measurement when necessary. However, they were driven by more than this. As already noted, they firmly believed that there was an observable and measurable underlying order, or *Mayet*, to the created universe and that this order was the original plan of God and perfect in all respects—an argument that I have advanced in Part I of this book. I believe that the ancient Egyptians were aware of this order and of its significance, and were deeply in awe of it. So much so, that they labored mightily to align their measures of time and distance to it as closely as possible, without unduly sacrificing accuracy. And I believe they were remarkably successful, because the systems of measure that they put in place so long ago survived for thousands of years, and many of them or their lineal descendants are still in use today.

Cultures, countries, empires all have come and gone throughout the ages, yet these measures, though corrupted at times, survived. When the Renaissance looked to the ancient world to retrieve much of the knowledge that had been lost over the intervening centuries, the ancient system of measurement was one of the most highly sought after. Scholars researched ancient manuscripts and investigated ancient ruins to ascertain the ancient the basis for the ancient Roman system, yet everything they found seemed to harken back still further in time to the measurements of ancient Egypt. It was the oddest thing. As Stecchini observed, "From the very beginning of literate cultures, documents indicate an extreme concern with the preservation of exact metric systems. The concern with precision seems to have lessened during the course of history. Early modern Europe was less careful than medieval Europe. The Greeks of the Hellenistic age were less careful than those of the classical age. Even though the Greeks of the classical age seem to have obsessed with the problem of correct standards,

---

[208] Plato; *The Dialogues of Plato*; Volume 7 of the Great Books of the Western World; Encyclopedia Britannica, Inc. William Benton, Publisher; University of Chicago, 1952; Timaeus 37-38, page 450d.
[209] Ibid, Timaeus 38, page 451a

they did not reach the subtlety of Egypt and Mesopotamia. One of the reasons why the study of the history of measures was actively pursued in the late Renaissance was that by that time standards had begun to waiver."[210]

The system of measures of time and distance that were used by the ancient Egyptians was too synchronized and too specific to have been the end result of a long period of evolutionary development, involving a lot of trial and error. This point has been made earlier, but bears repeating: when such singular numbers as 216 and 864 from the Pattern (*Mayet)* are found in such seemingly unconnected measures as the Earth's circumference in geographic miles (21,600) and the Moon's diameter in international miles (2,160), and in the length of the day in seconds (86,400) and the diameter of the Sun in international miles (864,000), it is hard not to conclude that this was done by purposeful design. It is also hard not to conclude that it must have all been done at precisely the same time, in order to have it all mesh so well. The whole system has an all-too-perfect feel about it for this to have occurred in any other manner. The ancient Egyptians hint at this, and also at the fact that they knew of the Pattern that is a part of their understanding of the nature of *Mayet*. It is appropriate to this issue that we repeat an earlier quote: "Great is *Mayet*, lasting and penetrating, it has not been disturbed since the time of Him who made it." [211]

All of this seems so satisfactory. Many of the most commonly used measures of time and distance were derived from and conform to the natural harmonic of the *Mayet*, and two of the most renowned monuments of the ancient Egyptians incorporate several of the more important of these measures as well as several of the most significant numbers from the progression. How tidy! And yet, what are we to conclude from all of this? The ramifications of any possible conclusions that can be drawn from it are so extraordinarily unsettling, so challenging, that the mind struggles desperately, grasping for traction. By whom, how, when, and why was all of this done? I don't know. Only one possible answer comes to my mind and that is that it was all done by the Hand of God at the act of creation, and then knowledge of it was given over to mankind for his benefit. This is, indeed, what many ancient legends aver—the legends of the Egyptians, the Mayans, the Incas, etc. Legends that all speak of a bearded man, who dwelt with them and taught them the arts of civilization, and who urged them to live in peace and harmony with one another. And who taught them how to count and measure by the motions of heaven, too. Perhaps, then, these legends are all true.[212]

It should be recognized, however, that even if the ancient Egyptians knew of the *Mayet* and of its significance to measurement, they could not have known all of its ramifications, such as the

---

[210] Tompkins, Peter; *Secrets of the Great Pyramid*, (Harper & Rowe Publishers, Inc. New York, NY; 1971); Appendix, *Notes on the Relation of Ancient Measures to the Great Pyramid*, by Livio Catullo Stecchini, page 308.

[211] Clark, R.T. Rundle; *Myth and Symbol in Ancient Egypt*; page 64

[212] Ibid, page 103. Also see: Hancock, Graham; *Fingerprints of the Gods* (Three Rivers Press, New York, NY; 1995) page 268

measures of the Astronomical Unit and the speed of light per second that derive from it. These are relatively recent advances in scientific knowledge. Surely there couldn't be anything in the Step Pyramid complex that would argue to the contrary. Or could there? Modern science has found that it takes approximately 500 light seconds for the Sun's light to reach the Earth, which would make the diameter of the Earth's orbit around the Sun roughly equal to 1,000 light seconds, measuring from the Earth's barycenter to the Sun's barycenter, and the circumference of its orbit roughly equal to 3,140 light seconds. These are familiar numbers. Recall that the perimeter of the temenos wall around the Step Pyramid complex is 3,140 royal cubits, which if rendered in the form of a circle, would create one with a diameter of 1,000 cubits. What do we make of this? Fascinating as the possibilities are, I'll leave the question to the reader to ponder it. I don't have any answers or explanations to offer.

As I noted in Part I, modern science is focused on the accuracy of measurement, which is entirely appropriate and has allowed for the great scientific advances made by our civilization. The ancient Egyptians were nowhere near as proficient and capable in this regard, but their level of competence was undoubtedly far greater than what they are generally given credit for currently. But their primary focus was not simply on accuracy, but on the interrelationship and synchronization of measurements, and their alignment with the underlying order of the universe—*Mayet*. It is these aspects of their science that most distinguishes it from modern science. Modern science can readily find the flaws and shortcomings in *Mayet*, but they would do well to appreciate the overall beauty of its harmonization with an underlying plan, a natural plan, of which the *Mayet*, in my opinion, is a very important aspect. *Mayet* was a central tenet of the ancient Egyptian religion, which enabled its civilization to grow and flourish for thousands of years. While modern science is certainly far superior to that of the ancient Egyptians in most respects, there is nothing comparable to *Mayet* in modern science, no sense of the unifying whole that underlies all visible form and order. In this respect, the ancient Egyptians achieved something truly magnificent with *Mayet*, something which has been unequalled to this day.

CHAPTER 42

THE COLONNADE AND THE SAROS CYCLE

In many respects the colonnade is among the most sophisticated architectural elements in the entire complex. In its stones, it embodies a great deal of knowledge about lunar cycles and eclipses, in addition to certain features that allow an adept user to forecast long range cyclical repetitions of several important astronomical events and their calendar dates. The colonnade's memory capacity and computational capabilities are such that drawing a comparison between it and a modern computer is well warranted.

While the colonnade was introduced earlier in this work, we will briefly review its main features again to refresh the reader, and provide additional detail, as well. There is a certain amount of repetition, but hopefully it will prove beneficial. In addition, we will introduce a number of eclipse and calendar cycles that will be germane to our analysis of the colonnade.

a. The Colonnade Described

   i. The only doorway to the complex is approximately 1 m. (1.9 cu.) in width and the door to it is made of stone, permanently set in the open position. Inside the doorway, there is a colonnade of parallel columns, which on first viewing appear to be fluted in the manner of Greek columns, but are in fact fasciculated or ribbed in a manner that is suggestive of bundled reeds. None of the columns are free standing; rather, they are engaged to the surrounding structure. The colonnade runs roughly from east to west. At its western end, it intersects with a transept or crossing structure that is composed of four pairs of similarly fasciculated columns oriented in a north-south direction. Each pair of the transept's columns is engaged structurally, but they are otherwise free standing. The colonnade is divided by a low gate between the 12th and 13th pair of columns (from the entrance), into two sections; the first has twelve pairs of columns and the second section, to the west of it has eight pairs, for a total of forty columns in the main colonnade. The transept is also set off from the main colonnade by a low stone gate and has a total of eight columns in it. A similar gate on the far side of the transept, leads into the south court of the complex. The height of the columns in the main colonnade is approximately 6 meters (20 Egyptian feet of 299.42mm each), not accounting for the added height of the capital, which will be described in detail later. The height of the columns in the transept is approximately 5 meters (16.69 Egyptian feet of 299.42mm each). Thus, the heights of the columns of the two sections MAY be in a 6/5 proportion to one another. (These measurements may not be definitive, because of the ruined state of the columns.)

ii. The main hallway of the colonnade narrows as one travels from east to west between the columns. It is also offset from the main axis of the complex (4.5° east of north) by approximately 1.5° south of west.[213] The walls adjoining the columns are not structurally tied into the walls behind the columns and are believed to have been designed to support the roof over the hallway. The undersides of the roofing blocks were cut in imitation of logs, laid perpendicular to the hallway's main axis. Extensive traces of red paint were found on the undersides of the roofing blocks, the column support walls, and the main structural walls behind the columns. Traces of red paint were also found on some of the columns.

iii. The columns of the hallway were cut with a noticeable taper, from 1.05m (2 royal cubits) diameter at the base to 0.785m (1 royal cubit) diameter at the capital[214], and were found to be some 6.2 meters (11.83 royal cubits) in height. The capitals were unusual in appearance, giving the impression that a sleeve or basket had been fitted over the reeds, with three of the reeds (four in a complete circle)—located at 90° angles from one another around the circumference of the capital—continuing upward on the exterior of the 'sleeve' to the roof. The rounded blocks or drums of the columns were cut to a fairly uniform height, each appearing to be about 0.25m. There were 21 drums in the columns up to the capital and four drums, or equivalents, in the capitals for a total of 25.[215]

b. Details of the Colonnade

i. Section I—columns in the eastern section of the main hallway, closest to the entrance. There are 12 pairs of columns in this section (24 total), each of which has **17** reeds, for a total of 408 reeds. (It is relevant to this discussion that there are **17-18** partial and total eclipses in the average exeligmos series. It is also relevant that the average arc measure of the Earth's shadow on the celestial sphere is 17°-18°.)

---

[213] Firth, and Quibell, Excavations of the Step Pyramid of Saqqara, plate 54

[214] Lauer, Jean Phillipe; *Saqqara, The Royal Cemetery of Memphis; Excavations and Discoveries since 1850*; page 92; Far more detail on the columns can be found in: Lauer, Jean-Phillipe, *Etude Complimentaires sur les Monuments du Roi Zoser a Saqqarah*, Annals du Service des Antiquities De L'Egypte, Cahir No.9, (Le Caire, MCMXLVIII)

[215] Firth, Cecil and Quibell, J.E., *Excavations at Saqqara, The Step Pyramid*, Volume I (Martino Publishing Company, Mansfield Connecticut, 2007) Pages 13-15 and 64-67.

ii. Section II—columns in the western section of the main hallway. There are 8 pairs of columns in this section (16 total). There are four columns on the south side of the western-most part of the section that have 19 reeds each. All of the rest of the columns in this section have 17 reeds. There is a total of 280 reeds in this section: 4, 19-reed columns, for a total of 76 reeds, and 8, 17-reed columns, for a total of 204 reeds. (Sections I and II have a total of 40 columns, 20 on the northside and 20 on the southside, which is the average number of Saros series active at any given period of time.)

iii. Section III—the transept. There are 4 pairs of columns in this section (8 total), each of which has 19 reeds, for a total of 152 reeds.

c. The Saros-Inex Panorama

At any given time, there are some 40 active Saros lunar eclipse series, with old series dying off and new ones beginning, continuously.[216] That the number of active Saros lunar eclipse series is exactly equal to the number of columns, 40, in the main hallway of the Step Pyramid is not coincidental; instead, it is a matter of specific design, as will be shown in this section, with each column representing a separate Saros series. Furthermore, it will be shown that the width of the colonnade between the columns represents an Inex cycle.

In 1955, G. van den Berg reviewed all of the some 5,200 lunar eclipses described in von Oppolozer's *Canon der Finsternisse* (1887) and then depicted them in a dramatic, two-dimensional matrix. Each column contained all of the various members of a particular Saros series arranged in chronological order, which were then staggered vertically so that each row represented a separate Inex series. The matrix, styled the "Saros-Inex Panorama of Oppolozer's Lunar Eclipses" is depicted in Chapter 10.

The Panorama's columns permit a ready comparison between the members of a particular Saros series, while its rows permit the same comparison between the members of a particular Inex series. These comparisons can be used to predict the circumstances of past and future lunar eclipses. The columns of the Step Pyramid's entryway performed the exact same function as van den Berg's Panorama, although their focus seems to have been primarily on partial and total eclipses, and not on penumbral eclipses. A more definitive statement regarding the penumbral eclipses cannot be made, primarily due to the ruined upper parts of the walls to which the columns are engaged. The columns, though, may have provided more information than the Panorama, because they are three-dimensional unlike the Panorama, which is two-dimensional.

[216] Espenak, Fred and Meeus, Jean; *Five Millennium Canon of Lunar Eclipses; -1999 to +3000 (2000 BCE to 3000 CE);* Section 5-3, page 59. http://eclipse.gsfc.nasa.gov/SEpubs/5MCLE.html

d. The Columns and the Saros Cycle

The dimensions of all the columns in the main colonnade, including the four with 19 ribs, are: height-11.83 royal cubits, width at base-2.0 r.c., width at top of capital 1.33 r.c., depth at base 1.5 r.c., and depth of column at the top of the crown-1.0 r.c.[217] (See the figure immediately following.)

Details and Measurements of Main Colonnade Column

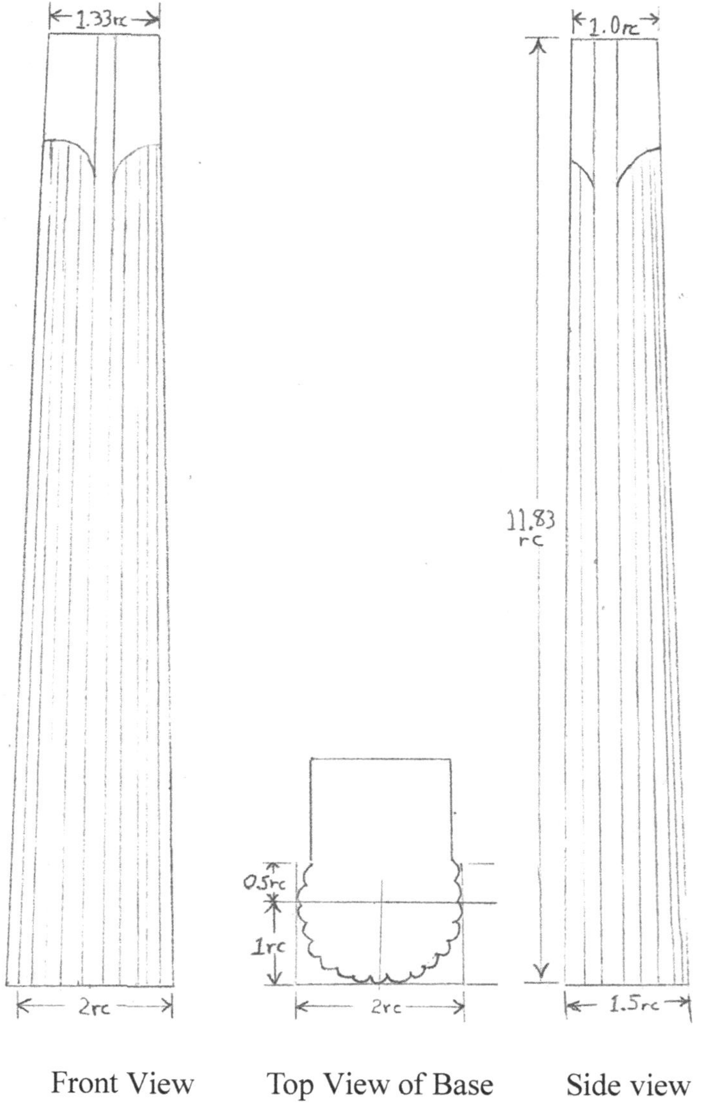

Front View Top View of Base Side view

(All measurements are in royal cubits of 524 mm.)

[217] The description and dimensions of the columns can be found at: 1) Lauer, Jean-Phillipe, Annales De Service Des Antiquites De L'Égypte, Cahier 9, "Etudes Complimentaires sur les Monuments Du Roi Zoser A Saqqarah, Response A Herbert Ricke" (Le Caire, Imprimerie De L'Institut Francais D'Archeologie Orientale, MCMXLVIII; page 39; and 2) Lauer, Jean Phillipe and Drioton, Etienne, *Sakkarah, The Monuments of Zoser* (Le Caire, Imprmerie De Institut Francais D'Archeologie Orientale, 1939) page 12.

Each column in the main hallway of the colonnade was designed to track a single Exeligmos cycle of a Saros series, most likely the one closest to Egypt. In this respect, the 25.5 reeds of each of the 17-reed columns along with their 18 grooved intervals represent the 34°-36° longitudinal window of the celestial sphere transited by the series during its life, from its birth to its death. The distance from the peak of each reed to the valley of each succeeding groove represents the approximately 1.5° longitudinal change on the celestial sphere (compare with the orientation of the hallway of the colonnade, which is also approximately 1.5°) in the location of the point of greatest eclipse that occurs from one exeligmos cycle to the next, during the life of the Saros series. On Earth's surface, this longitudinal shift, as represented by the reed-peak-to-groove-valley distance, is approximately 15° of longitude.[218]

It will be recalled that each member of a Saros series is geographically displaced westward by some 120°, because the Saros period is eight hours longer than an even number of days. As a result, a Saros series lunar eclipse will occur in the relatively same geographic region only once every 54.1 year, which, of course, is the exeligmos cycle.

The height and taper of each of the columns was designed to represent a hemisphere of the Earth, measured from the pole of the series origin to the equator and then to the opposite pole, which allowed the eclipse members of a Saros series to be tracked from the beginning of the series to its end. (The seeming impossibility of tracking the same series while using only one hemisphere was a problem that the ancient Egyptian elegantly solved, as will be demonstrated in a moment.) Every two drums of the 24-25 drums in the shaft of each column in the main hallway, was designed to represent approximately 900 km (540 international miles), or 7.5° of latitude. This is the degree of latitudinal change in the location of the point of greatest eclipse (center of eclipse shadow) between one exeligmos cycle (54.1 years) and the next in a Saros series.

It is also likely significant that the height of each column in the main colonnade, 11.83 royal cubits, is a number that approximates the arc range on the celestial sphere, 11.23°, within which umbral eclipses, both partial and full, can occur. (The measure 11.83 is predicated on Lauer's figure of 6.2 meters for the height of each column, but it is not known if his measure included the plinth or base of the column, which would significantly affect the stated measure.) Thus, it appears that all of the typical Exeligmos cycle's partial and full umbral eclipses (i.e. 17 of 24 total) were plotted vertically on the reeds of the visible column.

e. Plotting Saros Series and Exeligmos Series on the Columns

Once a Saros series has begun producing eclipses and its closest point of approach to the center of the Earth's umbral shadow, which occurs when the point of greatest eclipse is closest

[218] Espenak, Fred, NASA'a GSFC, *Periodicity of Lunar Eclipses*, https://eclipse.gsfc.nasa.gov/LEsaros/LEperiodicity.html Section 1.4, gives the actual figure as .48° per Saros cycle or 1.44° per exeligmos cycle.

to Earth's equator, has been estimated, the various members of the series can be plotted on a column. The plot begins with the capital of the column, as shown below. (INSERT DRAWING)

It will be noted that the plot involves the entire circumference of the Earth and not just the34-36° longitudinal eclipse window on the Earth's shadow on the celestial sphere, as described earlier. The columns were designed for both purposes, and there is no conflict between the two. A Saros series begins when the Sun and Moon enter the 34-35° longitudinal window for eclipses and it ends after the Sun and Moon pass beyond this window at its opposite end. And, as mentioned earlier, each of the three exeligmos cycles in a Saros series is separated by 120° from the other two. (This can be a difficult concept to grasp, so it is strongly recommended that readers who are not very knowledgeable of the geometry and mechanics of Saros eclipses review the linked NASA reference, taking particular note of figure 1, which shows the changes to the location of the point of greatest eclipse in each of the eclipses of the Saros series used in the example.)[219] The ancient Egyptians handling of this complex process can best be demonstrated and explained by rendering the plot from the column in a two-dimensional drawing, as shown below. (INSERT DRAWING)

On each column, the closest exeligmos cycle of a Saros series is tracked from the pole of its origin (represented by the capital of the column) to the equator (represented by the base of the column) and then back to the capital of the same column, which now represents the opposite pole of the series' origin. Thus, the entire Saros series is tracked from beginning to end. The plot depicts the transition of the series as it passes across the 34° celestial longitudinal range of the eclipse shadow on the celestial sphere. Individual exeligmos members of the series are plotted along its exeligmos path, moving two drums up or down and 1 reed left or right to the next member. Saros series at the ascending node are depicted on the north side columns of the hallway, with their exeligmos cycle plotted from west to east, or from left to right when facing the column, as the zone of eclipse visibility progresses from the North to the South pole and the point of greatest eclipse moves from west to east. Saros series occurring at the descending node are depicted on the south columns with their plots beginning at the South pole (capital of the column) and terminating at the North pole (capital of the column), and also run west to east. As remarked earlier, the space between the columns on the north side of the colonnade (Saros series occurring at the ascending node) and those on the south side (Saros series occurring at the descending node) is the inex cycle.

The plot on each column is V-shaped, beginning at the sheathed top of the column, to the base at the center of the column, and then returning to the opposite side of the sheathed capital. The individual exeligmos members occurring in the penumbra originate and terminate at the sheathed portion of the capital. Full eclipses occur toward the center of the column, within 4.5° of the column's centerline, and peak at or near its base, which represents the earth's equator.

[219] Espenak, Fred; Periodicity of Lunar Eclipses; NASA; http://eclipse.gsfc.nasa.gov/LEsaros/LEperiodicity.html; see figure 1, section 1.4. A more in-depth treatment of the subject can be found in the *Five Millennium Canon of Lunar Eclipses*.

f. An Alternative Plotting Procedure

There is an alternative, more complex way for plotting a single exeligmos cycle of a Saros series on a column, and that is to plot all three exeligmos cycles of a Saros series on the column, with the center facing reed of the column aligned on the prime meridian of Egypt. All three exeligmos cycles would be configured in the same manner on the column, i.e. starting from the top of the column with penumbral eclipses, then trending down toward the base of the column in a V-shape pattern, until the base is reached which marks the location of the equator and maximum full eclipses, then moving back toward the capital of the column, where the final penumbral eclipses of the cycles end at the opposite pole. The big difference between the plot of a single exeligmos cycle and that of all three cycles of the Saros series is that the plot of all three cycles begin, track and end in their true longitudinal locations, instead of having the one cycle peak at the base of the column on its center facing reed. This allows for easier reference of latitude and longitude, centered on the latitude of the observer and the prime meridian, which was then located in Egypt. (See later discussion of the Great Pyramid as datum for the prime meridian.)

The geographic plot on the columns most likely was a hybrid, azimuthal equidistant map, with its focal point on the pole, as represented by the capital of the column, and then with the locations of the hemisphere projected downward along the outward-flaring surface of the column to its base, which represents the equator, where it is roughly 1.5 X's the diameter of the capital. (The base of each column is 1.05meters or approximately 2 royal cubits in diameter.)[220] In a typical azimuthal equidistant projection[221], the projection of the hemisphere is downward onto a flat surface. In the case of the colonnade, the plot would be projected downward onto the circular plinth at the base of each column, but only for the longitude since the base of column occupies most of the available surface area on the plinth.

There are several advantages in the use of this alternative plot, some already discussed in the foregoing paragraphs. However, another advantage is that the longitude of the final penumbral eclipse of the Saros series can be noted and marked at the opposite pole of the next column, on the opposite side of the colonnade, proceeding eastward toward the entrance, which would denote the longitudinal location of the beginning of the next Saros series at the opposite node. North side columns represent Saros series occurring at the ascending node and south side columns represent Saros series occurring at the descending node. The intervening space between the two columns then would represent the Inex cycle between one Saros series and the next one at the opposite node.

---

[220] Lauer, Jean-Phillipe "Études Complémentaires sur les Monuments du Roi Zoser À Saqqarah" (Annales du Service des Antiquités de L'Égypte, Le Caire, Imprimerie de L'Institut Français D'Archeological Orientale, 1948) p. 39

[221] Wikipedia contributors, *Map Projection*, Wikipedia, the Free Encyclopedia, https://en.wikipedia.org/wiki/Map_projection (Accessed May 28, 2017)

g. Tracking Gamma and Eclipse Magnitude

Gamma, or the measure of the distance of the center of the Moon to the center of Earth's shadow, which is maximum at the furthest distance between the two and approaches or reaches zero, as they come into their closest proximity to one another for the entire series. Gamma could easily be tracked on the sides of each column, starting at the maximum measure toward the attachment point of the column to the wall, and then decreasing outward, with the value reaching zero or near zero near the face of the column at its base, the point of the greatest full eclipse for the series. As the magnitude of a series eclipses peaks and begins to decrease, Gamma can then be tracked as it increases on the opposite of the column. The depth of the column is 1.5 royal cubits, and the number of visible reads on each side is 8.5.

CHAPTER 43

IN THE MIDST OF COBRAS

Interestingly, the ancient Egyptian word for a hallway with columns was, *wꜣd* or *uatchit*, which is the same name they used for the cobra goddess, who graced the brow of every ancient Egyptian king.[222] The import of this dual meaning with respect to the colonnade of the Step Pyramid complex is immediately obvious: to the ancient Egyptians, its columns were cobras, the living cobras of the 40 active Saros lunar series. These 'cobras' are in every sense of the term, 'immortal', as the Saros series that they represent are born, die, and are then reborn again, ad infinitum. Thus, to walk along the colonnade between the 'cobras' is to literally walk through time itself. And the narrowing of the space between the two rows of columns, as one proceeds west along the length of the colonnade, can only but reinforce the sense of infinity that naturally arises from being in the midst of these 'immortal cobras.'

In light of the foregoing, it is easy to see why an earthly king would seek to associate himself with the Saros eclipse cycle, because it is such a compelling metaphor for divine power and immortality. Furthermore, it is just as easy to envision how earthly kings would then proceed to adopt that singular Egyptian symbol for the Saros eclipse cycle—the cobra—as a preeminent icon of their royal authority. In fact, there is very persuasive evidence that the kings of ancient Egypt did just this by gracing their crowns and various royal headdresses with the image of the rearing cobra, a symbol they used to convey their royal authority from the very earliest of times to the latest in the long history of the ancient Egyptian civilization.

An ancient Egyptian myth tells a story involving Atum, the creator god, and the first beings that he brought forth into existence as he drifted in the abyss: Tefnut, the first female, and Shu, the first male. Both came into being at the same instant, but after a time they became separated from Atum and were lost in the abyss. He grieved over their absence and sent his eye to look for them. In due course, the eye found them and brought them back to Atum, who rejoiced at their return, but the eye grew enraged when it found that another was in its place. Atum calmed the eye and placated it by binding it around his head in the form of a rearing cobra.[223]

[222] Gardiner, Sir Alan; *Egyptian Dictionary*; p 560; and Budge, E.A. Wallis; *An Egyptian Hieroglyphic Dictionary, Vol. I*, p 151 b; (Note: Gardiner uses the spelling *wꜣḏyt, while* Budge uses the spelling, *uatchit*)
[223] Clark, R. T. Rundle; page 90

The 'eye' that returns and finds another in its place seems to be an unequivocal reference to an eclipse event. And the description of it as being 'enraged' clearly conveys the thought of its being red—a color closely associated with lunar eclipses. There may be other possible explanations for this myth, but in my opinion, none of them come anywhere near to being quite as apt. And the transformation of the eye into a rearing cobra, Atum places about his head, would seem to bring this argument to its only possible conclusion: the rearing cobra is the symbol of the Saros eclipse cycle.

CHAPTER 44

THE COLONNADE; NUMBER AND *MAYET*

One of the central features of the *Heb-Sed* celebration was a ritual known as, 'encompassing the field.' Depictions of this ritual frequently show the king running between a set of markers similar to the pair of horseshoe-shaped cairns in the center of the Step Pyramid's south court. The significance of this ritual are not well understood, but it was believed to be similar to the king's 'circuit of the walls' run around the walls of Memphis, with its apparent emphasis on the king's health and stamina.[224]

The horseshoe-shaped cairns seem particularly well suited for a ritual run. There is no other structure in the south court, with the exception of an 'altar' located immediately adjacent to the base of the Pyramid, so the 'field' is wide open. If a ritual run was staged around these markers, the distance between them may be very significant. From marker to marker, the distance is 123 royal cubits, while a complete circuit is 246 royal cubits. Both of these numbers, particularly 123, present themselves several times in the complex.

a. Instances of the Sequences 1—2—3 and 2—4—6, in the Colonnade

The colonnade is 123 royal cubits in length, as was noted earlier. And, because the distance between its columns appears to narrow subtly, as one moves from the entrance toward the west and the south court beyond, the illusion created is one of parallel lines merging at infinity. If this illusion was intended, then dimensions of the two lines combined would be 246 royal cubits.

The total number of columns in the colonnade is **48**. There are **8** in section III (transept), **16** in section II (western part of the main hallway) and **24** in section I (eastern part of the main hallway). If these numbers are divided by 8, the results are: 6, 1, 2, and 3, respectively. The association of the number 6 with 1—2—3 is very significant with respect to the *Mayet.*

There are a total of 280 reeds in section II, 560 in sections I (408) and III (152) combined, and a total of 840 in all three sections. These three numbers—280, 560, and 840— are related as 1—2—3.

[224] Wilkinson, Toby A. H.; pages 213-214

b. *Phi* and 1—2—3

There are four columns in section II that have 19 reeds each as opposed to the 17 reeds found in the 12 other columns in the section. These four columns have a total of 76 reeds. If we multiply 76 by phi, 1.618, the product is, 122.968 or 123. Similarly, if the number of reeds in the columns of section III, 152, is multiplied by phi, the product is 246.

There is a direct connection between phi and the number sequence 1—2—3, which are the first three numbers in the Fibonacci series.

c. A Moveable Feast

There are several numbers in the colonnade that strongly suggest the ancient Egyptian astronomers who designed and built the Step Pyramid complex were using a solar calendar similar to the modern Gregorian calendar. In fact, they may have been using this same calendar or one substantially like it. For the sake of argument, I will assume that they were using the Gregorian calendar as the numbers in the colonnade support it.

Christian Easter and its Jewish antecedent, Passover, are both 'moveable feasts' in that they are dependent on astronomical phenomena, although not entirely so, thus the date of its observance moves in step with these phenomena. The principal phenomenon is the vernal equinox, the day that the Sun returns to the equator and reaches its zenith at an observer's meridian, and the second is the first full moon following the vernal equinox. These are astronomical events and the Gregorian calendar follows it as closely as possible, but affixes the date of the vernal equinox at the 21st of March for the purposes of determining the date for the feast of Easter. Similarly, the first full moon following the 21st of March is not the astronomical event, but one based on forecasting tools, such as the Metonic cycle of 235 synodic months or lunations, rather than actual observations. This calendar is accurate and reliable for both ecclesiastical and civil purposes, and reasonably follows the ephemerae of the astronomical phenomenon on which it is predicated. To the extent that it gets out of step with the actual astronomical phenomena, there are built in corrective measures, such as leap year, to adjust it. The calendar is based on a period of 5,700,000 years, after which all of the dates for Easter repeat again in exact order.[225]

The Metonic cycle is based on a cycle of 235 synodic months or 19 tropical years, after which the phase of the Moon will repeat on exactly the same date, within approximately 2 hours. Mathematically, the two periods are almost identical: 19 tropical years = 6,939.602 days and 235 synodic months = 6939.688 days. A related cycle is the Callippic cycle, which is equal

[225] Doggett, L. E. Reprinted from the Explanatory Supplement to the Astronomical Almanac, P. Kenneth Seidelmann, editor, with Permission from University Science Books, Sausalito, CA 94965; NASA Web Site, Calendars: http://eclipse.gsfc.nasa.gov/SEhelp/calendars.html

to four Metonic cycles. Interestingly, both the Metonic and Callippic cycles are reasonably accurate, short-term eclipse predictors.

As previously noted, there are four columns in the colonnade's section II, which have 19 reeds each, as opposed to the 17 reeds on the other 12 columns in this section. Quite clearly, these four columns compose one Callippic cycle of 76 tropical years (4 X 19) or four Metonic cycles of 19 tropical years each. If, for example, a full Moon was observed on a certain date, then exactly 19 tropical years or one Metonic cycle later a full moon will again occur on that date. The error rate is cumulative, though, and over time adjustments must be made to the calendar.

Also, the eight columns in section III, the transept, have a total of 152 reeds. This is a number that is a fairly accurate tool for forecasting Easter dates that repeat in the Gregorian calendar. For example, all 100 dates for Easter between 1948 and 2047 will repeat in the exact same order 152 years later, starting in 2100. Such repetitions on this 152-year cycle are rather frequent in the 5,700,000 years of the Gregorian calendar, but the number of dates that repeat after each cycle can contain 48, 52 or 100 dates.[226]

The columns in section II can provide one more number that is relevant to our immediate purpose. If the number of reeds in the 19-reed columns, 76, is multiplied by the number of reeds in the 17-reed columns, 204, the product is: 15,504. If we then divide this number by the number of days in the draconic month, 27.2, the result is, 570, which compares with the number of years of the Gregorian calendar's 5,700,000-year cycle.

Based on the foregoing, I think we can safely conclude that the ancient Egyptian astronomers were using a calendar very similar to the Gregorian calendar and attempting to forecast a moveable event similar to Easter. The question that naturally arises, though, is what was the nature of this event? It seems reasonable to assume that it was closely associated with astronomical phenomena, most likely with the vernal equinox and the first full moon following, but beyond this what are we to conclude? There is little to guide us forward, but there is some rather compelling evidence in the Step Pyramid complex that suggest the event may have been associated with the rituals and celebrations of the god, Osiris, the Egyptian god of the resurrection.

Osiris has a long and complicated history in ancient Egyptian religion, and the rights and rituals associated with him changed over this period. He is not specifically attested in Egyptian records until the pyramid age, where he was celebrated as one of the gods of the trinity identified with the city of Memphis: Ptah, Sokar, and Osiris, who's cosmic roles can be broadly understood as similar to those of the Christian holy trinity: father, spirit and son, respectively. There has been much written about the possible links between the ancient Egyptian religion and Christianity, and it is not my purpose here to delve into this issue in depth, but there are

---

226 Meeus, Jean; *Mathematical Morsels* (Willamann-Bell, Richmond, Va.) 1997, 2nd printing 2000; page 363 and following.

some aspects of the rights and rituals of Osiris that are relevant to our discussion, specifically his resurrection from the dead.

There is little written about the earlier celebrations associated with Osiris, but in what is today referred to as the New Kingdom, his festival was an annual one that lasted some eight days, including three days of lamentation, referred to as the three days of passion where the god lay helpless in his tomb. There were various rights and rituals associated with the festival, but the final concluding act was the 'raising up', which was performed by 'raising up' the '*djed* pillar', a pillar with four horizontal lines draw through its crown. Most Egyptologists believe this pillar is symbolic of a backbone, but, be that as it may, the *djed* pillar was always identified with Osiris and his resurrection, from the very earliest of times. In the Step Pyramid complex, there are numerous representations of the *djed* pillar, that are found both above and below ground. In my opinion, their connection with Osiris or with some earlier incarnation of his, as god of the resurrection, is obvious.[227]

It is tempting to conclude this subject here, but there is the matter of the date of the Osiris festival, and this is where matters get quite messy. Basically, the Egyptian civil calendar was composed of 360 days, three seasons, and 12 months of 30 days each, with five intercalary days added to make it 365 days in length.[228] In the New Kingdom, the festival of Osiris was celebrated in the fourth month of the flood season, which would place the event in our October, marking the onset of the planting season. Here the matter should end, but there is a problem and that is the Egyptian civil calendar was allowed to drift, so that the seasons became disassociated with it, because there were no intercalation provisions to correct it, as there are with the Julian and Gregorian calendars. The civil population was apparently quite content with this, as efforts to reform the calendar over the years were strongly resisted. The authorities recognized the problem and lamented it: "Winter is come in Summer, the months are reversed, the hours in confusion." But they were unable to reform the civil calendar until the Romans forced the issue and imposed the Julian calendar.[229] Even a festival as important as New Year's, which was celebrated around the 20th of June (Gregorian calendar) when the Nile began its annual flood, clearly an event critical for agriculture was allowed to drift and became uncoupled from the seasons, including the flood.

So, what are we to conclude at this point? Just this: there is no way to determine exactly when the date for the festival of Osiris was first set, or whether it had an astronomical connection or not. However, we may conclude by saying it is entirely possible that the original date for the Osiris festival was associated with the autumnal equinox and the first full Moon following, with the festival celebrations following these astronomical events. The Egyptian focus on the autumnal equinox for these celebrations, comports their growing season, while

[227] Clark, R. T. Rundle; chapters III and IV, but in particular page 132

[228] Egyptian Calendar; http://en.wikipedia.org/wiki/Egyptian_calendar

[229] Gardner, Sir Allen; *Egypt of the Pharaohs* (Oxford University Press) 1961; pages 64-65

the Judeo-Christian focus on the vernal equinox for the Easter celebration, comports with their growing season. Either the autumnal or the vernal equinox can be used to accurately reset the astronomical calendar and synchronize time with the seasons, and in this respect, the two events are interchangeable.

CHAPTER 45

VENUS TRANSITS AND THE PHOENIX CYCLE

The phoenix is the mythological firebird, a fabulous creature that appears only once every 500 to 1,000 years, returning to Earth after traveling to some far distant land. On its return, it renews the life force and is then consumed in a blazing fire. After the fire burns itself out, the offspring of the phoenix rises up from the ashes of its dead parent and begins its own long journey. [230]

This myth is a common one and versions of it can be found in many cultures around the world. In ancient Egypt, the phoenix was called the *benu* bird, which was associated with a peculiar pyramid-shaped perch, called the *ben-ben* stone that was located in the great temple of Atum in the city of Heliopolis—City of the Sun.[231] The Greek's may have received the phoenix myth directly from the ancient Egyptians or by way of the Phoenicians, who some believe may have taken their name from that of the bird. If the latter is true, then the Phoenicians most likely received the myth from the ancient Egyptians, who were close trading partners of theirs from time immemorial.

The first syllable of the word, phoenix, has a phonetic similarity to phi. The bird shown above is called the *neḥ* [232] bird, a word that may be the source for the second syllable in the word, phoe-nix. If this is accurate, then the Egyptian origins of this Greek word are evident. The absence of the syllable *phi* in the Egyptian word for the *neḥ* bird most likely reflects the reluctance of the Egyptians to speak or write of phi, except in cryptic or oblique terms. In any case, the *neḥ* and *benu* birds are most likely one and the same.

On the face of it, there is no apparent relationship between the Egyptian name for the *benu* bird and its Greek name, phoenix. However, as discussed above, the Greek name is phonetically related to the word phi, which in my opinion is the key to not only an understanding of the origins of the *benu* bird, but its religious significance to the ancient Egyptians, as well. The

[230] Wikipedia Contributors, "Phoenix (mythology)"; Wikipedia the Free Encyclopedia; http://en.wikipedia.org/wiki/Phoenix_(mythology)

[231] Lehner, Mark; *The Complete Pyramids, Solving the Ancient Mysteries*; page 34

[232] Budge, E.A. Wallis; An Egyptian Hieroglyphic Dictionary, Volume 1; page cxiv, glyph 7,8.

Egyptian words *benu* and *ben-ben* both have graphic sexual connotations, particularly *ben-ben*, which has a meaning closely connected to onanistic sexual arousal.[233] This is no doubt a crude connotation, but it is also an important one in the ancient Egyptian religious beliefs as to how God initiated creation. While this explicitly sexual reference may be lacking in discretion and modesty, it leaves no doubt that the ancient Egyptians believed that God alone was the source of creation. Accordingly, both words are also understood to mean procreation, and it is in this wider context that the meanings of the words *benu* and *ben-ben* are properly understood.

Earlier in this work, we discussed the religious significance of both the word and concept of 'phi' to the ancient Egyptians. It must be re-emphasized here that *phi* was a profoundly sacred concept to the ancient Egyptians and even the name could not be spoken or written. The Greeks, on the other hand and probably the Phoenicians as well, felt no such inhibitions and freely spoke of it, which is why there was no reluctance on the part of the Greeks to call the *benu* bird by its more appropriate name, phoenix, the bird of phi. The question then arises as to how and why a fiery bird came to be associated with phi? The fact that the *ben-ben* perch for the *benu* bird was found in the great temple of Atum at Heliopolis, the center for ancient Egypt's astronomers and from whence all of her measures of time and calendars derived, is critical to an understanding of how this association came about.

The ancient Egyptians had two ideas as to how and when life originated and how it continues to sustain itself. They are not the same; rather they are distinct efforts at explaining this deepest and most profound of mysteries. In their way of thinking, both of these stories or myths are accurate, but neither alone provides a complete understanding of this mystery. Underlying both myths is the ancient Egyptians belief that time and life are the same, and that all cycles of time originate with God and are recurring and self-perpetuating.[234]

According to the first myth, it was the *benu* bird that arose from the abyss on his perch and broke the silence of an eternal night with his call, thereby initiating life and the various cycles of time, and determining 'what is and what is not to be.' As such, the *benu* bird manifests the connection between the mind of God and creation, bringing to life with his call that which God wills into existence. Thus, the *benu* bird is the ancient Egyptian symbol for the divine word itself, the Greek *logos*.

In the second myth, the *benu* bird brought the vital essence of life into the world from a distant source that lay far beyond the world—the 'Isle of Fire', the place where the gods were born and from whence they arose into the world. In this myth, the bird is not only the bringer of life at the beginning of time, but periodically returns to the Earth, renewing the life force and the cycles of time initiated by God.

---

[233] Ibid; pp 34-35

[234] Clark, R.T. Rundle; *Myth and Symbol in Ancient Egypt*; pp 245-246

It is the second myth that is the basis for the mythical bird of fire, the phoenix. In the Middle Kingdom of ancient Egypt, the *benu* bird was identified with the soul of Osiris, the god of resurrection, and with the planet Venus. The association of the *benu* bird with the planet Venus is an important one for our purposes.[235] The time cycles of Venus provide us with a basis for an understanding of the deeper astronomical significance of the *benu* bird myth and of the importance that these cycles played in the Step Pyramid complex.

Most of these cycles will be found in the *Heb-Sed* court, but several important particulars from the planet Venus are also found elsewhere in the complex and will be introduced first, because they are important to the *Heb-Sed* discussions.

[235] Ibid; page 246

CHAPTER 46

THE PARTICULARS OF VENUS

a. *Temenos* Wall (Redan Count)

There are several particulars relating to Venus and its rotational and orbital periods in the *redan* count of the *temenos* wall, which, as we noted earlier, memorializes so many other cycles relating to time and the Moon. As noted earlier, when we first introduced the various major parts of the Step Pyramid complex, there are 192 *redans* along the *temenos* wall and 226 *redan* and *redan*-equivalents, counting the gates and bastions. These numbers correspond with the number of Venus solar days that it takes for Venus to orbit the Sun, 1.92, and the equivalent time period in Earth days, 224.7.[236] The one figure is a precise match, the other, is off by a day, unless the two redan equivalents from the entryway are excluded from the total which would make it 224 instead. Significantly, though, both figures are found on the *temenos* wall, which is where the numbers for the Earth and Moon orbital periods are found. This suggests that the ancient Egyptian astronomers were very familiar with the orbital and rotational periods of the planet.

b. Transept of the Entryway Colonnade

Another Venusian particular, which relates to the 243-year Venusian transit cycle, is memorialized in the colonnade. The 152 (8 X 19) reeds in the transept mark the number of Venusian synodic years in the Venusian transit cycle and its presence in the transept is appropriate enough, as the columns of the transept are used to track the eclipse members of the longer Saros series. The Venusian synodic period is equal to 584 Earth days and marks the time period between successive appearances of Venus at inferior conjunction with Earth. Five such successive appearances constitute an octaeteris cycle and defines a five-pointed star pattern inside the Venus orbital path, which is the reason that Venus is often identified with the five-pointed star. Meanwhile, 152 such appearances constitute the 243-year transit cycle: 152 X 584-days/synod = 88,768 days ÷ 365.242 days/year = 243.03 years.

---

[236] NASA web site. "Venus Fact Sheet"; https://nssdc.gsfc.nasa.gov/planetary/factsheet/venusfact.html

c. *Temenos* Wall (Panneaux Count)

This 243-year cycle may also be memorialized on the *temenos* wall, where its 4,383 panneauxs are almost equal to eighteen 243-year cycles: 18 X 243 = 4,374. However, if only the eastern segment of the wall— with its 1,458 panneauxs—was used for tracking the cycle, then there are exactly six (06) 243-year Venusian transit cycles represented: 6 X 243 = 1,458. There is an architectural feature on the eastern segment of the wall that appears to support this possibility.

In the court of the Chapel or Tomb of the North (Princess Int-ka-s Monument) there is a grouping of three, slender papyrus columns on its eastern wall that backs up against a mass of rubble fill and is directly opposite to the eastern segment of the *temenos* wall.[237] Lauer restored these columns and later rendered a drawing of how he envisioned their original appearance, which included an arched pediment above them.[238] These same elements—the three papyrus columns and the arched pediment—are also found on a hieroglyph that has as its root the word, *ṯḥn*, which means: 'sparkle' or 'dazzling', words that are typically associated with the planet Venus, although in the case of the hieroglyph the pediment, which is the hieroglyph for "sky", is superimposed on the three columns.[239] If this architectural feature does indeed represent the planet Venus, then its close structural association with the eastern segment of the *temenos* wall, with its accurate measure of the Venus transit cycle, seems both readily apparent and very appropriate.

It is also interesting that there are approximately 18.03 Venus transit cycles of 243 years in the entire *temenos* wall with its 4,383 panneauxs: 18.03 X 243 = 4,381.29 ≈ 4,383. The number 18.03 is, of course, the length of the Saros cycle, as measured in tropical years. Is this significant? Yes, very much so!

Venusian transits are relatively rare events Currently, four of them occur in the 243-year period of this cycle: 121.5 years without transits is followed by two transits separated by 8 years, then there is another period of 105.5 years without transits, which is followed by two more transits separated by 8 years, for a total of 4 transits in the 243 years of the cycle. However, there are long periods of time when there are only two or three transits in this 243-year period. Transits occur during the 'transit season' when the Sun is at one of the two Venusian nodes intersecting the ecliptic at the same time that Venus passes through the node. (Venus has both

---

[237] Firth and Quibell; plate 78

[238] Lauer, Jean-Philippe; CAHIER No. 9 ÉTUDES COMPLÉMENTAIRES SUR LES MONUMENTS DU ROI ZOSER Á SAQQARAH; Supplement aux Annales du Service des Antiquities de L'Egypte; LE CAIRE, IMPRIMERIE DU L'INSTITUTE FRANCAIS D'ARCHAEOLOGIE ORIENTALE ; MCMXLVIII; plate VI, no 1. 'Les colonnes papyrus de la Maison du Nord'

[239] Gardiner, Sir Alan; Egyptian Grammar; p 505, item S 15

an ascending and descending node where its orbit intersects the ecliptic.) The transit season lasts eight years during which one or two transits can occur, depending on the proximity of the first transit to the center of the Sun, i.e. the closer the first transit is to the Sun's center the less likely there will be a subsequent transit in the transit season.[240]

Earlier we made mention of the five-pointed star blocks found under the complex's Pyramid and South Tomb, and the connection of these blocks with the Venusian transit cycle. The fact that there are some blocks with two stars on only one face of the block and others with two stars on two, opposite faces is significant in this regard. There are long periods of time where there are only single transits (i.e. after a transit, there is no transit eight years later in the season) and other periods where there are always two transits in each transit season, i.e. a transit is followed eight years later by a second one in the transit season. Jean Meuss observed, "Between the years

-4000 and +8000, there are 27 single and 23 double transits at the ascending node, and 25 single and 24 double transits at the descending node. We are lucky to live now in a period of double transits: between the years 1500 and 3000, all transits are double. Between 3800 and 5200, all will be single, as they were between the years -500 and +500." [241] In my opinion, the star blocks with two stars on only one face represent single transit eras, while the blocks with two stars on two, opposite faces represent eras with double transits. The proportion of these aforementioned numbers, i.e. 27 single to 23 double, and 25 single to 24 double (combined, 52:47), is roughly comparable to the proportion of the blocks found under the South Tomb, i.e. 13 single face blocks to 12 double face blocks. The presence of this proportion in the number of star blocks would seem to further corroborate their intended significance and relationship to the Venusian transit cycle. I know of no definitive count of the number of star blocks under the Step Pyramid.

d. The Octaeteris Calendar Cycle

The octaeteris cycle is not an eclipse cycle, but a calendar one. It is a time period that is based on the fact that the Moon will appear in exactly the same phase on the same day of the year, within a day or two, every eight years. It is closely associated with and is nearly synchronous with the Venus appearance cycle of eight years. Venus, as it progresses along its near-circular orbit around the Sun, enters inferior conjunction with Earth five times in eight Earth years or thirteen times in Venusian years. The interval between successive inferior conjunction events is termed the Venusian synodic period and is some 584 Earth days in length (8/5 X 365 = 584). As Venus approaches or departs inferior conjunction with Earth, the Moon is close by in crescent phase, which creates a very spectacular and beautiful astronomical display. Venus is clearly visible as it approaches inferior conjunction in the evening twilight, when it

---

240 Meuss, Jean; Mathematical Astronomy Morsels III; pp 264-276; see also, Venus Transits: http://en.wikipedia.org/wiki/Transit_of_Venus

241 Ibid page 272

is referred to as the "evening star"; as it departs inferior conjunction, it is visible in the dawn twilight and is referred to as the "morning star." It takes approximately eight days for Venus to pass from an evening star to a morning star, a period of time when it is only faintly visible, if at all, in the daylight. The octaeteris cycle may have been of very special interest to the ancients, because it could be used to track the appearance of this display, which has fascinated cultures around the world since time immemorial. For the purposes of this book, whenever I refer to the octaeteris cycle it includes the Venus appearance cycle as well and the two cycles are considered to be one and the same.

The, 5—8—13 relationship (5 inferior-conjunctions—8 Earth-years—13 Venusian-years) of the octaeteris cycle traces a five-pointed star pattern, a pattern with an intricate relationship to *phi* (1.618), which is often referred to as the "golden mean." This 5—8—13 relationship itself is reflective of *phi*, as it is made up of the 4th, 5th and 6th numbers in the Fibonacci series, a series named after an Italian mathematician of the Middle Ages, who may have learned of it during his visit to Egypt. As this series progresses, it leads to ever finer measures of *phi* between any two successive member-numbers; e.g. 8/5 = 1.6, while 21/13 = 1.615. It is this 5—8—13 sequence from the Fibonacci series that is used to memorialize the octaeteris cycle in the Step Pyramid complex, where it is found in the *Heb-Sed* court.

e. The *Heb-Sed* Court

The *Heb-Sed* Court in the Step Pyramid complex is on the east side and just above the entryway colonnade. The only access to it was by means of a long passageway leading off from the colonnade, near its entrance in the *temenos* wall. The *Heb-Sed* court is one of the most intricate parts of the complex and, broadly described, is composed of a series of 'chapels', thirteen arrayed along the west side of the court and twelve along the east side, surrounding an open court dominated by a low stone dais with a twin stairway. All of the chapels are dummy structures with solid interiors, although many are provided with an elaborate entryway, including in some cases vestibules. Several have niches which are believed to have held statues, though this is not known for certain. The stone dais is found toward the southern end of the court. The twin stairways that provide access to the platform of the dais are both found on its eastern face.

It is from the shape and disposition of the thirteen shrines on the western side of the court that the 5—8—13 octaeteris sequence is found. In the following drawing[242], these shrines are seen arranged north to south. At the northern end, there is a group of three shrines, a flat-roof one followed by two round-roof ones. This group is followed by a gap and then by another group of ten shrines. In this second group there are three round-roof shrines, a flat-roof shrine, then five round-roof shrines, and finally by a flat-roof shrine. (See Appendix 3, Plate 67.)

[242] "Drawn by Philip Winton, © Thames & Hudson Ltd., London, from *The Complete Pyramids*, by Mark Lehner, Thames & Hudson".

Interestingly, four different number groups, each with a different three-number sequence from the first six numbers of the Fibonacci series, can be observed in the shape and disposition of these shrines. Taking the shrines as described earlier in their north-south disposition, the first flat-roofed shrine represents the number, one, the following two round-roof shrines represent the number, two, the three round-roofed shrines following the gap represent the number, three, and

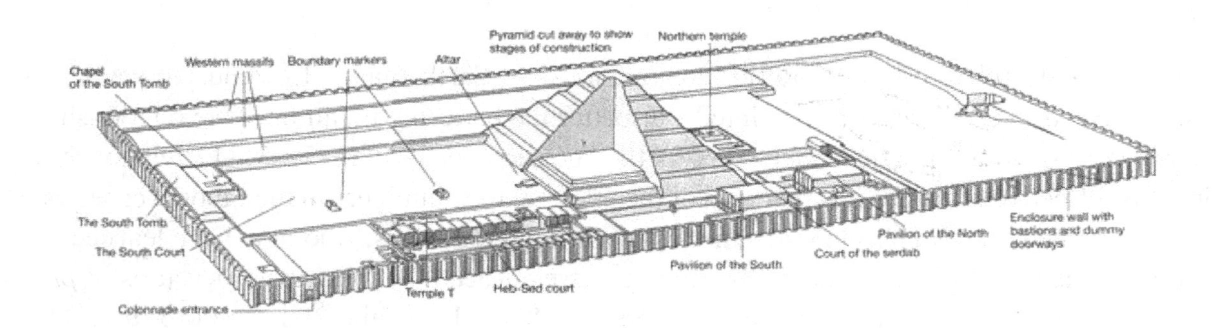

the five round-roof shrines following the intervening flat-roof one represent the number five. Then, reversing course and proceeding from south to north, all the round-roofed shrines found in the group of ten (eight) represent the number eight, and then all thirteen of the shrines on the west side of the court, taken as a whole, represent the number thirteen. These numbers, when grouped in the order presented, describe not only distinct, three-number sequences from the Fibonacci series, but four different progressions or cycles, as shown below.

1—2—3	First three whole integers
2—3—5	Metonic lunar cycle in synodic months (235)
3—5—8	Inex eclipse cycle in synodic months (358)
5—8—13	Venus inferior conjunction cycle and octaeteris calendar cycle, in 5—8—13 sequence, i.e. 5 Venus inferior conjunctions (octaeteris cycle), 8 earth orbits in tropical years, and 13 orbits of Venus measure in earth tropical years
1—2—3—5—8—13	First six numbers from Fibonacci series

The significance of each of these progressions and cycles should be immediately obvious, especially the Inex eclipse cycle, which I believe explains the meaning and purpose of the *sed* festival itself, as was detailed in an earlier chapter. However, there is another number that can be derived from the court's shrines that represents an important particular from the octaeteris cycle. The total number of shrines in the court is 25, which is the number of Venusian days in the octaeteris cycle: 13 Venusian years X 1.92 days/year = 24.96 Venusian days. There are only 1.92 Venusian days in the Venusian year, a number that is memorialized in the *temenos* wall as detailed earlier, because of the extremely slow rate of rotation of the planet. Thus, there are 4.99 days in the Venusian synodic period (24.96 ÷ 5 = 4.99 days).[243]

While each of these four, 3-number sequences is individually significant, the question arises as to whether or not there might be a broader, deeper significance in these four sequences when considered as four parts of a whole. It hardly seems likely that there could be a relationship or a common focus between an eclipse cycle (Inex), a lunar cycle (Metonic), and a calendar cycle (octaeteris), but there is and this relationship may reveal the underlying meaning of the phoenix myth.

There is one further area of the *Heb-Sed* court to be addressed. This is the "T-temple", which is located behind the shrines along the western side of the court and which can only be reached by a single passageway that gives access to it from the *Heb-Sed* court. The T-temple is a rather simple structure, but the exact purpose served by this structure is not known. However, among the more noteworthy features of the temple are the three fluted columns in its interior: one to the north, standing in isolation, and two to the west, standing as a pair. Do they signify the right triangle with sides equal to one and two, respectively, which was discussed earlier as a proxy for *phi*? If so, this would explain the nearby 'quarter turn' in the wall along the passageway that gives access to the temple. It is the quarter turn with radius equal to one, swung from the point of the angle formed by the shorter side and the hypotenuse, which is used to produce the *phi* proportion, as described earlier.

[243] Venus; Wikipedia; http://en.wikipedia.org/wiki/Venus

PART IV

EARTH AND THE GREAT PYRAMID

CHAPTER 47

MEASURES OF THE EARTH AND THE GREAT PYRAMID

The absence of any significant and specific reference to the Earth at the Step Pyramid complex is surprising, though there may be references that have eluded the author of this work. However, there is a nearby structure which incorporates a very significant amount of Earth measures, and appears to function as a scale model of the Earth. That structure is the Great Pyramid of Giza, a nearly contemporary structure and one located less than eight international miles north of the Step Pyramid complex. In addition to their being contemporaries in time and geographically close to one another, the knowledge incorporated into the two structures is complementary in many ways.

The Great Pyramid is without a doubt the most investigated and measured ancient monument in the world. It has occasioned a great deal of speculation, some of it well documented and well-reasoned, but also much that is baseless and even bizarre. This is unfortunate and has made serious discussions of the structure all but impossible and rendered most discussions of it, outside its widely-accepted role as a pharaoh's tomb, odious in the extreme. I seriously hesitated to even bring it up, lest it cast the rest of this work in an unfavorable light, but it must be examined, even if briefly, because of the demonstrated sophistication of the Step Pyramid complex and the lack of any significant references to the Earth in it. However, in our examination of the Great Pyramid we will avoid speculation as much as possible, and focus, instead, on the measures of the structure and its location, all of which pertain to the Earth. The connection of the Pyramid with the Earth has been recognized throughout history and it is one that is widely accepted, though not without significant disagreement.

a. Description and Significance of its Location

The Great Pyramid was built by Pharaoh Khufu, the second king of the 4th dynasty, who reigned for some 23-32 years, beginning around 2,551 BCE.[244] Until modern times, it was

[244] Mark Lehner, *The Complete Pyramids, Solving the Ancient Mysteries* (Thames and Hudson, London, 1997) pages 106-199.

the largest and tallest stone structure ever built. Its scale is massive—literally a man-made mountain—and its many superlatives do not begin to do it justice. It has inspired and awed men throughout the ages. It is also the only surviving member of the Seven Wonders of the ancient world.

The location and orientation of the Great Pyramid are among its most striking features. Its geographic location is: 29°51'45.03" N, latitude, 31°08'03.69" E, longitude, almost exactly on the 30th parallel of latitude. It's sides almost perfectly align to the cardinal points, with an average deviation of less than 3'6" of arc.[245] It's orientation was so perfect that the Pyramid was used to calibrate compasses and surveying instruments in earlier times.

The location of the Pyramid at the 30th parallel of latitude is significant, as it is located approximately 1,800 nautical miles from the equator (1,800 nautical miles X 6,000 nautical feet of 308mm each) or 1,796 nautical miles of 6,000 nautical feet of 308.66mm each. If the base diagonals of the Pyramid are extended form its location, along their respective 45° angle to the equator to the equator, then the arc of the equatorial circumference marked off by the diagonals is 3,600 geographic miles (1/6th of 21,600 geographic miles), or 60° of 360°, of the equatorial circumference.

Then, if the area of the circle made by the equatorial circumference of the Earth is divided into six equal segments of 3,600 geographic miles or 60° of arc each (the arc segment marked by the Pyramid's diagonals), creating a hexagon of six equilateral triangles, then the radius of this circle is 3,437.74 miles and each side of each of the six equilateral triangles is also 3,437.74 miles each.

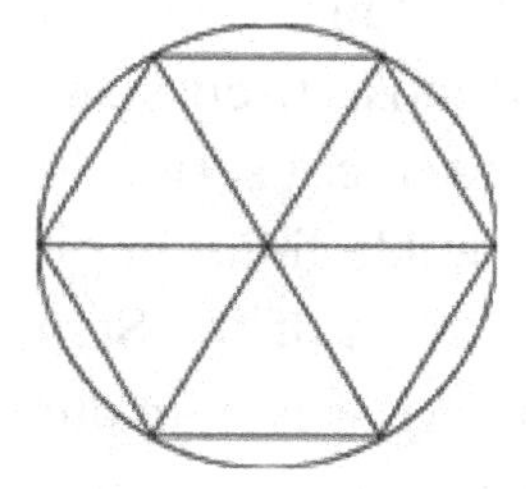

There is a further significance of the equilateral triangle in respect of the Great Pyramid's location at 30° N.

b. The Equilateral Triangle and the Equinoxes

On the equinoxes, both vernal and autumnal, the Sun stands directly above Earth's equator. At local apparent noon (LAN), which is the Sun's maximum observed height (altitude), on the meridian of the Pyramid's location, on the day of the equinox, the Sun is observed to be

[245] Wikipedia contributors, "Great Pyramid of Giza", https://en.wikipedia.org/wiki/Great_Pyramid_of_Giza, Wikipedia the Free Encyclopedia, accessed November 15, 2016.

60° above the horizon. (The Sun's height at LAN can be predicted as follows: 90° - 0° (Sun's declination at equinox) = 90°; then, 90° - 30° (latitude of observation at Great Pyramid) = 60°, which is the predicted altitude of the Sun at LAN on the equinox as observed from the Great Pyramid.) At this same instant, the Sun is also observed to be at an altitude of 60° at 30° S on the Pyramid's meridian in the southern hemisphere at LAN. An equilateral triangle forms at thisexact instant, with the Sun at the apex of the triangle and these two 60° angles of observation. This can be best visualized from the drawing shown below.

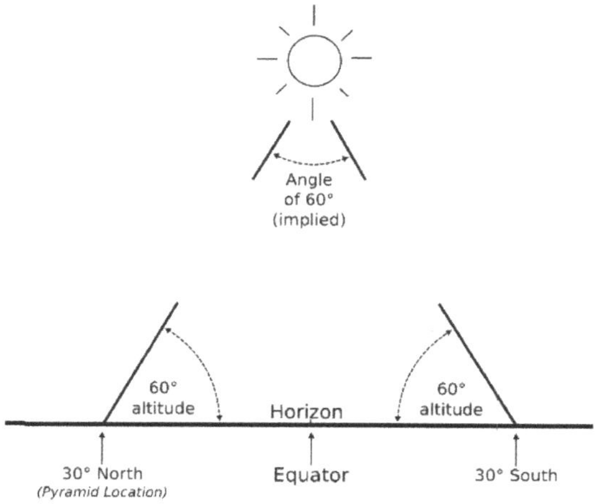

Angle
of 60°
(implied)
60°
altitude
Horizon
60°
altitude
30° North
(Pyramid Location)
Equator
30° South

The two foregoing sections demonstrate the significance of the Great Pyramid's location and its relationship to several manifestations of the equilateral triangle, as these arise from the measure of the Earth. The equilateral triangle is also at the heart of the measure of the Vesica Piscis, which is also related to the Great Pyramid, which will be discussed in detail in a later section.

c. The 36°—54°—90° Triangle and the North Ecliptic Pole

The 36°—54°—90° triangle is a close approximation of Pythagoras's 3—4—5. This triangle forms daily at the Great Pyramid's location with the North Ecliptic Pole, as the North Ecliptic Pole appears to circle the North Celestial Pole in the heavens, some 24° in celestial arc from the North Ecliptic Pole. (This circling motion is only an apparent motion, as in reality it is the North Celestial Pole that is slowly circling the North Ecliptic Pole in a slow, 25, 920-year long cycle. The location of the North Ecliptic Pole is and has always been stationary and unchanging.) The triangle is formed when the North Ecliptic Pole is directly above the North Celestial Pole, which places it at 54° celestial arc above the horizon at the Great Pyramid's location. This arc measure is the sum of the 30° celestial arc location of the North Celestial Pole and the 24° celestial arc location of the North Ecliptic Pole above it, for a total of 54° celestial arc. The triangle is completed by assuming that the angle on the distant horizon directly below the superimposed Poles is 90°, which would make the angle at the North Ecliptic Pole 36°. (See following figure. Also, note that the 30°—60°—90° is formed as well.)

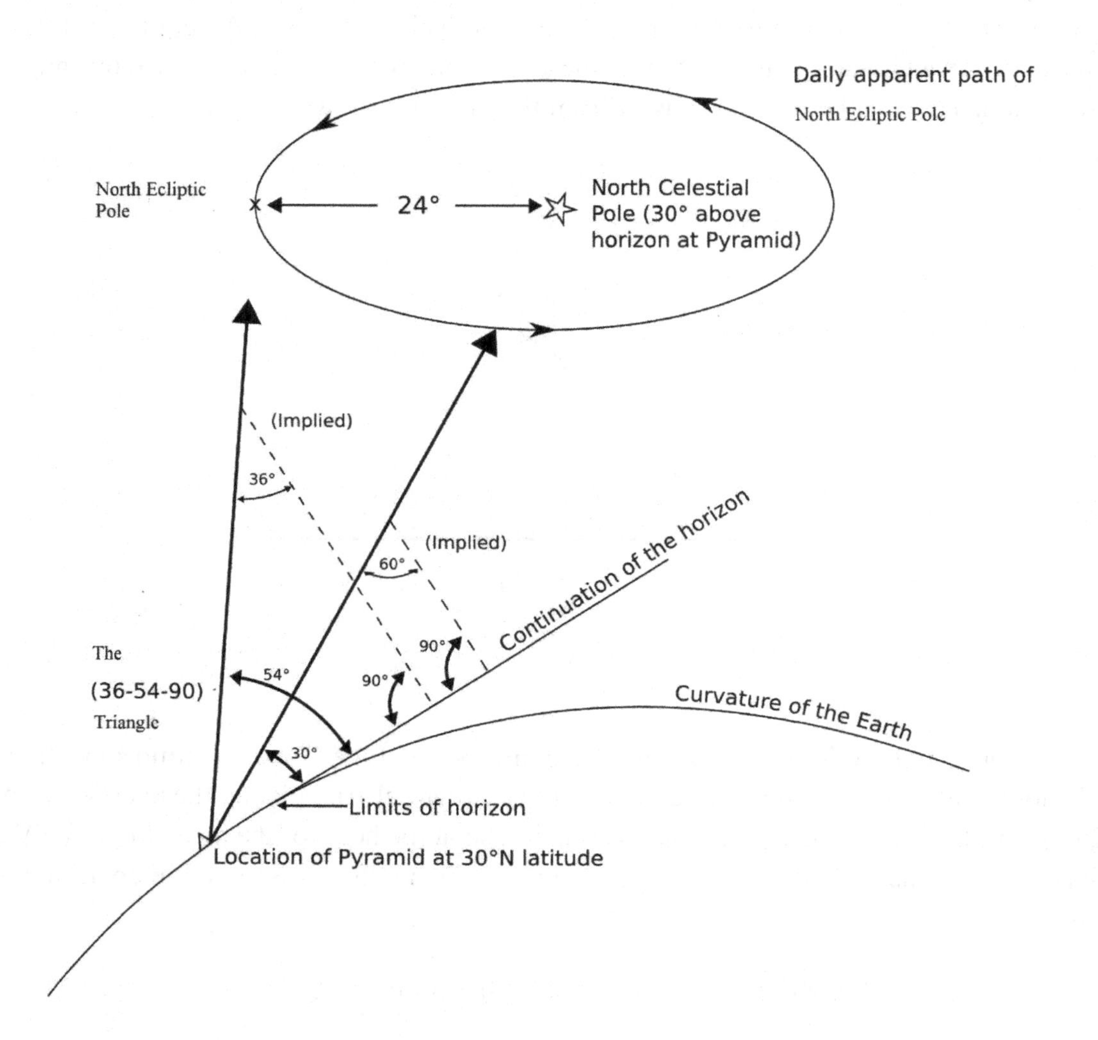

CHAPTER 48

MEASUREMENTS OF THE GREAT PYRAMID AND THEIR RELATION TO MEASUREMENTS OF THE EARTH

The Pyramid's dimensions are generally accepted to be: height, 146.59 meters and length of base, 230.4 meters. These figures are based on a royal cubit of 524 mm: height, 279.75 royal cubits, which is generally agreed to be 280 royal cubits, and length of one side of the base, 439.7 royal cubits, which is generally agreed to be 440 royal cubits. "Generally agreed to be" should be interpreted to mean the intended design dimensions, although even this understanding may not be entirely accurate.

Agatharchides of Cnidus, who lived around 100 BCE and spent a considerable time in Egypt, wrote that the base of the Pyramid was 1.25 stadia on a side or 5 stadia in perimeter. As a stadium is 600 feet, and the measure of one side of the base 770 Egyptian feet, as shown above, this does not seem to be accurate, but closer analysis proves he is correct.

The perimeter of the Pyramid is 4 X 440 royal cubits = 3,080 Egyptian feet. If we assume a common Egyptian foot of 300 mm in our calculations, a "nominal measure"[246], then there are 3,080 Egyptian feet in the perimeter of the base or 3,000 feet of 308 mm in the perimeter. (The foot of 308 mm is a good approximation of the modern measure of the nautical foot.) The figure of 3,000 feet is 4 X 750 Egyptian feet (base side), or 1.25 X 600 feet/stadium for a base side, thus Agatharchides is correct. The measure of the perimeter (3,080 Egyptian feet of 300mm, or 3,000 feet of 308mm) is also a good approximation for the measure of 30 arcseconds (0.5 arcminutes) of latitude, which would give a meridional circumference of 24,803 international miles (3,000 ft. X 2 X 308mm each X 21,600 ÷ 5,280 feet of 304.8mm each = 24,803 international miles). The modern figure for the Earth modeled as a spheroid is

[246] This "nominal measure" is at variance with the actual dimensions of the Great Pyramid, which are based on a royal cubit of 524 mm that defines the foot as 524 · 4/7 = 299.43 mm. Thus, the assumption of a foot of 300 mm, the usual measure of the Egyptian foot, also implies a "nominal measure" royal cubit of 525 mm. Are these "nominal measures" the intended or design measures of the Pyramid? Probably not; instead, they represent easy numbers to recall for an oral tradition, i.e. 3,000 feet or 5 stadia on the Pyramid's perimeter.

approximately 24,802 international miles. For the Earth modeled as an ellipsoid the figure is 24,860 international miles. (There is further discussion on this point in Chapters 55, 58, and 59.)

The perimeter of the base of the Pyramid is 3,000 feet of 308mm each or 30 arcseconds of latitude

For the equatorial circumference, which is 24,902 international miles, the difference is more significant when using this measure for 30 arcseconds of longitude. However, if we assume that the equatorial bulge is 1/280 of the height of the Pyramid (actual figure is approximately 1/298), then the measure for the foot becomes 308 mm + 1/280 X 308 mm = 309.1 mm, and the circumference is measured as: 21,600 miles X 6,000 feet of 309.1 mm ÷ 5,280 feet of 304.8 mm each = 24,891.7 international miles (the modern measure is 24,902 international miles).

a. Scale of the Pyramid to the Earth

If the height of the Pyramid 280 royal cubits, is taken as the scaled radius of the Earth, then the Pyramid's scale is approximately 43,200:1[247].

280 royal cubits X 524 mm/royal cubit ÷ 309 mm/geo. foot = 474.8 geo. feet in height
474.8 geo. feet X 43,200 ÷ 6,000 geo. feet/geo. mile = 3418.6 geo. miles in Earth's radius
3,418.6 geo. miles in Earth's radius x 2 = 6,837.2 geo. miles in Earth's diameter
6,837.2 geo. miles X π (3.1416) =21,479.7 geo. miles in Earth's circum. (actual: 21,600)

The Pyramid's scale to the Earth is 43,200:1, which is a number from the Pattern

[247] Tompkins, Peter; *Secrets of the Great Pyramid*, (Harper & Rowe Publishers, Inc. New York, NY; 1971); Appendix, *Notes on the Relation of Ancient Measures to the Great Pyramid*, by Livio Catullo Stecchini, Page 378

CHAPTER 49

IMAGE OF THE EARTH IN THE GREAT PYRAMID

The internal structures of the Great Pyramid generate the image of the Earth. This image arises from the fact that the height of the Pyramid is equal to the length of the diagonals of the base, each of which is 280 royal cubits in length. These radii generate a half sphere—Earth's northern hemisphere—which is represented by the Pyramid. (See figure following.) The structures referred to in the figure, and, in particular, the centerline of the ceiling of the fourth reliving chamber above the king's burial chamber, generate the Tropic of Cancer and the Polar or Arctic Circle.

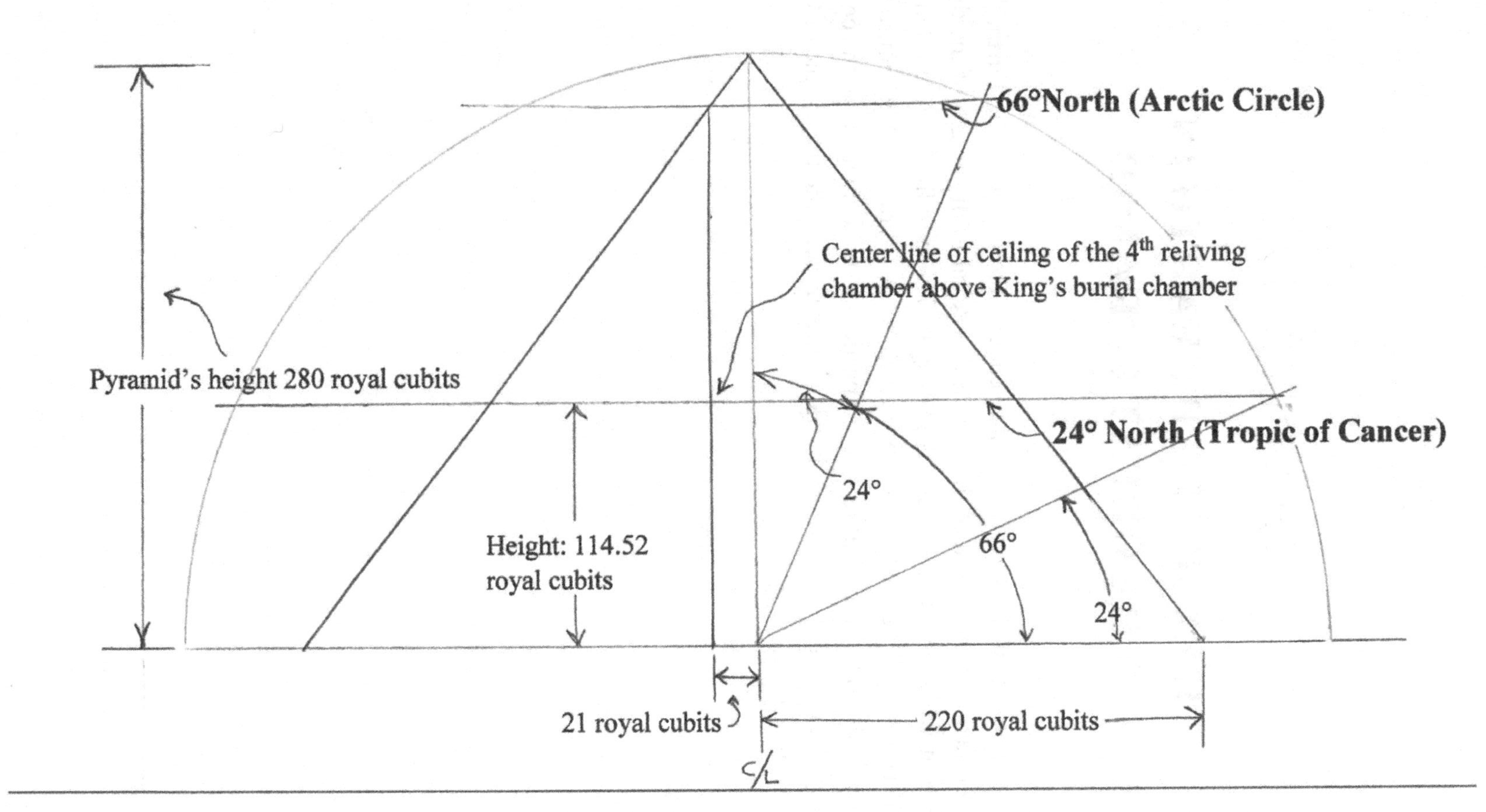

NOTES: 1) The scale of the drawing is ½ inch = 40 royal cubits of 524mm each. 2) All measures are taken from the surveys conducted by Maragioglio and Rinaldi, and/or Sir Flinders Petrie (see Appendix 4). 3) Currently, the Arctic Circle is at 66.34° North while the Tropic of Cancer is at 23.44° North.

CHAPTER 50

THE GREAT PYRAMID AND THE GOLDEN MEAN

The golden mean, or *phi*, is 1.61819 which is often represented by the symbol *phi*, and is one of the keys to understanding the Great Pyramid. This begins with the angle of the Pyramid's slope, which is 51°50'40", which equates to approximately 51.8°.[248] The trigonometric measures of this angle are:

Sine 51.8° = .785 ≈ $1/\sqrt{\varphi}$
Cosine 51.8° = .618 ≈ $1/\varphi$
Tangent 51.8° = 1.272 ≈ $\sqrt{\varphi}$

The Tangent measure of 51.8°, (1.272) relates the Pyramid's major dimensions to one another: The dimensions of the Pyramid's cross section triangle are: 220 royal cubits (base) X 280 royal cubits (height) X 356 royal cubits (hypotenuse). Starting with the base, 220 royal cubits, the height is 220 X 1.272 = 279.84 ≈ 280 and the hypotenuse (length of the slope of the Pyramid's faces) is 280 X 1.272 = 356.16 ≈ 356. Also, the hypotenuse (356) divided by the base (220) = 1.61819, the golden mean.

These measures and their relationship to *phi* would be interesting, but they have far greater significance. When the sine of 51.8° (.785) is multiplied by four, which represents the four faces of the Pyramid, the result is: 4 X .785 = 3.14 or pi, π, the key to measuring the circumference of a circle, π X d (diameter of circle), and the area of a circle, πr^2 (radius of a circle). Furthermore, when π is multiplied by four again, the result is: 4 X π = 4π, which is the key to measuring the surface area of a sphere, $4 \pi r^2$. The measure of the surface area of a sphere is very important in the creation of plane or flat surface maps from a sphere. In practical terms, the measure of the surface area of a sphere, $4 \pi r^2$, allows for the creation of equal area maps of the Earth and of the Celestial Sphere.

---

[248] Mark Lehner, *The Complete Pyramids; Solving the Ancient Mysteries* (London: Thames and Hudson, Ltd., 1997.) Page 108, caption.

The sine value of the Great Pyramid's slope angle, 51.8°, generates the measure of π.
Sin 51.8° = .785 X 4 (for the four sides) = 3.14

CHAPTER 51

THE GREAT PYRAMID AND MAPS OF THE EARTH

The Earth is a sphere, or close enough to a sphere, for most practical purposes. The Surface features of the Earth can be mapped out on the surface of a sphere of globe with reliable accuracy and little distortion. However, spheres depicting the Earth's surface features, while accurate, are difficult to use in many practical applications, such as computing angles and distance between locations, and calculating the square areas of features. Flat surface or plane surface maps are much easier to reference and use for these and other practical purposes. Unlike a sphere, they are easily drawn, very transportable, and they greatly facilitate reference and study. However, no matter the format chosen to depict the surface of a sphere on a plane surface (equal area, Mercator or conformal, azimuth or equidistant, etc.) there will always be distortion in the plane surface depiction, because the surface of a sphere simply cannot be accurately rendered onto a flat surface.[249] (It should be noted that all of the forgoing remarks also apply to plane surface maps of the celestial sphere, which are used for mapping the features of the heavens, and for tracking the movements of heavenly bodies and their rhythms.

a. The Significance of the Pyramid's Location and the Vesica Pisces

The significance of the geographic location of the Great Pyramid manifests itself in two important ways. First, its diagonals mark off 1/6th of the circumference of the Earth at the equator. And second, an equilateral triangle is created when the Sun stands directly above the equator on the equinox, with the Pyramid's location at 30°N and its corresponding location in the southern hemisphere at 30°S. These two factors have been described in foregoing sections, and both are important for generating plane surface maps of the Earth. The keys to this process are the Vesica Piscis, a familiar geometric figure shown below, and the formula for calculating the surface area of a sphere ($4\pi r^2$). It is no coincidence that the formula for calculating the surface area of a sphere is also found in the measures of the Great Pyramid. Recall that π is

[249] Wikipedia Contributors, "Map": https://en.wikipedia.org/wiki/Map, Wikipedia, the Free Encyclopedia; accessed January 17, 2017.

generated by multiplying sin 51.8°, .785, X 4 (for the four sides) = 3.14 X 4 (again for the four sides) = 4π or 4 X 3.14 = 12.56.

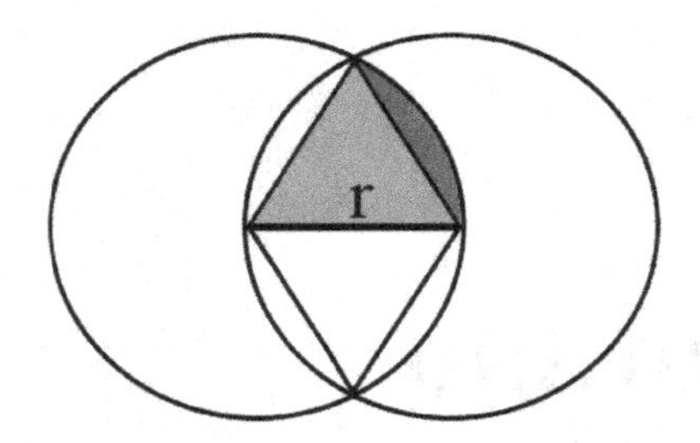

Source of figure [250]

In the drawing above, the image is created by taking two circles of equal radius, then overlapping them with one another, so that the circumference of each circle crosses the center of the other circle. The "Vesica Piscis" is the oval figure in the center of the two circles. The "r" in the drawing is the radius of both circles, as well as the base of the inscribed equilateral triangle. Furthermore, since the triangle is equilateral, "r" is also the dimension of each of the triangle's remaining two sides. The proof of this is that the sides of the triangle are also radii of the two circles. The proportional dimensions of an equilateral tringle are defined by three sides of equal length and a height of $\sqrt{3}$.

In the drawing shown above, the area of the triangle (shown in light grey) is defined by: base of .5r (radius) X height of ($\sqrt{3}$ X .5r) = $1.732/4r^2$ = $.433r^2$. The two triangles have a combined area of 2 X $.433r^2$ = $.866r^2$. The area of each of the four leaves (dark grey area) is calculated as follows: (θ/360° X r – sin θ/2) r^2, where θ is the angle opposite the leaf or 60°; then: 60°/360° or 1/6 X 3.1416 = .5236 – sin 60°, or .866/2, or .433 = .5236 - .433 = $.0906r^2$. The four leaves combined have an area of: $.3624r^2$. The area of the Vesica Piscis is the combined area of the two equilateral triangles, $.866r^2$ + the area of the four leaves, $.3624r^2$ = $1.2284r^2$. (The actual area of a Vesica Piscis depends on the dimension of the radius.)

b. Creating an Equal Area Map from the Vesica Pisces

To create an equal area map that is mathematically precise, start with the Vesica Pisces figure and swing two arcs, each centered on the outer circumference of its respective circle, on the centerline of the figure. Each arc is drawn with a radius equal to 2r, which is equal to twice the radius or "r" of the two circles comprising the Vesica Pisces. (All dimensions in the following discussions are based on "r" from the Vesica Pisces figure described in the foregoing sections. The actual value of "r" is not given.) This creates an "elongated fish" that has an area equal to $2.09r^2$, a figure that is exactly $1/6^{th}$ of the formula for the area of a circle, $4\pi r^2$, which is equal to $12.56r^2$ or 6 X $2.09r^2$. This is the reason that the Great Pyramid's diagonals and the

[250] Wikipedia Contributors, "Vesica Piscis", Wikipedia, the Free Encyclopedia, https://en.wikipedia.org/wiki/Vesica_piscis, accessed January 18, 2017. Drawing is from source: By שׁשׁב - Own work, CC BY-SA 4.0, https://commons.wikimedia.org/w/index.php?curid=45806912.

fact that they mark off 1/6th of the Earth's circumference at the equator, is so significant. (see figure following)

The proof that this "elongated fish" has an area equal to $2.09r^2$ is as follows, based on the same formula used in the foregoing to calculate the area of the Vesica Pisces: $(\theta/360° \times \pi - (\sin \theta/2)) \times (2r)^2$ = to the area of ½ of the "elongated fish", when divided in two lengthwise.

The angle of the foregoing triangle is equal to 86.2°. This is the value of θ for the purposes of area calculations. This value is derived from the actual measure of the height of a "elongated fish" triangle of 1.366r, which is equal to the height of the Vesica Pisces triangle, .866r + .5r. (The .5r addition is properly understood as the height of a rectangle, whose length is equal to 1.732 X .5r = .866r and whose diagonal is equal to r. This rectangle has the same area, .866r X .5r = $.433r^2$, as an equilateral triangle drawn on r, or .5r X 1.732 X .5 r = $.433r^2$.) The .866r height of the Vesica Pisces is derived as follows: $\sqrt{3}$ = 1.732 X .5r = .866r. The height, then, of the "elongated fish" is 2 X 1.366r = 2.732r. Given a hypotenuse equal to 2r, the sine value of ½ of the "elongated fish" (1.366r) is 1.366r/2r = .683, which corresponds to an angle of 43.1°, that would make the angle of the foregoing triangle equal to twice 43.1° or 86.2°.

Generating a Sphere from the Vesica Pisces

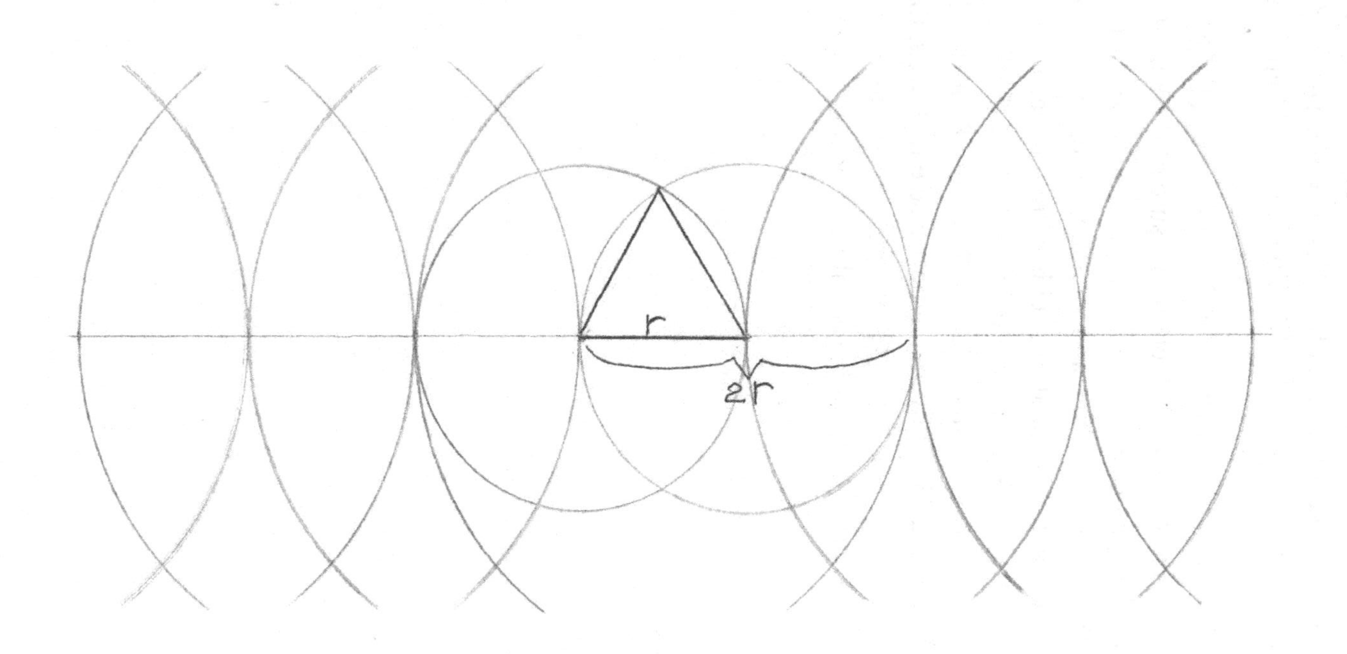

Using a radius of 2r, successive arcs are drawn, radiating outward from the central image of the Vesica Pisces. The area of one "elongated fish" is $(1.732 + .3624)\ r^2 = 2.0944\ r^2$. The area of six "elongated fish" is $6 \times (2.0944\ r^2) = 12.566\ r^2$. The area of a sphere is: $4 \times \pi r^2 = 4 \times (3.1416)\ r^2$, or 12.566. The area of a sphere, then, is $12.566\ r^2$, which is equal to the area of six "elongated fish" generated by the Vesica Pisces, or $12.566\ r^2$.

The area of ½ of an "elongated fish" is:
$(86.2°/360° \times \pi - (\sin 86.2°/2)) \times (2r)^2 =$
$(.24 \times 3.1416 - .99/2)\ 4r^2 = (.754 - .495)\ 4r^2 =$
$.26 \times 4r^2 = 1.04r^2$ (the area of half an "elongated fish").

The area of the "elongated fish", then, is twice the area of its half:
$2 \times 1.04r^2 = 2.08r^2 \approx 2.09r^2$ (the expected value).

c. Argument for the Area of the "Elongated Fish"

The argument for the area of the "elongated fish", as opposed to the proof laid out in the foregoing section, is very straight forward. There are two variations on the argument.

i. Given that the area of the equilateral triangle of the Vesica Pisces is $4.33r^2$ and that the area of the two equilateral triangles that compose the Vesica Pisces is $.866r^2$, then the area of the two triangles of the "elongated fish" is computed by simply **doubling the area of the two triangles of the Vesica Pisces**, $.866r^2$, which makes the area for the two triangles of the "elongated fish", $2 \times .866r^2$, or $1.732r^2$. Further, given that the area of the "4 leaves" from the Vesica Pisces is, $.3624r^2$, then adding the area of the "4 leaves" to the area of the triangles of the "elongated fish" $= 1.732r^2 + .3624r^2 = 2.0944r^2$, which is 1/6th of the area of a sphere: $12.56r^2$.

ii. Given that the height of the "elongated fish" is: 2.732r, (see foregoing section), then the area of the elongated fish is: $(1.366r \times .5r) = .683r^2$, which is the area of one triangle from the "elongated fish", $\times 2 = 1.366r^2$. Then, if this figure, $1.366r^2$, is added to **the double of the area of the "4 leaves" from the Vesica Pisces**, $.3624r^2 \times 2 = .7248r^2$, then the area of the "elongated fish" is: $1.366r^2 + .7248r^2 = 2.0908r^2$, which is also 1/6th of the area of a sphere: $12.56r^2$.

It is noteworthy that many of the foregoing numbers are very close approximations to numbers from the Pattern. These include:

1) The height of the equilateral triangle, which is $\sqrt{3}$ or 1.732r, which compares to the Pattern number 1,728.
2) The height of the Vesica Piscis is .866r, which compares to the Pattern number 864.
3) Half the height of a Vesica Piscis is .433r, which compares to the Pattern number 432.
4) The angle of a triangle formed by half the height of the segment of the "elongated fish" inscribed within a circle (the triangle's height) with the radius of this circle (the triangle's hypotenuse) is 43.1°, which compares to the Pattern number 432. The angle for the triangle drawn on the segment's full height is 86.2°, which compares to the Pattern number 864. (See foregoing discussion of these two triangles and their dimensions.)

These numbers are well within the accepted variation (equal to or less than .009) established for this purpose and are therefore considered to be manifestations of the Pattern in nature.

Note, that there is a necessary correction that must be made when using the "elongated fish" as a flat surface projection of the surface area of a sphere. The flat surfaces of the "elongated fish" are, in actuality, the surfaces of a hexagon-shaped sphere, rather than of a conventional sphere. Referring back to the figure of the "elongated fish", note that the horizontal or circumferential measure is 6 X r. As the circumference of a conventional sphere is computed as: πd or π2r, which coverts to 3.1416 X 2r or 6.2832r, a correction must be made. This correction is: 6.2832r ÷ 6 = 1.0472r. The measure of r, then, as depicted on the "elongated fish" projection must be multiplied by 1.0472 to correct the projection, to have it accurately reflect the spherical surface of a conventional sphere. This correction to r must be made both vertically and horizontally on the "elongated fish" projection on a flat surface.

d. Mercator Projection and Conformal Maps, and the Vesica Pisces

A Mercator or conformal projection can readily be created from the equal area map that is based on the six "elongated fish" that are derived from the Vesica Pisces. A uniform grid of latitude and longitude is simply laid on top of this equal area map to create a Mercator or conformal projection. The singular advantage of the Mercator projection is that all course lines drawn on it intersect the grid at precisely the same angles at every intersection. This makes such a projection invaluable for navigation and travel, and for orienting objects to one another for reference purposes, and Mercator projections are still in widespread use. Its disadvantage is that the distances are only correct near the equator and must be corrected elsewhere, with increasingly larger correction factors necessary for locations approaching the poles.[251] However, the underlying equal area map creates a correction key for this purpose. The space between the "elongated fish" readily provide this key.

e. The Great Pyramid, and Azimuthal Equidistance Projection Maps

A circle with radius drawn on the height of the Pyramid, 280 royal cubits, is equal to the circle whose perimeter is equal to the perimeter of the Pyramid's base or 440 royal cubits X 4 = 1,760 royal cubits (1,760/π = 560 royal cubits (the circle's diameter), which has radius 560/2 = 280 royal cubits. By design, then, the Pyramid represents half a sphere, or hemisphere, with radius 280 royal cubits, whose area is equal to ½ (4π X 280^2). From this hemisphere, an azimuth (arc direction) equidistant projection onto a plane or flat surface can readily be created. And, since the Pyramid is precisely oriented on true north, the point of the azimuthal projection could readily be the North celestial pole. However, the point of the azimuthal projection could also be the Pyramid itself, as its four faces are precisely aligned on the four cardinal directions: north, south, east, and west, and its four diagonals are also precisely aligned on

[251] Wikipedia contributors, "Mercator projection", Wikipedia, the free encyclopedia, https://en.wikipedia.org/wiki/Mercator_projection (accessed January 26, 2017)

the four intermediate directions: northeast, southeast, southwest, and northwest. From these points, all other compass directions can readily be derived. All lines drawn to objects on an azimuthal projection originate at the point of projection and are correct in their direction. Azimuthal maps can be drawn from any sphere, including the Earth and the celestial sphere of the heavens.[252]

An azimuthal equidistant projection can be produced by using the scale of the Great Pyramid to the Earth, which as previously determined in an earlier section, is, 1: 43,200. Using this scale, all objects or locations on the Earth can be accurately plotted on an azimuthal equidistant map from the point of projection. However, a problem with any azimuthal equidistant projection is that it has substantial area distortion, which increases with distance from the point of projection.

A fine example of an azimuthal equidistant projection map is depicted below. This projection uses the North celestial pole as its projection point.[253] (Note that area distortion increase with distance from the North celestial pole, reaching an extreme toward the regions near the South celestial pole.) A similar azimuthal equidistant projection, with the North celestial pole as its projection point, is on the official emblem of the United Nations. Again, all objects or locations depicted along the meridians of an azimuthal equidistant projection of an entire sphere are accurate in direction and distance from the point of such a projection.

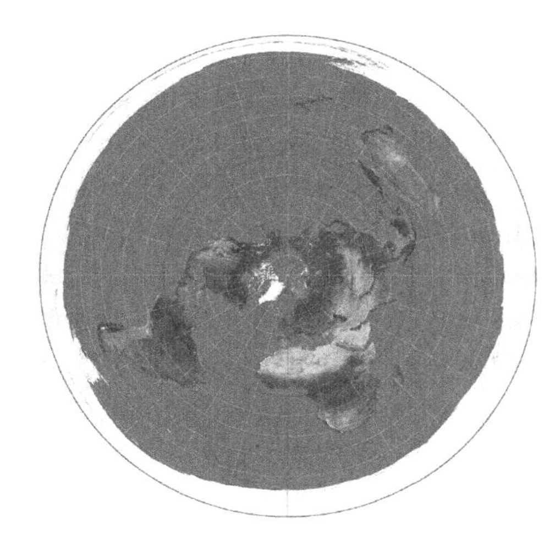

To make a flat surface map from an azimuthal equidistant projection of the Earth, it is only necessary to recognize that the quarter arc of a circle is 1.57 X its radius, or 1.57 radians.

[252] Wikipedia contributors, "Map projection", Wikipedia, the free encyclopedia, https://en.wikipedia.org/wiki/Map_projection, (accessed January 26, 2017)

[253] Wikipedia contributors, "Azimuthal equidistant projection", Wikipedia, the free encyclopedia, https://en.wikipedia.org/wiki/Azimuthal_equidistant_projection, (accessed February 1, 2017). The map depicted was drawn by Strebe, on August 15, 2011, who describes the work as, "The world on azimuthal equidistant projection. 15° graticule, polar aspect. Imagery is a derivative of NASA's Blue Marble summer month composite with oceans lightened to enhance legibility and contrast. Image created with the Geocart map projection software."

Thus, the radius of the flat surface map is 1.57 X the scale-of-the-projection for the radius. If the Great Pyramid is used this scale is 1: 43,200, which defines a hemisphere with a radius of 280 royal cubits. The flat surface of an azimuthal equidistant projection from the Pyramid's hemisphere is a circle with a radius of 1.57 X 280 = 439.6 royal cubits. If this distance is doubled to approximately 880 royal cubits, the entire sphere can be depicted on a flat surface.

The surface area of the Pyramid supports this. The dimensions of each triangular face of the Pyramid, as previously noted, are: base 440 royal cubits and height 356 royal cubits. The area, then, of each face is 220 X 356 royal cubits = 78,320 square royal cubits, with the total surface area of the Pyramid being 4 X 78,320 = 313,280 square royal cubits. The square root of this figure, $\sqrt{313,280}$, is 560 royal cubits, which, if taken as the radius for a circle, is equal to twice the radius of the circles drawn from the Pyramid's height and from the perimeter of its base. The significance of this larger circle, which has a radius that is twice the radius of the other circles, 2 X 280 = 560, is that it accommodates the projection of the southern hemisphere onto the same flat surface used for the northern hemisphere, thus depicting the whole sphere of the Earth onto the projection's flat surface. Again, the flat surface used for the projection is a circle with a radius that is 1.57 X 560 = 879.2 $\approx$ 880.

f. Likely Origin of the Ancient World Maps of Charles Hapgood

A number of maps have been found that proved to be of surprising accuracy and that depicted areas of the Earth, such as the Antarctic, which were not known until the late nineteenth and early twentieth centuries, when explorers first arrived and catalogued their findings. A number of these maps were in azimuthal equidistant projection format, as detailed by Charles Hapgood in his book, *Maps of the Ancient Sea Kings,* which he published in 1953. The most famous of these maps was the Piri Reis azimuthal equidistant projection map of the Atlantic coastline of South America and Antarctica created in Constantinople in 1513. An analysis of the Piri Reis map by the U.S. Air Force indicated that the point of projection for the Piri Reis map was located in the vicinity of modern Cairo, a location in the immediate vicinity of the Great Pyramid. Could this and other maps analyzed by Charles Hapgood have originated in ancient Egypt? A very intriguing question and one worth pursuing, but one that will not be pursued further here.[254]

[254] Graham Hancock, The Evidence of Earth's Lost Civilization, Fingerprints of the Gods, (Three Rivers Press, New York, 1995) A good summary of the issue of ancient maps and Chares Hapgood's analysis of the Piri Reis map in particular can be found in Chapters 1 and 2, of Part 1.

CHAPTER 52

THE PYRAMID'S SCALE TO THE EARTH REVISITED

The Pyramid's height is 280 royal cubits of 524mm each, as was detailed earlier in this chapter. From this dimension, along with its slope angle of, 51.83°, the other major dimensions of the Pyramid can be derived. Tan 51.83° is 1.272, which is the square root of *phi*; 280 ÷ 1.272 = 220, which is the half base of the Pyramid; 2 X 220 is the base dimension of the Pyramid; and sin 51.83° is .7862 (.7862 is = 1/√*phi*); 280 ÷ .7862 = 356, which is the slant height of the structure. However, a height half of 280 would have sufficed to give the Pyramid the same shape and dimensional relationships, and saved the builders enormous labor and materials. Why wasn't the height of the Pyramid set at 140 royal cubits instead? The answer seems to be that 280 incorporates *phi*, the golden mean and the key measure of the equilateral triangle, the √3, or 1.732. *Phi*, or 1.618, X √3 = 1.618 X 1.732 = 2.80, which has the same root, 28, as 280. This number, 28, further ties the Pyramid to the equilateral triangle formed by the Pyramid's location and the equinox, as well as the 36°—54°—90° triangle formed with the North Ecliptic Pole and the 30°—60°—90° triangle formed with the North Celestial Pole. (See Chapter 47.) Thus, *phi* or one of its cognates generates all the dimensions of the Pyramid, and it may be confidently stated that *phi* underlies the basic design of the structure.

a. <u>"mr" Pyramid</u>

There is a very strong argument to be made that the Great Pyramid is the *mr* pyramid (see drawing below). Recall from earlier Chapters in Part III that the *mr* pyramid is represented by a glyph that contains two triangles: 30°—60°—90° and 54°—36°—90°. As the key measures incorporated in these two triangles—1.732 and 1.618, respectively—when multiplied together produces the number for the height of the Great Pyramid, as detailed in the foregoing paragraph, then both triangles can be said to have generated the Pyramid. No other pyramid has these characteristics. Therefore, the Great Pyramid must be the *mr* pyramid.

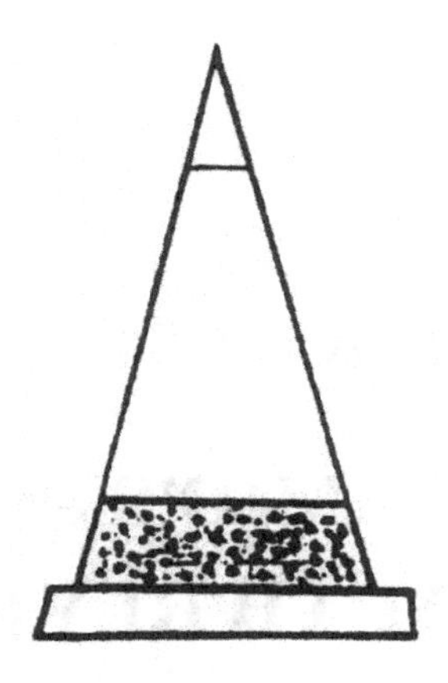

b. Phi, and the measure of a Saros series and the Platonic Year

If *phi* is the half base of an equilateral triangle whose height is 2.80 (1.732 X 1.618), then the sides of the triangle, including its base, are each 2 X *phi*, or .3236, on a side. If four of these equilateral triangles are combined to form the four faces of a pyramid, with each side being an equilateral triangle with side equal to .3236, then the perimeter of the pyramid is 4 X .3236, which is 1.2944, a number which is approximately the length of the average Saros series and a number from the Pattern, 1.296. The root of this number 1,296, is ½ the root of the number of the Platonic Year of precession, 25,920 years, which is 2,592, (2,592 ÷ 2 = 1,296). This is by design, as the Pyramid's diagonals also mark off 1/6th of the band of the ecliptic at the celestial equator, or a double great age (4,320 years). The number 28 also closely approximates a number from the pattern, 279,936, which is found at Pattern pair (2,187/128). It is noteworthy that, like the number 6, 28 is also a perfect number.

Defined Value	Pattern Number	Difference Δ	Dev. ± from Pattern
280 (Pyramid's height)	279.936 ($6^7 \cdot 10^{-3}$)	.064	+.00023

The Pyramid's height is a reflection of the Pattern.

It should also be pointed out that both *phi* (1.618) and the square root of three (1.732) are very close approximations of numbers from the Pattern, which are, respectively, 162 and 1,728; 162 has a pattern pair of: (2/81), and 1,728 has a pattern pair of: (64/27).

Defined Value	Pattern Number	Difference Δ	Dev. ± from Pattern
1.618 (*phi*)	1.62 ($3^4 \cdot 2 \cdot 10^{-2}$)	.002	-.0012
1.732 (√3)	1.728 ($6^3 \cdot 8 \cdot 10^{-3}$)	.0004	+.0023

Both *phi* and the √3 are manifestations of the Pattern in nature.

A final question: Since the Great Pyramid is a scale model of the Earth, based on the height of the Pyramid (280 cubits) and the approximate ratio, 43,200:1 (radius (polar) of the Earth to 280 royal cubits), as was demonstrated earlier, might the Earth itself also be a manifestation of the product of *phi* (1.618) and the height 30°—60°—90° triangle (1.732)? The answer is, yes, and there is evidence that God specifically designed the Earth with *phi.*

c. The Creation of the Earth and *Phi*

In Plato's Timaeus, there is a section which Timaeus unequivocally states that God Created the universe with *phi.*

> "Wherefore also God in the beginning of creation made the body of the universe to consist of fire and earth. But two things cannot be rightly put together without a third; there must be some bond of union between them. And the fairest bond is that which makes the most complete fusion of itself and the things which it combines; and proportion is best adapted to effect such union. For whenever in any three numbers, whether cube or square, there is a mean, which is to the last term what the first term is to it; and again, when the mean is to the first term as the last term is to the mean—then the mean becoming first and last, and the first and last both becoming means, they will all of them of necessity come to be the same, and having become the same with one another will be all one." (Author's emphasis added.)

Subsequently, in the same paragraph, Timaeus further states:

> "And for these reasons, and out of such elements which are in number four, the body of the world was created, and it was harmonized by proportion, and therefore has the spirit of friendship; and having been reconciled to itself, it was indissoluble by the hand of any other than the framer."[255] (Author's emphasis added.)

There is, then, a remarkable and seemingly definitive consonance of the available evidence that God created the Earth with *phi.* Summarizing and recapitulating this evidence, as presented in Chapters 39, 47, 50, as well as in the instant Chapter, then:

i. At the Great Pyramid's location at 30° North, there is a direct connection between the Pyramid and the North Celestial Pole, by means of an implied 30°—60°—90° triangle, as well as the North Ecliptic Pole, by means of an implied 36°—54°—90° triangle—which, as was shown, is connected to the generation of *phi* (Chapter 47, Section c).

[255] Plato, *The Dialogues of Plato*, translated by Benjamin Jowett (Encyclopedia Britannica, Inc., published with permission by The University of Chicago, The Great Books, 1952) *Timaeus* 31-32, page 448, column 2 top.

ii. Anticipating an issue that will be discussed in detail in Chapter 60, the North Ecliptic Pole, with respect to the Earth, is the one location in the heavens that has never moved and is, therefore, beyond time, which is one of the defining attributes of God, i.e. He is beyond or outside of time. God, then, can be identified with the North Ecliptic Pole, and hence with the 36°—54°—90° triangle with its association with *phi*.

iii. The Great Pyramid is the pyramid depicted in the "mr pyramid" hieroglyph, which depicts both the 30°—60°—90° and the 36°—54°—90° triangles, as mentioned in the foregoing paragraph; the second triangle is connected to *phi*, as stated earlier (Chapter 39, Section d). These two triangles, when their respective measures of 1.618 (*phi*) and 1.732 (height of 30°—60°—90° triangle) are multiplied by one another, produce the root number, 2.8, of the height of the Great Pyramid, which is 280 royal cubits, a measure that in turns scales to the dimensions (specifically, the radius) of the Earth, as 1:43,200—a Pattern number. All of the Pyramid's dimensions are generated by *phi* or one of its cognates. And, since the Pyramid is a scaled to the Earth's dimensions, the Earth, too, was generated by *phi*.

iv. Timaeus's statement regarding *phi* and the Creation of the Earth, as was presented in the foregoing quoted material, clearly states that God used the golden mean or *phi* to "harmonize' his Creation.

CHAPTER 53

PRECESSION AND THE GREAT PYRAMID

In an earlier section, it was shown that the Pyramid's diagonals when extended to the equator, from the Pyramid's location, creates an equilateral triangle on the day of the equinox. This triangle is defined by the Pyramid's location at 30°N with a corresponding location at 30°S and the position of the Sun itself at local apparent noon. At the same instant, an equilateral triangle is also formed above and along the equator, between the two points where the Pyramid's diagonals intersect the equator. These two equilateral triangles are identical in dimensions.

This second equilateral triangle also marks off an arc equal to 1/6th of the length of the celestial equator, and, because it is a circle of similar apparent dimensions (great circle) on the celestial sphere, this distance is also equal to an arc of 1/6th of the band of the zodiac. This second equilateral triangle has a direct connection to the phenomenon of the precession of the equinoxes, because in terms of the time dimension of precession, the Sun at local apparent noon on the day of the equinox, marks off 4,320 years on the band of the zodiac, with a distance equal to half of this time-period (2,160 years) to the east and also to the west of the Sun at that instant. The significance of this is that on the day of the equinox, with the Sun above the equator, it marks a precise location on the band of the zodiac, which is, for all practical purposes, the exact same location for sunrise, local apparent noon, and sunset on the day of the equinox. Accordingly, the Pyramid's diagonals mark precise locations **on the horizon** where the band of the zodiac intersects the Earth's equator on the day of the equinox, as the Sun travels above the equator. Specifically, the diagonals intersect the zodiac on the equator at 9 a.m. and 3 p.m. local time, from the Pyramid's location, on the day of the equinox.[256]

With accurate star charts, oriented on the North ecliptic pole, the location of these points of intersection can be marked and the charts adjusted for precession, at the rate of approximately 20 minutes of time per year, one day every 72 years, and 1° of arc across the band of the ecliptic every 72 years. This is the significance of the Pyramid's diagonals and probably explains why

[256] Peter Tompkins, *Secrets of the Great Pyramid* (Harper and Row, New York, NY; 1971) Chapter XI, Almanac of the Ages; see the bottom illustration on page 122, which shows the Pyramid's shadow at various times and days throughout the year.

the Great Pyramid was named, "The horizon of Khufu", the name of the Pharaoh historically credited with its construction. However, to confirm the rate of precession, periodic position fixes of the Great Pyramid must be taken, using the stars to affix the Pyramid's exact location. Position fixes using the stars were most likely performed just before sunrise and just after sunset on the day of the equinox, during twilight when the horizon can be scene, as has always been the practice when affixing position with the stars, since time immemorial. **Once the Pyramid's position is ascertained, it is then plotted on a star chart—not a terrestrial chart—which allows for the direct measure of the rate of precession.**

The Pyramid obviously remains in its same terrestrial position, but its apparent position with respect to the stars, as does every other position on Earth, changes with precession. It is the position of the stars that appear to change with time, due to precession. To correct for precession, the position of the Pyramid in the stars would have to be returned to its Prime Meridian 0°-30°North position. Drawing a line from the North ecliptic pole through the newly affixed position of the Pyramid would then intersect the band of the zodiac at its precession-adjusted location, i.e. the band would drift westward on the celestial sphere with its orientation on the North ecliptic pole. Shifting this line to restore the Pyramid to its constant location, i.e. 0° longitude, 30° latitude, causes the entire map of the heavens to rotate westward with respect to this line. The best map on which to plot the changes in the position of the stars would be an azimuthal equidistant map with the North ecliptic pole as the projection point and the band of the ecliptic depicted along the circumference of the projection. This type of map readily tracks the changes to band of the ecliptic, but also tracks the changes to the location of the North celestial pole, which also moves due to precession. Although the azimuthal equidistant map with its projection point on the North ecliptic pole is ideal for tracking precession and its effect on the apparent location of the stars, the best map for using stars for plotting terrestrial locations would be an azimuthal equidistant map with its focus point on the North celestial pole and the celestial equator along the circumference of the projection.

a. The Great Pyramid as Datum

For map making and measuring time, a landmark with a known location is necessary. Today, this function is performed at Greenwich, England, which marks O° longitude (Prime Meridian), and 12:00pm noon, at the same instant when it is 12:00am midnight, **at the beginning of the next day**, on the international dateline, at position 180°East/West. To reemphasize, the datum is also used for laying out the geographic grid system of latitude and longitude of the Earth. If the Pyramid was used to track and measure precession, then the Pyramid was used as datum for time measure, and the day likely began and ended at noon, at the Pyramid's meridian. The Pyramid was also likely used to mark 0° longitude for laying out the grid system of the Earth. Proof for this latter contention is found in the fact that the southwest-northeast diagonal of the Great Pyramid, when extended out to the northeast, in the direction of Heliopolis, directly intersects an ancient obelisk that was erected around 1,970 BCE by Pharaoh Senusret I in

Heliopolis. This ancient obelisk is the only surviving major monument of ancient Heliopolis. The measure of the line drawn between the Great Pyramid and this obelisk is very significant.

The coordinates of the Great Pyramid are: 29° 58' 45" North (29.979°North), 31° 08' 03" East (31.134°East).[257] The coordinates of the Heliopolis obelisk are: 30.129°North, 31.3075°East.[258] The differences between these measures are:

Obelisk:	Latitude:	30.129°N	Longitude: 31.3075°E
Pyramid:	Latitude:	29.979°N	Longitude: 31.1340°E
		0.15°	0.1735° (corrected for Lat. = 0.15°) *

(* The measure of longitude varies as the cosine of latitude. Cosine 30° = 0.866 · 0.1735° = .1502° ≈ 0.15°)

These coordinates mark a square, 0.15°, or 9 arcminutes or miles on a side (one arcminute of latitude is equal to one nautical mile of 6,000 feet of 308mm). This square has a diagonal equal to 1.414 (the measure of the diagonal of a square) · 9 = 1.272. This number is very significant, as it is equal to the √*phi* or √1.618 · 10. And, as has been shown earlier, *phi* is a number that is embedded in both the design and dimensions of the Great Pyramid. If the dimensions of this square are increased by a factor of 2/3, then the dimensions are: 15 arcminutes X 15 arcminutes X 21.21arcminutes (diagonal) or miles. This square and the diagonal between the Pyramid and the obelisk most likely constituted the baseline measure for laying out the Earth's grid and the grid of Egypt, although the measure would have to be corrected for the discrepancy between the Pyramid's latitude (29.979° and the 30° North latitude location of the grid. Corrected for the discrepancy, the baseline between the Pyramid and the obelisk likely served as the baseline measure for surveying Egypt to mark boundary and property lines. This may explain while this obelisk is the sole surviving one from Heliopolis, of the many that are believed to have once stood there. Times may change but boundary and property lines usually remain stable, as they determine ownership, and fees and taxes etc. due to the government, regardless of who is in charge!

The square of 15 arcminutes · 15 arcminutes is a base unit for the measure of both time and the latitude/longitude grid of the earth. The longitudinal measure of this square (15 arcminutes) is equal to one minute of time. (60 arcminutes (1° of arc) = 4 minutes of time. 15° of arc = 1 hour of time. 360° of arc = 24 hours of time.) This type of grid system would best be depicted on a conformal map, where directions can be shown cutting the lines of longitude and latitude at the same angles. However, the weakness with such maps is the very significant distortion in the areas depicted as one moves toward the polar regions. If the Pyramid was used as datum

---

[257] Wikipedia contributors, "Great Pyramid of Giza", Wikipedia, the Free Encyclopedia, https://en.wikipedia.org/wiki/Great_Pyramid_of_Giza

[258] Wikipedia contributors, "Heliopolis (ancient Egypt)", Wikipedia, the Free Encyclopedia, https://en.wikipedia.org/wiki/Heliopolis_(ancient_Egypt)

for laying out the grid system of the Earth, the grid would have to start at 30°North and not at the latitude of the Pyramid, which is slightly below this latitude.

The obelisk's location in Heliopolis is further noteworthy, as Heliopolis is the location of the temple dedicated to the worship of Re, the god of the Sun, and specifically Re in his form of *Re-Horakhty*, **Re, as Horus-of-the-two-horizons**. Heliopolis is Greek for the "Sun City", which the Egyptians called, *Iunu*, the "Pillar."[259] The connection of the Heliopolis obelisk to the southwest-northeast diagonal of the Great Pyramid, the "Horizon of Khufu", is well documented and, as demonstrated in the foregoing paragraphs, a matter of fact.[260] It is further significant that the "baseline" that derives from this diagonal likely began the grid system of the Earth that delineates both geographic measure as well as its relation to the measure of time. Is this obelisk the "pillar" that gave the ancient city its name? This is a distinct possibility, and one that is supported by the fact that Heliopolis was famous throughout its long history as a center for astronomy. More tantalizing yet, could this obelisk be an incarnation of the legendary *Benben* stone[261] of Heliopolis, the "Pillar of the First Time"? There is no reason that it couldn't be! Especially in view of the possibility that the term "first time" may refer to the fact that if the Pyramid was the datum for measuring time and distance, and as such served as the "International Dateline" of the ancient world, it would be the place where the measure of time, literally, first began. And it is the "baseline" of the obelisk that gives dimension to the measure of time, i.e. 15 minutes of geographic arc equals one minute of time.

There is one further point to consider with respect to this obelisk. If the location of the Great Pyramid with its near-perfect orientation on true North and its proximity to 30°North latitude—a position likely determined by the physical limitations in the Cairo area of placing the structure any closer to 30°North than this—is a stunning feat of astronomical capability, then the placement of the obelisk is even more stunning, as its location is precisely aligned with the Pyramid's southwest-northeast diagonal and in a position so as to precisely define a square 9 nautical miles on a side with a diagonal of 12.72 nautical miles, a figure that has been shown to be related to *phi*, the golden mean. It needs to be recalled that the obelisk is located some 24 kilometers, or 14.615 international miles, from the Pyramid. Such precision in siting the obelisk and erecting it in this location would have to be done with rather sophisticated surveying techniques, something clearly beyond the known capabilities of the ancient Egyptians. This is yet one more example of a long list of anachronisms found in the monuments of ancient Egypt!

---

[259] Wikipedia contributors, "Heliopolis (Ancient Egypt), Wikipedia, the Free Encyclopedia, https://en.wikipedia.org/wiki/Heliopolis_(ancient_Egypt), accessed February 24, 2017

[260] Robert G. Bauval; "The Giza diagonal" and the "Horizon of Khufu"; article found at Unexplained Mysteries web site, http://www.unexplained-mysteries.com/column.php?id=233055 ; Mr. Bauval gives a good summary of the history of the noted connection between the Great Pyramid and the obelisk of Senusret I in Heliopolis.

[261] Wikipedia contributors, "Benben", Wikipedia, the free encyclopedia, https://en.wikipedia.org/wiki/Benben , accessed February 24, 2017

CHAPTER 54

A GIZA PLAN

There have been numerous theories that the Pyramids on the Giza Plateau were designed and sited according to an over-arcing or master plan. Some of these theories are easily discounted, but others are very credible and rooted in reliable survey data. Probably the best argument for a master plan is the one presented by J.A.R. Legon, The Plan of the Gia Pyramids.[262]

Among the more significant findings by Mr. Legon are that the Giza Pyramids were alligned and enclosed within a ground plan that is a rectange with side dimensions: 1,417.5 royal cubits (523.75mm) on the east-west (width) side by, 1,732 royal cubits (523.75mm) on the north-south (length) side. (See drawing shown below.) These numbers are very significant as they are the square root of two, $\sqrt{2}$, (approximately) and the square root of three, $\sqrt{3}$. (The $\sqrt{2}$ is actually, 1,414, but the measurement of the width may have been intended to incorporate the slightly longer, common royal cubit of 525.065625, instead of the royal cubit of 523.75 used with the lengthwise dimension. This would make the measure of the width 1,413.948, which is much closer to the $\sqrt{2}$. That this was the cubit actually used in the width will become evident in a later chapter, where the derivation of Earth's dimensions from *phi* are discussed.) These two square roots are also the key dimensions of the equilateral triangle and of the isoceles triangle, respectively. The reader will no doubt recall the importance of these two triangles in the design of the Great Pyramid and its relationship to map making, and to the measures of time and distance. Recalling but two examples from the earlier discussions: 1) the diagonals of the Great Pyramid mark of the sides of an isoceles triangle that intersects the equator and marks off 1/6th of the Earth's circumference (3,600 geographic miles), while an equilateral triangle is formed on the Pyramid's meridian at local apparent noon on the equinox, and is the key measure to the making of equal-area maps.

Are the dimensions of this Giza rectangle merely confirmation of the importance of the $\sqrt{2}$ and the $\sqrt{3}$, as has been demonstrated in the foregoing sections, or do they have some larger significance? Certainly it would be reasonable to conclude that they merely confirm the importance of these two square roots. But I hesitate to close this argument on such an

[262] The Plan of the Giza Pyramids, J.A.R. Legon: http://www.legon.demon.co.uk/gizaplan.htm Illustration is provided by kind permission of J.A.R. Legon.

assumption, particularly as there is no clear purpose served by the two other pyramids on the Giza Plateau, with any larger meaning as there is for the Great Pyramid. Or is there? As it turns out, there is an outsized importance to the siting and orientation of the second Giza pyramid, which is universally attributed to Pharaoh Kharfre, 3rd king of the Fourth Dynasty. This pyramid measures time with a precision that is breath-taking even by modern standards.

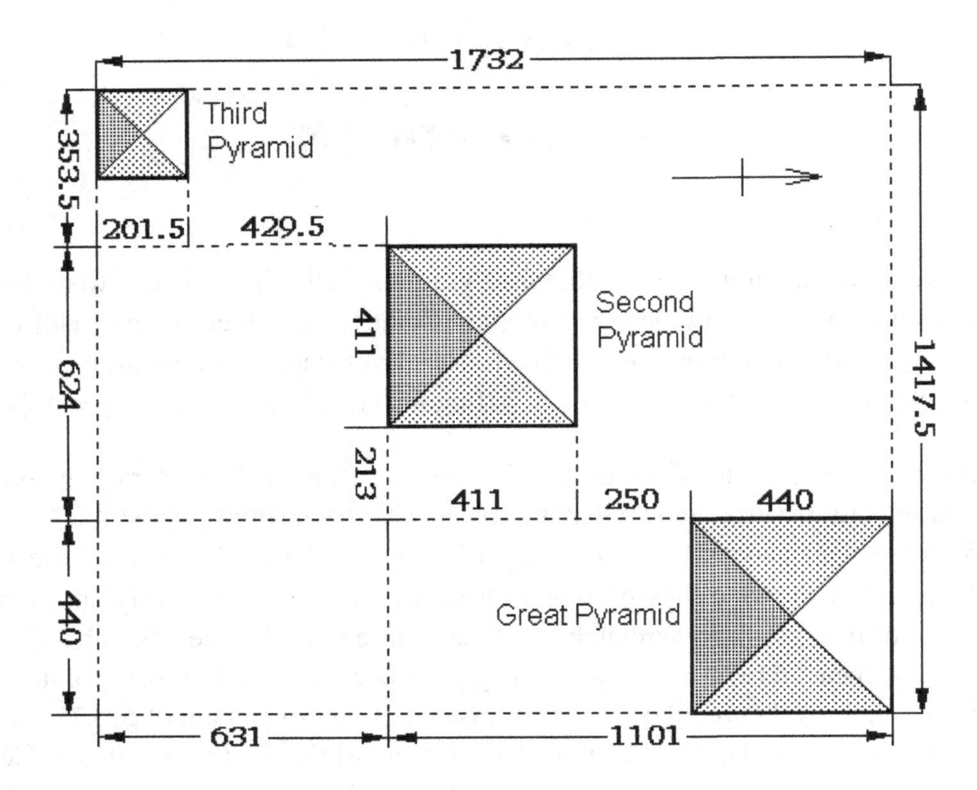

Dimensions of the Giza Site Plan in Royal Egyptian Cubits **© J.A.R. Legon, 2000**

CHAPTER 55

GIZA AND THE MEASURE OF A MINUTE OF TIME

One of the most impressive aspects of the Great Pyramid of Giza is that it is precisely aligned, diagonally, with an obelisk to its northeast at Heliopolis that was erected by Senusret I around 1900 BCE. Even more impressive is that this relationship between the two structures almost defines the distance covered by the Earth in its rotation in one minute of time. This was not a curious coincidence, but a matter of specific design. Even more impressive is that other, nearby structures located at Giza (Khafre Pyramid) and Heliopolis (2nd but now missing obelisk of Senusret I), and one or more structures at Letopolis (now lost or missing) form a related, isosceles triangle of 45°—90°—45° that is directly to the measure of a minute of time.

Letopolis

The distance between Heliopolis and Letopolis is the distance covered by the rotation of the Earth in 1/√2 minute of time.

Heliopolis

The distance between Letopolis and Giza is the distance covered by the rotation of the Earth in 1/√2 minute of time.

Giza

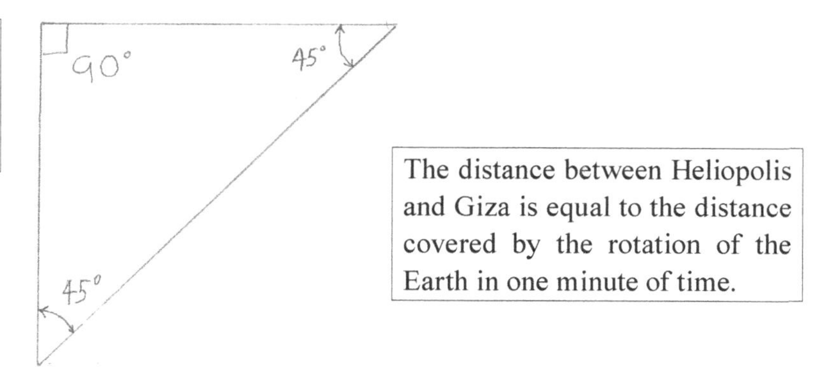

The Heliopolis—Giza—Letopolis Isosceles Triangle (45°—90°—45°) and Time

a. Background of the locations.

Heliopolis is one of the oldest cities in Egypt. It is the site of the great temple of Atum-Re, where Egypt's astronomers practiced their art, and where the cycles of time and calendars originated. It is the site of the "primeval mound" where time began. **Letopolis** was the site of an observatory and a temple dedicated to Horus.[263] Letopolis was also mentioned by Strabo as being the site of a tower that was used for astronomical observations by Eudoxus of Cnidus, a 4th century BCE Greek mathematician-astronomer, who studied astronomy for two years in Heliopolis.[264] The "tower" mentioned by Strabo is now lost.[265] **Giza** is one of the most renowned archaeological sites in the world, and here is found the Pyramids of Giza, including the Great Pyramid of Khufu, and the Pyramid of Khafre.

b. Arc measures of key Egyptian astronomical structures.

Geographic Coordinates of Egyptian Astronomical Structures

Monument / Location:	Khafre Pyramid, Giza	Heliopolis Obelisk	Great Pyramid (Khufu Pyramid), Giza	Assumed, Letopolis Observatory
Latitude	29.9761N	30.1290N	29.979N	30.1290N
Longitude	31.1308E	31.3075E	31.1340E	31.1308E

General Comments on the accuracy of measures:

i. The location of the second obelisk erected by Senusret I at Heliopolis is unknown, but likely a critical component of these measures;
ii. The precise location of the astronomical structure(s) at Letopolis—modern Ausim— is unknown; however, it is believed to have been at the same latitude as the Heliopolis obelisk (30.1290N) and the same longitude as the Khafre Pyramid (031.1308E), but this is only speculation.
iii. The Khafre Pyramid is not oriented precisely to the cardinal points, as is the Great Pyramid; thus, its SW-NE diagonal does not create a precise 45° NE angle with the Heliopolis Obelisk. This is why the missing second obelisk at Heliopolis is critical, as it may lie precisely along the Khafre Pyramid's diagonal.
iv. The margin of error in the methods and devices used by the Egyptians in making their measurements are unknown. The margin of error in the as-built measures of the various structures referenced here is also unknown.

263 Toby A. H. Wilkinson, *Early Dynastic Egypt* (New York, NY, Routledge; 1999) pp.273 and 287
264 Wikipedia contributors, "Eudoxus of Cnidus", Wikipedia the Free Encyclopedia; https://en.wikipedia.org/wiki/Eudoxus_of_Cnidus Accessed September 30,2018.
265 Strabo, *Geographica*, Volume XVII, Chapter I, 30.

The arc distances between the geographic coordinates of the locations in the foregoing table converted to international miles (assumes 60 geographic miles per degree, 6,000 geographic feet per minute, 308mm per geographic foot):

c. Heliopolis Obelisk to Khafre Pyramid.

	Latitude	Longitude
Heliopolis Obelisk:	30.1290N	31.3075E
Khafre Pyramid:	29.9761N	31.1308E
	0.1529*	0.1767 (corrected for cosine 30°N = .1528) *

* Creates square of ~ 0.153° on a side

0.153° · (60 miles · 6000 · 308) / (5,280 · 304.8) · √2 = 14.90769 miles.

d. Heliopolis Obelisk to Letopolis Observatory.

Heliopolis Obelisk:	30.1290N	31.3075E
Letopolis:	30.1290N	31.1308E
	0.0	.1767 (corr. for cos 30.129° = .1528 ~ .153)

0.153° · (60 miles · 6000 · 308) / (5,280 · 304.8) = 10.5413 miles.

e. Heliopolis Obelisk to Great Pyramid.

Heliopolis Obelisk:	30.1290N	31.3075E
Great Pyramid:	29.9790N	31.1340E
	.15	.1735 (corrected for cosine 30° = .15)

* Creates square of ~ 0.15° on a side

0.15° · (60 miles · 6000 · 308) / (5,280 · 304.8) · √2 = 14.615 miles.
(Interestingly, the geographic measure of this distance is: .15° · (60 miles · 6000 · 308) · √2 = 12.7279 ~ 12.72 geographic miles. This is √*phi* (√1.618) = 1.272 · 10.)

f. Distance measures and a minute of time.

Distance measures on the surface of the Earth that a fixed point on the celestial sphere crosses in one minute of time—given Earth as a sphere—due to Earth's rotation

Reference for measure / Distance measured	(1) **WGS 84 Measures** Earth as sphere (authalic measure); circumference: 24,873 miles	(2) **WGS 84 Calculator[266] Measures** (By Calculator, using the geographic coordinates from the foregoing section)	(3) **Arc Measures** Arc distances converted to inter. miles from calculations in the foregoing section	(4) Egyptian Measures 24,803.15 miles, (Earth as sidereal sphere.)
Measure of distance covered in one minute of **solar/sidereal** time	14.9507 mi.*in one minute of **solar** time. (24,873/1,440 mins-day = 17.2728 miles, corrected for Lat.: 30.0525°N)			14.9084 mi.*in one minute of **sidereal** time. (24,803.15/1,440 mins-day = 17.224 miles corrected for Lat.: 30.0525°N)
Khafre P. to Heliopolis obelisk		14.934 miles	14.90769 miles	
Heliopolis obelisk to Letopolis	10.5718 miles* (Derivation: above measure 14.9507/√2 or 1.4142)	10.58 miles	10.5413 miles	10.5421 mi.** (Derivation: above measure 14.90867/√2 or 1.4142)
Great P. to Heliopolis obelisk		14.657 miles	14.615 miles	

NOTES for table:

*Distances are corrected for Latitude based on an average latitude for the astronomical monuments of 30.0525°N, the cosine of which is 0.86556.

[266] WGS-84 Geoid Distance Calculator. Published by CQSRG. Copyright, 2010 by Mike Turnbull. http://cqsrg.org/tools/GCDistance/

**The distance between the Heliopolis obelisk and Letopolis is corrected based on the latitude of both monuments, 30.1290°N, the cosine of which is 0.86489.

General comment. All figures depicted in the table are in international miles of 5,280 feet of 304.8mm each.

Column (1). This is the measure of distance using the WGS 84 authalic (equal area) measure for a sphere, with radius of 3,958.76 miles, which has the same area as the WGS 84 reference ellipsoid. A comparable measure can be arrived at by averaging the Egyptian measures for the radii of Earth's equatorial and meridional circumferences, 3,963.4 miles and 3,953.12, respectively, which creates a sphere with a radius of 3,958.26 miles and a circumference of 24,870.48 miles. Compare the circumference of this sphere, 24,870.48 miles to the WGS-84 authalic sphere, 24,873 miles.

Column (2). This is the measure of distance between selected points obtained by entering the geographic coordinates of the points in a distance calculator.

Column (3). This is the measure of the distances from the arc measures in the foregoing section.

Column (4). This is the measure of distance in **sidereal** time, using a sidereal sphere. The measure of the equatorial and meridional dimensions of this sphere were obtained by using the 'nominal', not actual, dimensions of the perimeter of the base of the Great Pyramid:

- 308mm/nautical foot ·3,000 feet/one-half minute of latitude or longitude · 120 half minutes in a degree of latitude or longitude · 360° = 39,916,800,000mm = 24,803.15 miles.
- Alternatively, 525mm/royal cubit · 1,760 cubits/one half minute of latitude or longitude · 120 half minutes of latitude or longitude in a degree of latitude or longitude · 360° = 39.916,800,000mm = 24,803.15 miles.)

g. Comments and conclusions.

The difference between the WGS-84 calculated value obtained for the distance measure of a **minute of solar time** at the geographic location of these astronomical structures (Column 1), 14.9507 miles, compared to the measure of actual distance measure between the Khafre Pyramid and the Heliopolis Obelisk (Column 2), 14.934 miles, is .0167 miles. Converted to time, this difference is: .0167/14.9507 · 60 sec./min. = .067 seconds. Thus, these structures are accurate to within .067 seconds or 59.933 seconds in measuring a minute of solar time. The related measure of the minute of time between the Heliopolis Obelisk and the Letopolis structure(s) is likely of comparable accuracy.

The accuracy of the measure of a **minute of sidereal time** cannot be stated with certainty, even though the calculated values for it (Columns 3 and 4), respectively, 14.9084 miles and 14.90769 miles, which are almost identical, are known. The location of the second, but now missing, Senusret I obelisk at Heliopolis is critical in this regard. The difference between the WGS-84 calculated value of the distance measure of a **minute of solar time** (Column 1), 14.9507 miles, and the calculated value of the measure of a **minute of sidereal time** (Columns 3 and 4), rounded to 14.908 miles, is .0427 miles, which converted to feet, is 225.456 feet. This difference could well represent the distance of the location of the second obelisk from that of the surviving one. The difference in the distances covered by a **minute of solar time** and that of the **minute of sidereal time** is an extremely fine measure, given that the distance measure of a **minute of sidereal time** derived from a **minute of solar time** at this latitude is: 365.24/366.24 · 14.9507 miles = 14.909 miles, a difference of .0417 · 5,280 = 220 feet. This is almost an imperceptible difference, especially if measuring a minute of time, as it would represent a difference of approximately .06 seconds. As a practical matter, then, even though the Egyptian astronomers likely used sidereal time, and their astronomical structures reflect this, these structures, even if precisely positioned, likely could not be used by themselves to measure such fine differences. What other devices or techniques may have been used in conjunction with them are not known. Likely, though, devices containing liquid mercury were used to record the exact instant of a star's zenith.

Solar time was also observed and measured, but it is only reliably accurate during certain times of the year. During other times, it must be corrected for analemmas that arise from the Earth's elliptical orbit around the Sun.[267]

The structures cited in this section functioned as a clock(s) to measure a minute of time, both solar and sidereal. However, even though the Egyptian astronomers were aware of the difference between solar and sidereal measures, for day-to-day practical purposes, this distinction was not important. In this regard, the measure of a minute of time was likely observed and recorded several times daily, both day and night, and then transmitted via light signals to other astronomers and observers throughout Egypt to set their clocks and keep them accurate. These structures, then, served as a national time observatory, among their other astronomical purposes. In its day, it functioned much as the Royal Observatory of Greenwich in Great Britain did in measuring time.

 [268]

Interestingly, one of two hieroglyphs for Letopolis is the symbol depicted above, the other being the hieroglyph for town. If the symbol represents the disc of the Sun in the center, and

[267] Wikipedia contributors, "Analemmas", Wikipedia the Free Encyclopedia, https://en.wikipedia.org/wiki/Analemma accessed September 30, 2018

[268] Wikipedia contributors, Gardiner's Sign List, Wikipedia, the Free Encyclopedia, https://en.wikipedia.org/wiki/Gardiner%27s_sign_list (Accessed June 1, 2017) Also see, Sir Alan Gardiner, Egyptian Grammar, Third Edition, (Griffith Institute, Oxford, 1957) page 503, entries R 22 and R 23.

the two parallel lines to either side (the two closest lines are tangent to the circle) of it represent the left and right limbs of the Sun, then the implication is that a measure was taken of the Sun's passage on each of the limbs. It is not hard to extend this logic to infer that the first measure was taken of the left or leading limb of the Sun at Heliopolis and the second measure was subsequently taken of the same limb at Letopolis. A second pair of readings were then taken at each location of the right or trailing limb. All observations were likely done at Local Apparent Noon (LAN) at both locations. This procedure would accurately measure the time elapsed between the two locations, which, as we have demonstrated earlier, is approximately .707 of a minute of time at the 30th parallel of latitude. However, unless definitive proof can be found of the exact location of the observation tower at Letopolis, nothing further can be stated with any degree of confidence.

CHAPTER 56

THE GOLDEN MEAN; HELIOPOLIS, GIZA AND SAQQARA

The location of the Great Pyramid at Giza and the Step Pyramid complex of Saqqara also are tied to *phi*. The distance between the two structures in nautical feet (308mm) and nautical miles of 6,000 nautical feet is computed as follows:

Great Pyramid:	Latitude:	29.979°N	Longitude:	31.1314°E
Step Pyramid:	Latitude:	29.870°N	Longitude:	31.2160°E[269]
		0.109°		0.0846° (corr. = 0.0732°) *

(* The measure of longitude varies as the cosine of latitude. Cosine 30° = 0.866 · 0.0846° = .0732°)

.109 degrees of latitude converts to 6.54 nautical miles (60 nautical miles/degree of latitude), and .0732 degrees of longitude converts to 4.39 nautical miles (60 nautical miles/degree of longitude. The rectangle with height 6.54 nautical miles and width 4.39 nautical miles has as its diagonal a distance equal to 7.83 nautical miles. If the geographic distance between the Great Pyramid and Heliopolis, which as calculated earlier is 12.72 nautical miles, is multiplied by 1/*phi*, or .618, the product is 12.72 · .618 = 7.86 which is approximately equal to the distance between the Great Pyramid and the Step Pyramid of Saqqara: 7.83 nautical miles. Moreover, 7.83^2 is 61.3 or approximately $.618 \cdot 10^2$; or, even more interestingly, $\sqrt{7.83} = 2.798 \approx 2.80$. This is the exact same number generated earlier that from the product of $\sqrt{3}$ or 1.732 and *phi* or 1.618 = 1.732 · 1.618 = 2.80, which is related to height of the Great Pyramid, 280 cubits.

a. Heliopolis and Saqqara

The distance between the Heliopolis obelisk and the Step Pyramid of Saqqara is also noteworthy, which is calculated as follows.

[269] Wikipedia contributors, "Saqqara", Wikipedia the Free Encyclopedia, https://en.wikipedia.org/wiki/Saqqara (Accessed May 17, 2017). Note that the location in the complex from where the survey was taken is not provided, but is assumed to be the pyramid.)

Obelisk:	Latitude: 30.129°N	Longitude: 31.3075°E
Step Pyramid:	Latitude: 29.870°N	Longitude: 31.2160°E
	0.259°	0.0915° (corrected = 0.079°) *

(* The measure of longitude varies as the cosine of latitude. Cosine 30° = 0.866 · 0.0915° = .079°)

Multiplying both dimensions by 60 nautical miles (6,000 feet of 308 mm each) yields: Latitude: 0.259° · 60 = 15.54 nautical miles; and, Longitude: .079° · 60 nautical miles = 4.74 nautical miles. The diagonal of this rectangle, with dimensions 15.54 X 4.74, which is the actual distance between the two locations, is: cos. 17.03° (this is the angle opposite 4.74) is .95615; then 15.54 ÷ .95615 = 16.253 nautical miles, which is the distance between the two sites. This figure, 16.253 compares to *phi* · 10 = 16.18, with a difference of .073, or .0045 percent. Again, as was the case with the Great Pyramid and the Heliopolis obelisk, there is an extraordinary correlation between *phi* and the distance between the Heliopolis obelisk and the Step Pyramid of Saqqara.

CHAPTER 57

THE GIZA PLAN REVISITED

J.A.R. Legon's finding that the Pyramids of Giza form a great rectangle, 1,732 royal cubits X 1,417 royal cubits, may support the foregoing argument that the Giza (Khafre Pyramid) / Heliopolis diagonal defines the dimensions of the geographic square at the 30th parallel of latitude, as well as the measure of the minute of time at that latitude. Multiplying 1,732 ($\sqrt{3} \cdot 1{,}000$) by the cosine of latitude 30° (.866) = 1500 and then repeating the process a second time, $.866 \cdot 1500 = 1{,}298.92$. These numbers, 1,500 and 1,298.92, when divided by 100, provide the dimensions of the geographic square of 15 arcminutes at the 30th parallel of latitude, which are 15 miles X 12.9892 miles. These numbers almost exactly match the actual numbers, 15 miles X 12.9866 miles. This was by design. The second dimension, $1{,}417 \approx 1{,}414$ ($\sqrt{2} \cdot 1{,}000$), reflects the geometric basis of this diagonal, which is the 45° right triangle.

As noted in the foregoing paragraph, the dimensions of the Giza rectangle, 1,732 royal cubits X 1,414 royal cubits, are, respectively, equivalent to: $\sqrt{3}$ X 1,000 royal cubits by $\sqrt{2}$ X 1,000 royal cubits. The $\sqrt{3}$ and the $\sqrt{2}$ are interesting in that they are instrumental in creating a table of "special trigonometric values", as illustrated below

Degrees	*Radians*	*sin θ*	*cos θ*	tan θ	csc θ	sec θ	cot θ
0°	0	0	1	0	–	1	–
30°	$\frac{\pi}{6}$	$\frac{1}{2}$	$\frac{\sqrt{3}}{2}$	$\frac{\sqrt{3}}{3}$	2	$\frac{2\sqrt{3}}{3}$	$\sqrt{3}$
45°	$\frac{\pi}{4}$	$\frac{\sqrt{2}}{2}$	$\frac{\sqrt{2}}{2}$	1	$\sqrt{2}$	$\sqrt{2}$	1
60°	$\frac{\pi}{3}$	$\frac{\sqrt{3}}{2}$	$\frac{1}{2}$	$\sqrt{3}$	$\frac{2\sqrt{3}}{3}$	2	$\frac{\sqrt{3}}{3}$
90°	$\frac{\pi}{2}$	1	0	–	1	–	0

These special values include trigonometric measures for the specific angles 30°, 45°, 60°, and 90°, as well as their radian (rad) equivalents.[270] Radian measures are based on the radius of a

[270] Illustration source is found at: http://study.com/cimages/multimages/16/imagetrigfunctions9.jpg

circle and π, where 1 π rad = 180° [271], and it is not a coincidence that the dimensions √3 X 1,000 by √2 X 1,000, which, when added together, 1,732 + 1,414 (½ of the rectangle's perimeter, or ½ (180°) of the circumference of a circle with equivalent dimensions) equals 3,146, a factor of 3.146 and a close approximation of π (3.141592), with a difference of +.0044. Furthermore, it is not a coincidence that when the dimension's primary factors are subtracted from one another as follows, √ 3 - √2 or, 1.732 - 1.414 equals .318, which when multiplied by 4 equals 1.272, the square root of the golden mean, or √1.618. Recall from the earlier discussion of the Giza (Great Pyramid)-Heliopolis diagonal that the distance between the Great Pyramid and the Heliopolis obelisk was found to be 12.72 nautical miles, which is a factor of √1.618 or 1.272. And finally, when √3 (1.732) is divided by the √2 (1.414) the result is 1.22489, a number that is almost exactly equal to 1.2248, the area-measure of the Vesica Pisces, which was discussed in an earlier section and found to be an essential measure in constructing equal area maps. In view of their broader significance and importance as discussed above, none of these measures are likely to be coincidences; instead, they are most likely critical parts of an overall design in the construction of the Giza monuments.

It may also be significant that the length of the entry-way colonnade of the Step pyramid complex is 122.9 royal cubits—a number with a very similar root to the area-measure of the Vesica Pisces (i.e. 1.22) and one that is within .0034% of the area-measure of the Vesica Pisces. Does this Step Pyramid complex number also represent the √3 ÷ √2 = 1.22.489? If so, does it have some larger meaning that is unique to the site of the complex? These questions will have to remain unanswered.

[271] Wiki Contributors, "Radian", Wikipedia the Free Encyclopedia, https://en.wikipedia.org/wiki/Radian (Accessed April 26, 2017)

CHAPTER 58

THE GOLDEN MEAN AND MEASURES OF THE EARTH

The golden mean. Also: *phi*, the golden ratio, divine proportion, $(1 + \sqrt{5}) \div 2 = 1.6180339$. The golden mean, or *phi* as it will be hereafter referred to, has been a great source of wonder and fascination for mankind. It has been widely celebrated and sometimes derided over the centuries. Johann Kepler called it the jewel of geometry. And Plato's "Timaeus" claimed that it was used in the creation of the Earth itself.[272] Timaeus may be entirely correct, as will be demonstrated in a series of concise formulas laid out in the sections that follow, which pertain to the measures of the Earth, where *phi* is a primary factor. All are Earth-commensurate measures. All are orders of magnitude of 10. The other primary factor is the number 6, or its factors 2 and 3. (In both factors, their arithmetic cognates are also included, i.e. squares, square roots, reciprocals, etc.; ex. *phi* incudes, its square: 2.6180339, square root: 1.27202, and 1/phi: .6180339.) These formulas lead to very exacting measures, comparable to the very finest of those of modern science.

Phi $\cdot\sqrt{6}$**, R⊕**

a. Earth's equatorial radius.

The above formula gives the measure of the radius of the Earth on the equator, which in turn readily gives the equatorial circumference of the Earth. *Phi*, or 1.6180339, which when multiplied by 2.449489—the $\sqrt{6}$ or $(\sqrt{3} \cdot \sqrt{2})$—yields 3.963357, that when multiplied by 10^3, gives the equatorial radius of the Earth in international (the SI designation) or statute miles (5,280 statute feet of 304.8mm), i.e. 3,963.357 miles. This figure for the radius yields the circumference of the Earth; $(3{,}963.357 \cdot 2 \cdot \pi) = 24{,}902.5$ miles, which represents a difference of one mile from the World Geodetic System 84 (WGS-84) reference ellipsoid's measure of

[272] Plato," Timaeus"; Stephanus reference numbers 31 and 32. There is another possibility in that Timaeus may be referring to a combination of the two triangles he previously examined. In any case, the golden mean is also used

24,901.461 miles.[273] The Earth radius calculated by this formula may be accurate enough to be identified, as it is here, with the astronomical symbol, $R_{\oplus}$.

$$\sqrt{Phi} \cdot 6 \cdot 525.065625\text{mm}$$

b. Earth's equatorial circumference.

The above formula gives the equatorial circumference of the Earth in common Egyptian royal cubits of 525.065625mm[274]. The √phi, or 1.27202, multiplied by 6 then multiplied by 525.065625mm, then multiplied by 10^7 yields, 24,900.6 international miles. Compare this figure with the WGS-84 figure for the circumference of the Earth cited above, 24,901.461 miles, a difference of less than one mile. The derivation of the common Egyptian royal cubit is a fascinating story, and it is related below because of its relevancy to this subject. (The royal cubit of 525mm is of further significance, as it precisely relates distance on Earth to sidereal, or star time.)

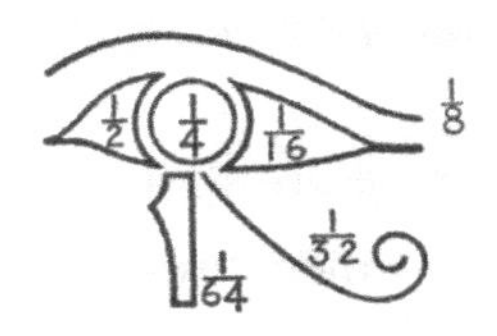

c. The Eye of Horus.

One of the most widely recognized symbols of ancient Egypt is the Eye of Horus, or *wedjat* eye. It is often depicted in fractional parts as shown above. These fractions add up to 63/64. Sir Alan Gardiner calls it the "corn measure" and comments on the missing 1/64th part as being "...supplied magically by Thoth." Gardiner identifies these fractions with the *ḥeḳat*-measure of volumes.[275] Gardiner's association, in this respect, is extremely relevant, because as metrologists understand, almost all volume measures have their origin in linear measures. Considering, then, the *wedjat* eye, this leads to the most fundamental of linear measure relationships: 63/64 of the international foot of 304.8mm = 300.0375mm. This is the measure of the common Egyptian foot, which, in turn, leads to the common Egyptian royal cubit of 525.065625mm.[276] There are four hands in an Egyptian foot and seven hands in a royal cubit; thus, 300.0375mm ÷ 4 = .75009375mm and .75009375mm · 7 = 525.065625mm. There are other measures of the Egyptian royal cubit that are relevant to this subject, but more on this

[273] Wikipedia Contributors, "Earth" Wikipedia the Free Encyclopedia, https://en.wikipedia.org/wiki/Earth, see the column to the right, which lists Earth's particulars. (Accessed August 29, 2018)

[274] Roger S. Bagnall, *The Oxford Handbook of Papyrology* (Oxford University Press, 2009) pp 185-186. Bagnall uses the term "sacred cubit" to refer to the cubit of 525mm. However, since it was in such widespread use, I prefer to call it the common royal cubit.

[275] Sir Alan Gardiner, *Egyptian Grammar; Being an Introduction to the Study of Hieroglyphs;* Third Edition Revised 2005 (Griffith Institute, Oxford; University Press Cambridge) § 266, pages 197-198

[276] All measures are reduced to metric measures for ease of reference.

later. Which of these measures came first, the Egyptian foot or the international foot? Likely both measures came into existence at the same time.

√*Phi* · 6 · 523.75mm

d. Earth's meridional circumference (1).

√*phi*, or 1.27202, multiplied by 6 then multiplied by 523.75mm[277], then multiplied by 10^7 yields, 24,838.21miles. This is a reasonable figure, but it is interesting in that the radius it yields is 3,953.123 miles. Compare this figure to the fixed polar radius cited by WGS-84, which is 3,949.9 miles[278], a difference of a little over 3 miles. (The WGS-84 measure assumes that the Earth is a uniform ellipsoid and disregards local variations for purposes of calculating this radius measure.)

√*Phi* · 6 · 524.21mm

e. Earth's meridional circumference (2).

√*phi*, or 1.27202, multiplied by 6 then multiplied by 524.2176mm, then multiplied by 10^7 yields, 24,860.388 miles.[279] (This is not a measure of circumference, because the Earth is an ellipsoid and not a sphere; rather it is a straight measure of the perimeter of the meridional distance around the Earth, through the poles.) This measure compares to the WGS-84 figure of 24,859.73 miles, a difference of less than a mile.,[280] (The Earth's meridional distance is the same dimension that the measure of the meter was originally predicated upon.) The royal cubit of 524.2176mm is a derived measure using the measure of the height of the Great Pyramid. The Pyramid's height is determined by p*hi*, 1.6180339, · √3, or, 1.73205, = 2.8025. This is the significance of the *mr* pyramid.[281] This figure, 2.8025, is roughly equal to the widely accepted height of the Pyramid, 280 royal cubits. Some sources equate the height of the Pyramid with the polar radius of the Earth, as the Pyramid itself, in their opinion, is a representation of the Earth.[282] If the theoretical height of Pyramid, 280.25 royal cubits, is multiplied by Petrie's royal cubit, 523.75mm, and then divided by the actual height of the Pyramid, 280 royal cubits, the

[277] This is Sir W. M. Flinders Petrie's royal cubit of 20.62 inches or 523.75mm, which he deduced from his measures at Giza. See his survey report, "The Pyramids and Temples of Giza", which can be found on line at: http://www.ronaldbirdsall.com/gizeh/ . See Chapter 20, "Values of the Cubit and Digit". Petrie's cubit of 20.62 inches is a widely cited value for the royal cubit.

[278] Wikipedia Contributors, "Earth Radius" Wikipedia the Free Encyclopedia, https://en.wikipedia.org/wiki/Earth_radius See "Fixed Radius" section. (Accessed August 30, 2018)

[279]

[280] Wikipedia Contributors, "Earth" Wikipedia the Free Encyclopedia, https://en.wikipedia.org/wiki/Earth , see the column to the right, which lists Earth's particulars. (Accessed August 29, 2018)

[281] Gardiner, *Egyptian Grammar;* sign O 24.

[282] Peter Tompkins, *Secrets of the Great Pyramid* (Harper and Rowe, Publishers, New York)

result is 524.21mm, the value of the cubit cited in the formula. The height of the Pyramid thus incorporates two measures for the meridional circumference of the Earth.[283]

Phi

f. <u>The bond of creation.</u>

"Now that which is created is of necessity corporeal, and also visible and tangible. And nothing is visible where there is no fire, or tangible which has no solidity, and nothing is solid without earth. Wherefore also God in the beginning of creation made the body of the universe of fire and earth. But two things cannot be rightly put together without a third; there must be some bond between the two of them. And the fairest bond is that which makes the most complete fusion of itself and the things which it combines; and proportion is best adapted to affect such a union. For whenever in any three numbers, whether cube or square, there is a mean, which is to the last term what the first term is to it; and again when the mean is to the first term as the last term is to the mean—then the mean becoming first and last, and the first and last becoming means, they will all of them of necessity come to be the same, and having become the same with one another will be all one…And for these reasons, and out of such elements which are in number four, the body of the world was created and it was harmonized by proportion, and therefore has the spirit of friendship; and having been reconciled to itself, it was indissoluble by the hand of any other than the framer."[284] (Author's ellipsis added.)

g. <u>Origin of *phi*-based measures of the Earth.</u>

There are many questions that arise from this presentation, but only one answer: God designed the Earth and created it with dimensions and measures reflecting their underlying origin in *phi* and number. And it was He that implanted in mankind the ability through reason to perceive and understand His creation. The measures cited in this presentation were not gradually worked out over time, as civilization progressed; instead, they were a priceless gift given to mankind at the dawning of his awareness.

Time and geographic distance

h. <u>Sidereal time and Earth commensurate measures.</u>

The cubit of 525mm is directly and precisely linked to the measure of sidereal or star time, which is often used by astronomers in making celestial observations. This linkage in time to

[283] It is noteworthy that the Great Pyramid also derives its measurements from *phi.* Given a height of 280 royal cubits—itself derived from, $phi \cdot \sqrt{3}$—the perimeter of its base is, $8 \cdot (280 \div \sqrt{phi})$, or $8 \cdot (280 \div 1.27202)$, or $8 \cdot 220$,the measure of the half side on the base, = 1761 (actual measure is 1760), and its apothem or slant height is, $220 \cdot phi = 356.165$ (actual measure is 356).) This should not be surprising if, indeed, the Pyramid represents the Earth.

[284] Plato," Timaeus"; Stephanus reference numbers 31 and 32

measured geographic distances on Earth is extremely useful in making such observations. In one sidereal second, the surface of the Earth as a sphere at the equator rotates 880 cubits of 525mm past a given celestial body on the celestial sphere. Sidereal time relates to solar time as 366.24/365.24. Sidereal time flows more evenly than solar time, which is measured from local-apparent-noon to local-apparent-noon and can vary significantly throughout the year. (See Chapter 59.)

There is one other measure that relates time to distance by means of the height of a 30°—60°—90° triangle, which is √3 or ≈1.732. If this number is multiplied by 10 ≈ equal 17.32, and then divided into the Earth's circumference, 24,901.461 international miles, the result is 1,437.7864, which compares with the number of minutes in a day, 1,440—a difference of 2.22 minutes. If, instead, the Earth's circumference, 24,901.461 miles is divided by 1,440 minutes in a day, the result is, 17.29268 or .0278, a difference from 17.32 of .02782. Likely, though, the time measure for Earth's daily rotation of 1,440 minutes is the primary measure, and the less precise measure of the rotational rate, 17.29268 miles/minute, while almost perfect, too, is secondary.

Pi, π

i. Derivation of pi.

There is another measure that can be derived from the product of √6 · *phi*, or √3 · √2 · *phi*, which equals 3.963357, as was detailed earlier in Section a. Earth's equatorial radius. If this measure, 3.963357, is squared, which equals 15.708203, then divided by 10, which equals 1.5708203, and then multiplied by 2, the final product equals 3.4164078. This figure, 3.14164078, is ≈ pi or 3.14159265 and is accurate to within 5 decimal positions, .00004813 (3.14164078 – 3.14159265 = .00004813). A coincidence or another example of a divine plan?

CHAPTER 59

SIDEREAL TIME AND EARTH MEASURES

a. The Great Pyramid and Earth measures

The measures of time are dependent on uniformity, as the Earth rotates on its axis and orbits the Sun. There can be no differences in these measures to accommodate variations in linear distances cross the face of the Earth as it moves, like there are for local variations when measuring local latitude and longitude. Therefore, the Earth must be understood as a perfect sphere to accommodate uniformity in the measures of time throughout the World, no matter the location. It is in this regard that the oft-repeated stories in history of the measure of the perimeter of the base of the Great Pyramid being equal to five stadia (600 feet) or one-half minute of latitude or, alternatively, of longitude, are very relevant. It is often assumed that these measures must pertain to the equatorial circumference of the Earth or the measure of latitude above or below the equator, but in these instances the measures of the Pyramid do not agree with actual Earth measures found on the equator. Instead, these measures express a mathematical relationship.

The measure of a degree of latitude at the Great Pyramid's location is 110,852 meters.[285] Dividing this figure by 60, one minute of latitude at this location is 1,847.533 meters, which when divided by the 6,000 feet in a nautical mile (a time-honored method for measuring latitudinal distances) yields a foot of 307.9222mm or approximately 308mm. Thus, a half minute of latitude at this location is 6,000/2 or 3,000 · 308mm = 924,000mm. If this figure is divided by the nominal measure of the perimeter of the Pyramid's base, i.e. 1,760, the result is 525mm, which is equal to the measure of the common Egyptian royal cubit. (Neither of these measures accords well with the actual measure of the perimeter of the base, which from Cole's survey is: 921,453mm, and is equivalent to 1) 1,755.15 royal cubits of 525; or 2) 2,991.7 feet of 308mm.) However, in this regard the purpose of the nominal measure of the Pyramid's base are to express a mathematical relationship, and there are major ramifications in the symmetry of these two measures.

[285] Wikipedia Contributors, "Latitude" Wikipedia, the Free Encyclopedia, https://en.wikipedia.org/wiki/Latitude

Since the foot of 308mm is a measure of latitude and the royal cubit of 525mm is a measure of longitude (see chapter 59, section), the implication is that these are the measures of a sphere as both measures on this sphere are interchangeable, i.e. any circle drawn on the surface of the sphere, which has as its center the center of the sphere, is equal in measure (commonly termed a great circle). If the two measures in the base of the Pyramid are equal, then 308mm/foot · 3,000 (feet in one-half minute of latitude) · 120 (half-minutes in a degree of latitude) · 360 degrees in a circle (This is equivalent to a scale of 43,200:1.) = 39,916,800,000mm = 24,803.15 miles; and 1,760 · 525mm = 924,000mm, in one-half minute of latitude, which, when multiplied by 120 (half-minutes in a degree of longitude) and then by 360 degrees in a circle (This is also equivalent to a scale of 43,200:1.) equals 39,916,800,000mm = 24,803.15 miles. The equatorial circumference of this sphere is thus equal to the meridional circumference.

b. Correlation of Earth measures with sidereal time

If the foregoing measure of the circumference of the sphere, 39,916,800,000, is divided as follows, the following time/distance correlations are obtained:

1) by the number of sidereal seconds in a sidereal day, 86,400, the result is 462,000mm, which when divided by 308mm = 1,500feet/sidereal second, or
2) when divided by 525mm = 880 royal cubits/sidereal second, and
3) when divided by 1,440 sidereal minutes in a day, yields, 27,720,000mm that when divided by 308mm/foot = 90,000 feet and then divided by 600 feet in a stadium = 150 stadia/sidereal minute.

However, the foregoing correlations of time/distance are only applicable on the equator; elsewhere, they must be corrected by the cosine of the latitude. Accordingly, at the latitude of the Great Pyramid, cosine 30° = 0.866, and the time/distance correlations for 30°N are as follows:

1) 1,500 feet of 308mm/sidereal sec. · .866 ~ 1,300 feet of 308mm/sidereal sec., or
2) 880 royal cubits of 525 mm/sidereal sec. · .866 ~ 762 royal cubits of 525mm/sidereal sec., and
3) 150 stadia (600 feet of 308mm)/sidereal minute · .866 ~ 130 stadia/sidereal minute.

c. Earth as a sphere and the measures of time.

As symmetrical as all of this is, though, the measures do not accord with the actual measures of the Earth. The Earth is not 39,916,800,000mm in circumference. However, if one is measuring time by the stars—sidereal time—then these measures of time/distance are entirely correct, as the sidereal day of 23.9344 hours is shorter than the solar day of 24 hours, which makes one solar day 1.00274 longer than a sidereal day. The radius of the sphere with circumference of 24,803.15 miles is 3,947.5439 miles, which when multiplied by 1.00274, is 3,958.36 miles.

This is approximately the authalic, or equal area, measure of the Earth's radius, when the area of the WGS-84 reference ellipsoid is considered as a sphere, is equal to 3,958.76 miles. Sidereal measures of time are all shortened versions of the solar measures, i.e. each sidereal day is composed of: 1) 86,400 sidereal seconds = 365.24/366.24[286] solar seconds, 2) 1,440 sidereal minutes = 365.24/366.24 solar minutes, and 3) 24 sidereal hours = 365.24/366.24 solar hours, etc.

The following table compares the measure of the surface distance covered in one second of rotation across the authalic and sidereal spheres, which are derived from, respectively: 1) the WGS-84 reference ellipsoid, and 2) the mathematical relationships of latitude and longitude from the base of the Great Pyramid. In both cases, the "sidereal sphere" is merely a convenient construct to demonstrate that sidereal time is also circular and based on Earth's rotation. In all cases time, including for both the solar and sidereal spheres, distances are measured on the actual face of the Earth, i.e. the WGS-84 reference ellipsoid; however, the WGS-84 derived authalic sphere is a convenient and accurate way to track and measure time, which is invariably circular and based on Earth's constant rotation.

Authalic and Sidereal Sphere Time/Distance Measures

Sphere and second used. / Distance measure	WGS 84 authalic sphere, with circ. of 24,873.62 miles. (Distance covered in 1 **solar** second)	Great Pyramid authalic sphere, with circ. of 24,870.37miles. (Distance covered in 1 **solar** second)	WGS 84 sidereal sphere, with circ. of 24,805.7 miles. (Distance covered in 1 **sidereal** second)	Great Pyramid sidereal sphere, with circ. of 24,803.15 miles. (Distance covered in 1 **sidereal** second)
Royal cubits of 525mm	882.5*	882.385*	880.09 ~ 880*	880*
Feet of 308mm	1,504.262*	1,504.065*	1,500.15 ~1,500*	1,500*

* However, the foregoing correlations of time/distance are only applicable on the equator; elsewhere, they must be corrected for latitude.

d. Conclusion:

Sidereal time is the most accurate expression of Earth time and the most reliable, which is why astronomers use it. It is also the time clock of the ancient Egyptian astronomers and is the basis for their synchronization of time/distance measures. However, the Egyptian astronomers were well versed in the use of solar time, as well, and could apply the necessary corrections to it to make it usable.

[286] The fraction 365.24/366.24 arises from the fact that there are 365.24 days in a solar year, which is equal to the 366.24 days in a sidereal year, thus solar time relates to sidereal time as 365.24/366.24.

CHAPTER 60

MAPPING THE CELESTIAL SPHERE

Star charts using the azimuthal equidistant format can be readily created, using either the North celestial pole or the Great Pyramid as the point of projection. The internal structures of the Pyramid most probably functioned as a sophisticated observatory at some point after its construction. This was the basic concept in a book published in the nineteenth century by an eminent British astronomer, Richard Anthony Proctor.[287]

The Descending Shaft of the Pyramid is focused just below the North celestial pole, and directly on the arc of the circle traveled by the pole as it slowly circles the North ecliptic pole, in response to precession over some 25,920 years. At various times, stars on or near this circle have shown directly down the shaft, where they could be used to mark the exact meridian of the Pyramid, at the exact instant of its "southing" or intersection with the circle of the North celestial pole, directly below the pole. This can be witnessed by reflecting the southing of the star in a pool of mercury or water at the intersection of the Descending and Ascending Shafts, up the Ascending Shaft, where the time of the southing would be noted. Thereafter, all stars that can be observed through the open slit of the corbelled roof of the Grand Gallery can be timed as they cross this meridian, and their respective declination and hour angle (celestial latitude and longitude) noted. (The location of the celestial equator, from which declination and hour angles are calculated, is readily fixed at 90° below the Celestial north pole.) The edges of the slit of the Gallery would allow for the placement of measuring devices as the star neared the meridian. Using the Slit in the roof of the gallery, approximately 80° of the celestial sphere could be observed and the exact positions of all stars on it could be recorded. If the top end of the Ascending Shaft opened onto the flat platform of the truncated Pyramid at that level, as Proctor believed, then virtually all the bodies of the Northern celestial sphere and many in the Southern celestial sphere could be observed and their position on the sphere calculated. The platform at this level of the Pyramid is a square that is exactly half the area of the base of the Pyramid.

[287] Richard Anthony Proctor, *The Great Pyramid, Observatory, Tomb and Temple.* This book is likely not readily available, but an excellent summary of its findings can be found in: Peter Tompkins, *Secrets of the Great Pyramid*, (New York, NY, Harper & Row Publishers, Inc., 1971) First Edition; see chapter 12, pp 147-158.

The cross-section dimensions of the Descending and Ascending shafts are roughly 2 royal cubits (1.07 meters) in width and 2.28 royal cubits (1.2 meters) in height. The length of the Descending Shaft from the Pyramid's opening to its intersection with the Ascending Shaft is 53.9 royal cubits (28.25 meters).[288] The remarkable precision of the stonework of the Descending shaft is noteworthy: the width of the entire passageway varies in azimuth (side-to-side) by approximately .25 inches and in altitude (up-and-down) by approximately 0.1 inches throughout its entire 350 foot length. The variation for azimuth and altitude nearest the entrance of the Descending Shaft is even more exacting: 0.02 inches.[289] Such exacting precision would be unnecessary for a mere entryway, but essential for an astronomical observatory.

Once the stars that could be observed from the Pyramid were affixed in declination and hour angle or right ascension, then azimuthal equidistant charts of the celestial sphere could readily be drawn. Thereafter, long term observations of the movements of the Sun, Moon and planets would allow for accurate measurements of their orbits and appearances. Similarly, with long term observation, the phenomenon of precession could be observed and measured, which would allow star charts to be updated to account for it. From this data, astronomical forecasting can be done and astronomical calendars can readily be drawn up. Forecasting solar and lunar eclipses, and planetary transits and occultations could also be done.

However, regardless of the foregoing, the design, construction, and long-term operation of the Great Pyramid as a precision astronomical observatory is an anachronism of the most extreme sort. Virtually none of this comports with the known levels of scientific and technological understanding and capabilities of the ancient Egyptians of the time. And it certainly doesn't comport with the evolution of civilization as generally accepted. Again, as was the case with the knowledge and use of the numbers from the Pattern for distance, mass, and time measure, the understanding and use of astronomy was not something gradually acquired by humanity; instead, it was taught to us, just as the Pattern and the measures derived from it, was taught to us.

As to when the Great Pyramid was constructed, it can be the generally accepted date of around 2,550 BCE, but if this date is accurate then the sophisticated and mature of the science that was built into it must have been acquired many centuries earlier, to allow time for the science to develop and mature through careful observation and experimentation. However, given the generally accepted scientific knowledge of the ancient Egyptian civilization at the time, such a process would have required many long centuries—which, in and of itself, would create an anachronym, because going back much further than 3,000 BCE is simply not supported by the generally accepted record of ancient Egyptian civilization and its progress. This, not the date of the Pyramid's construction, is the more important issue. That the science was taught to us, would seem to be the only viable answer.

---

[288] Maragioglio & Rinaldi, drawings from their survey of the Great Pyramid; available on Graham Hancock's web site, at: http://grahamhancock.com/phorum/read.php?1,331194,331194

[289] Peter Tompkins, *Secrets of the Great Pyramid*, (New York, NY, Harper & Row Publishers, Inc., 1971) First Edition, page 151.

CHAPTER 61

VESICA PISCES AND THE MOTIONS OF THE EARTH, AS TRACED IN THE HEAVENS

a. Vesica Pisces, and the North Celestial and North Ecliptic Poles

The Vesica Pisces and its connection to the drawing of maps has been covered in an earlier section. However, this simple geometric figure, which is first recorded in Euclid's First Book as Proposition 1, has a far greater significance, one that has profound ramifications, as it depicts the relationship of the North Celestial Pole of the Earth to the North Ecliptic Pole. The North Ecliptic Pole appears to orbit the North Celestial Pole, but the North Ecliptic Pole is stationary and constant, with the North Celestial Pole slowly orbiting it, counter clockwise, in a Platonic Great Year cycle of 25,920 years. This motion is caused by a wobble in the Earth's axial rotation, or precession, that also causes the precession of the equinoxes. Daily, though, it is the North Ecliptic Pole that appears to orbit the North Celestial Pole of the Earth.

The location of the North Ecliptic Pole, in its apparent motion about the North Celestial Pole, can be affixed with certainty when the location of the ecliptic is known, as it is 90° north of the ecliptic. At the Summer Solstice, it is located 24° below the North Celestial Pole. (The measure of the angular distance is the same as the measure of the axial tilt of the Earth to the ecliptic, or Earth's orbital plane, which varies between 22.1° and 24.5°, but for the sake of simplicity is assumed to be 24°.) At the Winter solstice, it is located 24° above the North Celestial Pole. At the Summer Solstice, it is located 24° south of the North Celestial Pole. At the Vernal Equinox, it is located 24° west of the North Celestial Pole and at the Autumnal Equinox it is located 24° east of the North Celestial Pole. For the following purposes, the North Ecliptic Pole is assumed to be at the Vernal Equinox location, or 24° east.

From the Great Pyramid's location at 30°N, the North Celestial Pole is located at 30° above the horizon, when facing north. (The position of the North Celestial Pole is currently marked by the star Polaris, but in the past, it was often located in the empty spaces along its orbital about the North Ecliptic Pole.) At the vernal equinox, the North Ecliptic Pole is located at 30° above the horizon, and 24° to the west of the North Celestial Pole. At the Autumnal Equinox, it is located at 30° above the horizon, and 24° to the east of the North Celestial Pole. At the

Summer Solstice, it is located 24° directly below the North Celestial Pole, at a height of 6° above the horizon. At the Winter Solstice, it is located 24° directly above the North Celestial Pole, at a height of 54° above the horizon. These numbers are merely estimates of the North Celestial Pole's exact location at specific astronomical events that mark the location of the ecliptic. It should be remembered, though, that the location of the North Ecliptic Pole makes a daily 360°, counter clockwise, circuit of the North Celestial Pole, due to its apparent motion; therefore, the North Ecliptic Pole reaches these locations every day of the year, but calculating the North Ecliptic Pole's specific location each day requires exact knowledge of the ecliptic's location at the time of the day the observation is made.

With the North Ecliptic Pole located due east of the North Celestial Pole at the Vernal Equinox, the North Ecliptic Pole is at the center of a circle, which is the first circle of the Vesica Pisces. At the same moment, the North Celestial Pole is at the center of a second circle of equal dimensions, which is the second circle of the Vesica Pisces. The circumference of the circle of the North Ecliptic Pole (circle A in the illustration shown below[290]) is defined by the Earth's precession, during the Platonic Great Year. The circumference of the circle of the North Celestial Pole (circle B below) is defined by the daily, apparent orbital motion of the North Ecliptic Pole, which is caused by the Earth's rotation on its axis. Both circles are equal in measure, with radii of 24° each, and both circles are the result of the Earth's motions. The center of circle A is the North Ecliptic Pole; the center of circle B is the North Celestial Pole. The North Ecliptic Pole is stationary and unvarying, while the North Celestial Pole is in constant motion. The motions of the North Celestial Pole are the very essence of time, which define not only the Platonic great year and its divisions, but also the various measures of the day (hour, minute, and second). The overall width of the two overlapping circles of the Vesica Pisces, as defined by the motions of the Earth about the North Celestial Pole, is approximately 72°.

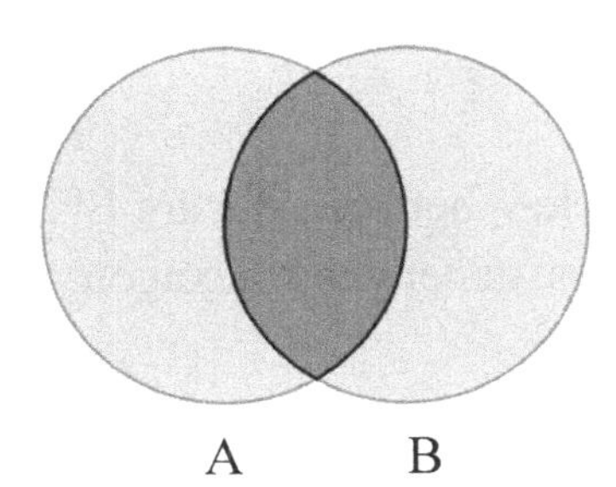

A B

b. <u>Vesica Pisces and Triangles</u>

The Vesica Pisces can be used to define the isosceles triangle and the √2 (1.414), the equilateral triangle and the √3 (1.732) and the 1 X 2 rectangle having as its hypotenuse the √5 (2.23). The 1 X 2 rectangle is formed by the width of the overlapping, almond-shaped center, 1, and the

[290] Image is reproduced from copy shown at: Wikipedia contributors, "Vesica Pisces", https://en.wikipedia.org/wiki/Vesica_piscis (accessed June 11, 2017). The author is Tomruen, who created it on November 7, 2015.

length (diameter) of the circle(s) (2). The √5 is the diagonal of this rectangle. The isosceles triangle is half of this rectangle, which is a square of 1 X 1, divided along its diagonal. The equilateral triangle is formed when a horizontal line is drawn through the centers of both circles and through the overlapping, almond-shaped center. The straight cord through the overlapping, almond shaped center is the base of the triangle, from the ends of which two cords of equal dimension are drawn to meet at the apex. See the illustration below.[291]

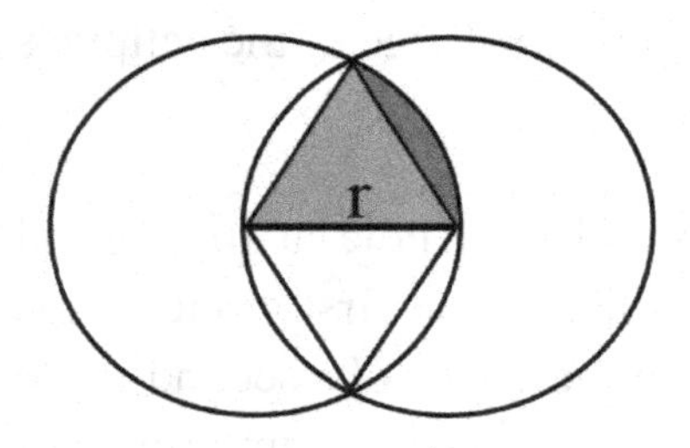

c. Vesica Pisces and *Phi*

The equilateral triangle when enclosed in a circle can be used to define the golden mean (phi, or 1618). See the illustration below[292]

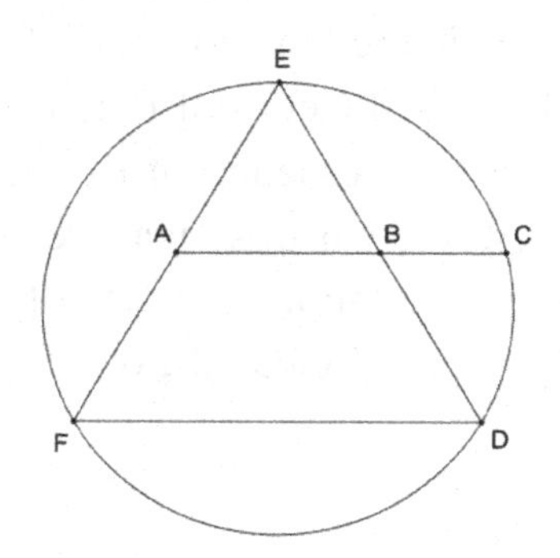

The cord, ABC, when extended to meet the opposite side of the circle—between the arc EF, and then replicated and arrayed about a circle, marked off in 60° arcs, generates a hexagon.

[291] Wikipedia contributors, "Golden Ratio", Wikipedia the Free Encyclopedia, https://en.wikipedia.org/wiki/Golden_ratio (accessed June 12, 2017); see Section titled, "Odom's construction." Drawing was published by בששׁ.

[292] Wikipedia contributors, "Golden Ratio", Wikipedia the Free Encyclopedia, https://en.wikipedia.org/wiki/Golden_ratio (accessed June 12, 2017); see Section titled, "Odom's construction." Drawing was published by Kmhkmh.

d. The Gods of Egypt and the North Ecliptic Pole

The ancient Egyptian god Amun, who was the preeminent god of the Egyptian pantheon in the Middle and New Kingdoms, was known as the "hidden one."[293] It is tempting to identify Amun with the North Ecliptic Pole, which is in an empty, or hidden, portion of the heavens, but this is problematic. However, the North Ecliptic Pole, as it orbits about the North Ecliptic Pole, frequently occupies empty space also, thus making an identification of the epithet, "hidden one" with the North Ecliptic Pole an ambiguous one at best. Still, it is an attractive identification, and it may well merit further analysis. However, there is a much firmer analogy between the North Ecliptic Pole and the Memphite god, Ptah, who is often referred to with a curious epithet, "South-of-His-Wall."

Ptah, South-of-His-Wall, was often referred to as the "Living One", or the "Lord of Eternity." His theological origins are fortunately preserved in an important document that was copied by the Pharaoh Shabako (c. 700 BCE) from an ancient text preserved in the temple dedicated to Ptah in Memphis. In this document, there is no doubt that Ptah is the ultimate source for the created world.

> "In the form of Atum (the creator god of Heliopolis) there came into being heart and there came into being tongue. But the supreme god is Ptah, who has endowed all the gods and their *Ka's* (spirits) through that heart [of his] which appeared in the form of Horus (the elder) and through that tongue [of his] which appeared in the form of Thoth, both of which were forms of Ptah." (Author's parenthetical statements added for clarity.)[294]

This text quite clearly anticipates the divine-word-made-manifest doctrine of the Greeks, known as *logos*. And it very neatly ties in with the duality between God and the "creator god" of Plato's Timaeus, as quoted below in the following section. Ptah, then, is God, and the "creator god" is Atum. It follows, then, that Ptah, "who is upon the Great Place" can be identified in the heavens with the North Ecliptic Pole and Atum can be identified with the North Celestial Pole. As to Ptah's epithet, "South-of-his-Wall", it seems only logical that it refers to the observation of the North Ecliptic Pole, at the summer solstice from Giza/ Heliopolis, when it is, indeed, "south" of the apparent "wall" that is defined by the actual movements of the North Celestial Pole about the unchanging and eternal North Ecliptic Pole.

---

[293] Manfred Lurker, *The Illustrated Dictionary of the Gods and Symbols of Ancient Egypt*, (Thames and Hudson, London, English Language Edition 1980) pages 25-26.

[294] R.T Rundle Clark, *Myth and Symbol in Ancient Egypt* (Thames and Hudson, London, England, 1959; reprinted 1995) pages 58-67; quote is found on page 61.

e. God and the Created Image of Eternity

The constant and unchanging nature of the North Ecliptic Pole calls into mind the essence of God, which is also eternal and unchanging, and His relationship to the created universe, which is in motion and time, as is the North Celestial Pole. The nature of this relationship perhaps has never been better described than it was in Plato's *Timaeus,* in two passages that were cited earlier. (Note that the Creator God in these passages is separate and distinct from the Divine, Eternal Nature. As explained in an earlier passage of *Timaeus*, the Divine, Eternal Nature handed off the actual act of creation to a Creator God, lest His Divine, Eternal Nature be divided or copied in the act of creation, thus becoming a part of Him or like Him.)

> "When the father and creator saw the creature which he had made moving and living, the created image of the eternal gods, he rejoiced, and in his joy determined to make the copy still more like the original; and, as this was eternal, he sought to make the universe eternal, so far as might be. Now the nature of the ideal being was everlasting, but to bestow this attribute in its fullness upon a creature was impossible. **Wherefore he resolved to have a moving image of eternity, and when he set in order the heaven, he made this image eternal but moving according to number, while eternity itself rests in unity; and this image we call time**. For there were no days and nights and months and years before heaven was created, but when he constructed the heaven he created them also. They are all parts of time, and the past and future are created species of time, which we unconsciously but wrongly transfer to the eternal essence; for we say that he "was," he "is," or he "will be," but the truth is that "is" alone is properly attributed to him, and that "was" and "will be" are only to be spoken of becoming in time, for they are motions, but that which is immovably the same cannot become older or younger by time, nor ever did or has become, or hereafter will be, older or younger, nor is subject at all to any of these states which affects moving and sensible thing and of which generation is a cause. **These are the forms of time, which imitates eternity and revolves according to a law of number."** (Author's emphasis added.)

Further:

> "Time, then, and the heaven came into being at the same instant in order that, having been created together, if ever there was to be a dissolution of them, they might be dissolved together. It was framed after the pattern of the eternal nature, that it might resemble this as far as possible; for the pattern exists from eternity, and the created heaven has been, and is, and will be, in all time. Such was the mind and thought of God in the creation of time."[295]

[295] *The Dialogues of Plato,* translated by Benjamin Jowett (Encyclopedia Britannica, Inc. University of Chicago, Great Books, 1952) for first quotation: Timaeus 37-38, page 450, column 2, first opening paragraph;

(It should be noted that Timaeus was an astronomer, and that the tale he told was related to him by Critias, who learned of it from Solon, who heard it from an Egyptian priest.)

f. Vesica Pisces and the Motions of the Earth

A Vesica Pisces also defines the Earth's motions with respect to the Sun.

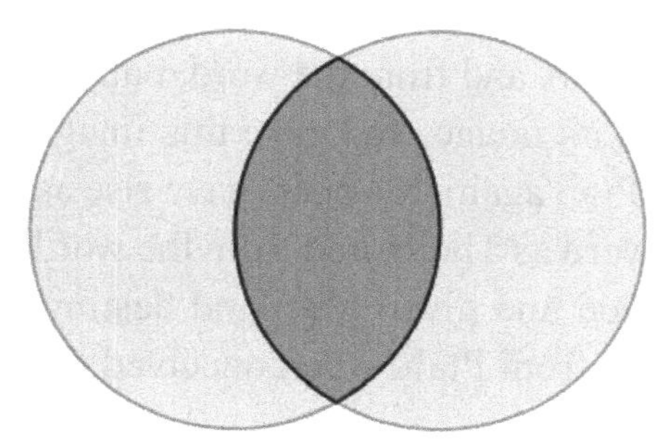

Sun at Center Earth at Center
A B

Circle A, with the Sun at its center, is stationary and unchanging. Circle B, with the Earth at its center, is in constant, counter-clockwise, movement with respect to the Sun. The circumference of Circle B defines the Sun's daily, apparent motion, as it circles the Earth, defining the length of the day. The circumference of Circle A defines the Earth's yearly orbit about the Sun, as well as the Great or Platonic year that results from precession, defining the length of each. Both circles define the ecliptic or path of the Sun.

This image also derives from the diagonals of the Great Pyramid at the equinox, when an equilateral triangle is formed along $1/6^{th}$ of the equator, at local apparent noon on the day of the equinox, on the Pyramid's meridian. A corresponding equilateral triangle derives from this event, with its apex at the center of the Earth, and its sides extended out at a 60° angle to the circumference of Circle B, which is the ecliptic.

The Sun is at the center of the plane of the ecliptic, which is centered on the North Ecliptic Pole. While the Sun is clearly at a substantial distance, 93 million miles which is defined as the Astronomical Unit (au), from the Earth, and, by inference, by the same distance from the Earth's orientation to the North Ecliptic Pole. However, the North Celestial Pole of the Sun and the North Ecliptic Pole of the Earth are parallel lines, which meet at infinity, and in this narrow sense, then, it can be said that the Sun is also oriented on the North Ecliptic Pole. This is important for interpreting the associated cosmology of the ancient Egyptians. Specifically, the Sun is identified as Atum-Re, who is a form of Atum, and thus also issues from Ptah, the Eternal Nature, or God. And Atum-Re, like Atum, is also a creator god, and all Earthly creation originates with Atum and Atum-Re. "**Wherefore he resolved to have a moving image of eternity, and when he set in order the heaven, he made this image eternal but**

and for second quotation: Timaeus 37-38, page 451, column 1, first opening paragraph.

moving according to number, while eternity itself rests in unity; and this image we call time." The Sun exists in time and is, therefore, a part of this "moving image", though it may appear to be otherwise.

Paraphrasing and synthetizing from the words of the unknown theologian of Ptah in Memphis and from Timaeus, the following is offered in summation of the thoughts expressed therein: From the heart of Ptah, thought in the form of Horus (hawk) arose, and from thought the tongue of Ptah uttered the word in the form of Thoth (ape or ibis), and from the word, pattern was created. Then, from this pattern or ideal, Atum created its image, and gave this image motion and revolution which are forms of time. Then from Ptah again, his heart gave rise as Horus to thought, and from thought his tongue gave rise to word as Thoth, and from the word Atum-Re was created. From Atum-Re, the world was created and given form and destiny, and from Atum it was given motion and time, but both arose from Ptah, who conceived the thought and gave it voice.

CHAPTER 62

THE VESICA PISCES OF TIMAEUS AND ITS SIGNIFICANCE

There is a passage in Timaeus 36 that describes the Vesica Pisces and implies that it is at the heart of the design of the Solar System. In part, this text reads as follows:

> "This entire compound he divided lengthwise into two parts, which he joined to one another at the center like the letter X, and bent them into a circular form, connecting them with themselves and each other at the point opposite to their original meeting point; and comprehending them in a uniform revolution upon the same axis, he made the one the outer and the other the inner circle. Now the motion of the outer circle he called the motion of the same, and the motion of the inner circle the motion of the other or diverse. The motion of the same he carried round by the side to the right, and the motion of the diverse diagonally to the left. And he gave dominion to the motion of the same and like, for that he left singe and undivided; but the inner motion he divided in six places and made seven unequal circles having their intervals of ratios of two and three, three of each, (author added underscoring for emphasis) and bade the orbits proceed in a direction opposite to one another; and three [Sun, Mercury, Venus] he made to move with equal swiftness, and the remaining four [Moon, Saturn, Mars, Jupiter] to move with unequal swiftness to the three and to one another, but in due proportion."

The first part of the text, where in the disposition of the two parts are described, "... joined to one another at the center like the letter X..." quite clearly describes the upper intersection of the two equal circles that comprise the Vesica Pisces, while the lower intersection of the Vesica Pisces is clearly implied in the ensuing text, "...connecting themselves and each other at the point opposite to their original meeting point..." (author's underscoring). (See following figure.)

There is a slight problem, though, in reconciling the ensuing text, "...and comprehending them in a uniform revolution upon the same axis, he made one the outer circle and the other the inner circle." However, this is readily resolvable if another circle is swung around the entire figure, touching both circles on their outer perimeter. Outside of this circle is the "motion of the same", while inside of it is the "motion of the other or diverse." That this, indeed, is the meaning of the text will become more apparent later in this section, when the intricacies of the Vesica Pisces are described in detail. Actually, the inner circle is derived from the equilateral triangle formed by the Vesica Pisces, which is detailed in the following figure.

The Vesica Pisces of Timaeus

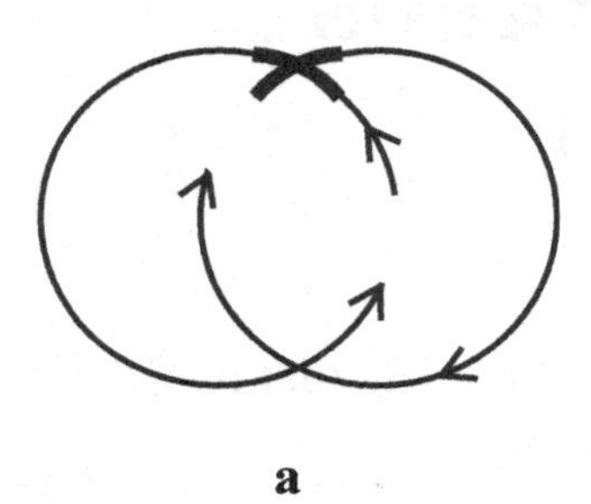

a

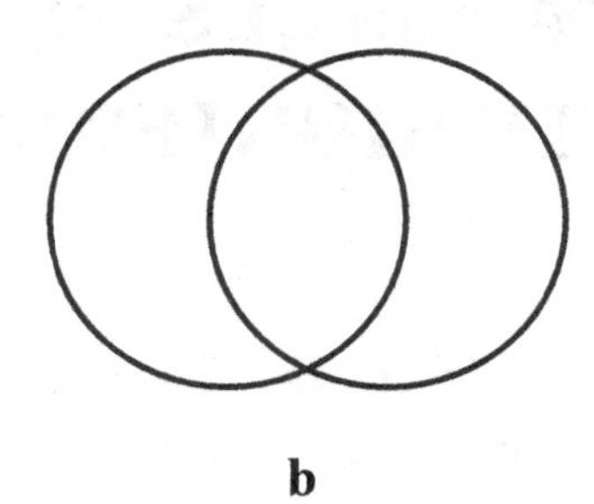

b

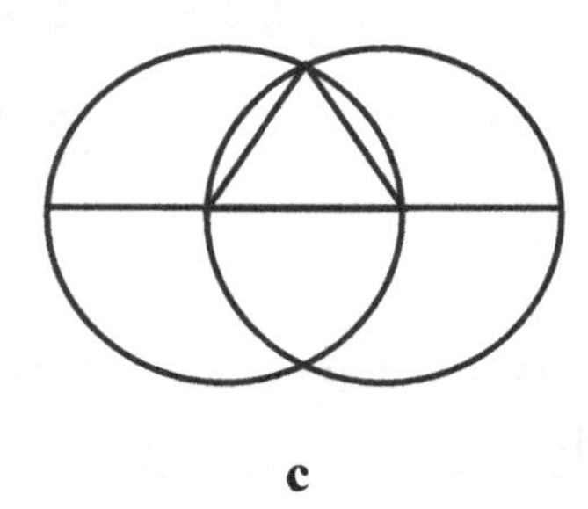

c

<u>Notes on images:</u>

a) The evolution of Vesica Pisces starting from the "X" at the upper crossing to "the point opposite at the lower crossing.
b) The completed Vesica Pisces formed by two intersecting circles, with each circle crossing the center of the other.
c) The equilateral triangle formed in the Vesica Pisces starting at the common radius of the intersecting circles. The center of the base of the triangle is the point from which two concentric circles are derived from rotating the entire figure. **"...and, comprehending them in a uniform revolution upon the same axis, he made the one the outer and the other the inner circle. Now the motion of the outer circle he called the motion of the same, and the motion of the inner circle the motion of the other or diverse."** Rotating the entire figure about the center point yields the "outer circle", while rotating the inner almond-shape yields the "inner circle."

a. <u>The Seven Orbiting Bodies</u>

There are seven orbiting bodies that describe the "unequal circles" referred to in annotations to the quoted text, which are arranged in two groupings; they are: 1) Sun, Mercury, and Venus, and 2) Moon, Mars, Jupiter, and Saturn. In this format, which is properly described as a geocentric or relative motion perspective, the Earth is fixed in its position and all of the orbiting bodies appear to orbit about it. A relative motion perspective is useful for a number of purposes, such as time keeping, compiling celestial maps and navigation. It's biggest drawback, though, is that it does not properly portray the true relationships of the bodies with respect to one another, particularly the Sun and the planets Mercury and Venus. A very important consideration, though, with respect to the quoted text, is that the orbiting bodies are listed by annotations, which may or may not reflect Plato's intent or actual meaning.

Given this, there is a distinct possibility that what was intended by the text was not a relative motion perspective, but a true perspective, with the Sun stationary and all the other bodies, including the Earth, orbiting about it. The Moon, from a true perspective, cannot be an orbiting body of the Sun, since it orbits the Earth. But not including the Moon in the count of orbiting bodies creates a serious textual problem, which specifies that "...seven unequal circles..." were drawn from the orbiting bodies, because from a true perspective there are only six: 1) Mercury, 2) Venus, 3) Earth, 4) Mars, 5) Jupiter, and 6) Saturn. (The planets of Uranus, Neptune and Pluto, a dwarf planet, were not known in classical times, therefore, they are not considered here.)

On its face, this problem would seem to preclude the possibility that what is contemplated by the wording of the text is a true perspective. However, there is another factor that may come into play, which would not only resolve this problem, but also explain the meaning of the text, where it is specified that, "...seven unequal circles having their intervals of ratios of two and three, three of each..." There is a dwarf planet in the asteroid belt between Mars and Jupiter, which is named, Ceres, and it is this planet that constitutes a seventh circle. Setting aside how it was possible for any earlier civilization to know of the existence of Ceres, its existence fits the meaning of the text quite satisfactorily, particularly when the meaning and significance of the text "intervals of ratios of two and three, three of each..." are discussed. However, it must be recognized that Earth's circle is not considered in defining these six intervals. Earth is the seventh circle, not Ceres which is one of the six circles with intervals of two and three. Earth does not factor into these six divisions, except to define the measure of the orbital periods in days and mean orbital radii in miles of the other six. Earth is the benchmark for these measures.

b. The Six Divisions of the Vesica Pisces and the Astronomical Unit (AU)

The three sides of the equilateral triangle formed by the Vesica Pisces are three of these six divisions. (See above figure.) The other three are comprised of the three intersecting lines, each of which originates in an angle and extends such that it divides the side opposite in two equal lengths. The intersection of the three lines locates the position of the Earth with respect to the Sun, which is located at the very center of the Vesica Pisces. This is the basic design of the Solar System, as it also shows the Astronomical Unit (AU), which is the mean distance between the Earth and the Sun. at the location of the Earth. The intervals between the Sun and the other planets are all defined in terms of the AU.

The significance of the Vesica Pisces' equilateral triangle in the design of the Solar System is mirrored in nature by the equilateral triangle formed by the great conjunction series of Jupiter and Saturn, which has a period of roughly 60 years, with the three individual conjunctions in each series occurring roughly 18-20 years apart. In fact, the Solar System from the Sun outward to Saturn can be drawn on the equilateral tringle of the Vesica Pisces. The apex of the triangle represents the position of Saturn and the position of the dwarf planet Ceres at the intersection of the three lines and 1/3 of Saturn's height, and the position of the Earth at a height equal to 1/3 of Ceres' height. The respective AU measures are: Saturn, 9.58 AU; Ceres, 2.77 AU; and Earth 1 AU, which are approximately 9, 3, and 1. However, for clarity only the basic design of the Solar System is shown on the following figure.

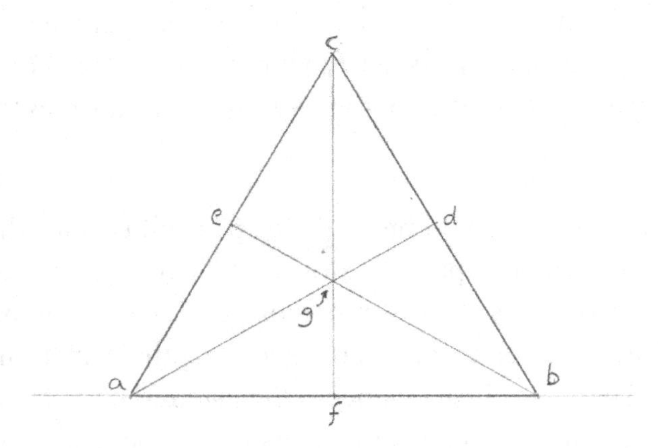

The Equilateral Triangle of the Vesica Pisces, and the Six Divisions of Timaeus

Notes: 1) The six divisions are created by the following cords: ac, cb, ba, cf, da, and eb. These six divisions create a number of right triangles, each of which is a 30°—60°—90° triangle. The hypotenuse of each such triangle is equal to twice its lesser side and the greater side is √3 times the lesser side. This relationship is the same no matter the scale of the triangle. Thus cords cf and af are both √3, even though their respective scale differs.

2) Cord cg is twice the length of cord gf; thus, point g is 1/3 of the height cord cf, or 1 of 3. This is a critical relationship, as it determines one of the basic measures of the solar system, the Astronomical Unit (AU), which is defined as the mean distance of the Earth to the Sun (approx. 93 million miles). <u>Accordingly, point f depicts the relative location of the Sun, point g depicts the relative position of the Earth—thus, depicting the AU measure as one—and point c depicts the relative position of the dwarf planet Ceres, or approximately 3 AU.</u>

c. <u>The Planetary Intervals and Their Correction Factors</u>

This section begins with another quote from Timaeus 36, which immediately precedes the one already cited, ad includes a part of Timaeus 35, as well:

"After this he filled up the double intervals [i.e. between 1,2,4,8] and the triple [i.e. between 1,3,9,27], cutting off yet other portions from the mixture and placing them in the intervals, so that in each interval there were two kinds of means, the one exceeding and exceeded by equal parts of its extremes [as for example 1, 4/3, 2 in which the mean 4/3 is one-third of 1 more than 1, and one-third of 2 less than 2], the other being that kind of mean which exceeds and is exceeded by an equal number. Where there were intervals of 3/2 and of 4/3 and of 9/8, made by the connecting terms in the former intervals, he filled up all the intervals of 4/3 with the interval 9/8, leaving a fraction over; and this fraction expressed was in the ratio of 256 to 243 (author's underscoring added for emphasis). And thus the whole mixture out of which he cut these portions was exhausted by him."

The first mean [1, 4/3, 2] from the foregoing, defines the position of the Earth relative to Venus and Mars in the Solar System. Venus is approximately .75 AU from the Sun, which, when multiplied by the mean, 4/3, yields the correct position of Earth at 1 AU; while twice the Venus AU is 1.5 (.75 X 2 = 1.5), which is the correct position of Mars at 1.5 AU, relative to which, the Earth's position at 1 AU, is one-third less than that of Mars, 1.5 AU, (1.5 – (1/3 X 1.5 = .5) or 1 AU. This is the only known purpose served by this mean in the design of the Solar System, though it is obviously a very important one.

The second mean involves an interval that is an "equal number." This is a bit harder to define from the information in Timaeus, which notes that its intervals are: 3/2, 4/3 and 9/8. The intervals between these three numbers, which are equal to: 36/24, 32/24, and 27/24, respectfully, are 4/24 and 5/24, numbers that are clearly not equal. However, if these three intervals, 3/2, 4/3, and 9/8, are multiplied together, the product is: 1.5 X 1.33 X 1.125 = 2.25, the square root of which is 1.5; further, the second and third fractions in the series when multiplied together, 1.33 X 1.125 = 1.5. Clearly, when the fractions are multiplied together there is a noteworthy presence of the fraction 3/2 or 1.5 in the series. Other numbers, though, can be produced:

1.5 X 1.333 = **2**;
$\sqrt{1.5}$ or 1.2248, + (4/3 or 1.3333)^2 or 1.7777 = **3**;
3/2 + 4/3 + 9/8 + 256/243 = **5**.

These three numbers, 2, 3, and 5, comprise the first three numbers of the Fibonacci series. This series generates ever finer measures of the golden mean, 1.681. The presence of this series in the fractions is also consistent with Timaeus, as found in sections 31 and 32, wherein the golden mean is described as an inherent element used by God in the creation of the universe, imparting to it harmony.

The first two numbers, 2 and 3, while their presence is not readily apparent in the fraction series are the intended intervals or "equal numbers." The numbers 2 and 3 are also the intervals between the planetary orbits as stated by Timaeus. (See the underscored text in the prior or first Timaeus quote.) There is one final consideration, though, and that is that derivative forms of these two numbers, 2 and 3, are the actual intervals, which are ½ and √3, respectively. That √3 is one of the two intended intervals is further supported in the following paragraph.

There are two intervals cited in Timaeus 36 that "...he filled up all the intervals of 4/3...", with 9/8 and 256/243. (See the underscored text in the foregoing or second Timaeus quote.) The 4/3 interval, when squared, = 1.777, which is roughly equal to 1.732 or √3. Was this fraction, 4/3, selected because it is the simplest fraction that leads to 1.732 or √3? This may be the case. (It is interesting that almost the exact √3, or 1.732, can be derived from slight variations of the second means' fraction series when multiplied together as follows: 1.5 X 1.3 (vs. 1.333) X 8/9 (9/8 inverted) = 1.733.)

It is also significant that the first three fractions—3/2, 4/3, and 9/8—all appear in the Vesica Pisces itself. If the two intersecting circles of Vesica Pisces are divided horizontally along their diameters, this line is intersected by the arcs of the circles twice, such that the entire line is divided into halves of a circle, or three halves, i.e. 3/2. Each circle of the Vesica Pisces is divided into thirds, such that if the center 'almond figure' is disregarded, the remainder of each circle is 2/3 or 4/3 if the remainders of both circles are combined. The third fraction, 9/8, is the factor that defines the relationship between the first two fractions: 4/3 X 9/8 = 3/2, or 3/2 ÷ 4/3 = 9/8.

d. Estimating the AU of the Planets, using Timaeus

Calculating the AU of each of the seven planets of Timaeus (i.e. Mercury, Venus, Earth, Mars, Ceres, Jupiter, and Saturn), using the numbers from the foregoing section, can now proceed. The respective AU distance of each of these bodies are depicted in the following table:

Planet	(1) Calculation of Timaeus Astronomical Unit (TAU) by intervals: ½ or √3	(2) TAU from column (1) corrected by Timaeus fractions: 9/8, 8/9, 256/243, or 243/256 ***	(3) TAU Adjusted for Corrections From column (2)	(4) Actual Astronomical Unit (AU), by Current Measure
Mercury	.375 (1/2 of Venus TAU)	.375 X 256/243	.395 TAU	.387 AU
Venus	.75 (1/2 of Mars TAU)	.75 X 243/256	.712 TAU	.723 AU
Earth	1 (TAU)	None	1	1
Mars	1.5 (1/2 of Ceres, 3 TAU)	None	1.5 TAU	1.52 AU

Ceres (dwarf planet)*	3 TAU **(1.732 X √3 = 3 TAU)**	3 X 8/9 X 256/243	2.8 TAU	2.77 AU
Jupiter	3 X √3 = 5.2 TAU	None	5.2 TAU	5.2 AU
Saturn	5.2 X √3 = 9.00 TAU	9.00 X 9/8 X 243/256	9.61 TAU	9.58 AU
Uranus**	9 X √3 = 15.59 TAU	15.59 X 9/8 X 256/243	18.48 TAU	19.23 AU
Neptune**	15.59 X √3 = 27 TAU	27 X 9/8	30.375 TAU	30.05 AU
Pluto** (dwarf planet)	27 X √3 = 46.76 TAU	46.76 X 8/9 X 243/256	39.45 TAU	39.5 AU

Solar System AU Calculations, from Timaeus

Notes accompanying the table:

* **Ceres is the point from which the intervals originate,** with the interval of ½ being applied to the planets inward toward the Sun, starting with Ceres, and the interval of √3 being applied to Ceres and the planets outward. Ceres's position is the first interval of √3, which originates in the geometry of the equilateral triangle, as depicted in the earlier figure in this section, "The Equilateral Triangle of the Vesica Pisces, and the Six Divisions of Timaeus." In this figure, the cords "af" and "fc" are both √3, which then define the height of the triangle as, **√3 X √3 = 3 TAU**.

** Uranus, Neptune and Pluto were not included in the seven orbiting bodies of Timaeus; however, they are included here to demonstrate the continuation of the √3 interval and the correction factors 9/8 and 256/243 that Timaeus says was used to define the intervals of the Solar System's outer orbiting bodies. This raises an interesting question: Did Timaeus stop at Saturn because of technical limitations that prevented his knowledge of these three planets, or because Saturn's orbit, along with that of Jupiter, defines the equilateral triangle of the Great Conjunction that can be implied in the equilateral triangle depicted in the earlier figure in this section, "The Equilateral Triangle of the Vesica Pisces, and the Six Divisions of Timaeus?"

*** There is no pattern or inherent logic behind the application of the corrections, which is a decided weakness in my argument. In the table, the corrections were simply applied until the TAU closely approximated the modern AU of each planet. This is admittedly rather arbitrary, but then Timaeus knew the underlying geometry of the Solar System and the measure of the precise corrections that need to be applied to have their estimated AU reflect actual reality, so there is justification for it. There are two possible explanations: 1) Timaeus knew the precise AU measure of each planet and applied precise corrections to their **ideal location** (i.e. as determined by the geometry of the equilateral triangle—which defines the value of the AU and locates Ceres—and by the intervals of ½ and √3, which broadly places each planet in its location) to reflect their **actual placement**, as was done in the instant table; or 2) the divine architect designed the Solar System used **ideal locations,** and then measured the imperfections in the created system with these fractions to reflect the reality of the **actual locations**. There is another possibility: Timaeus knew of and anticipated Kepler's third law of planetary orbits, where the square of the orbital period is proportional to the cube of the orbital radius.

Timaeus speaks of a seventh circle, in this same part of his treatise on the vesica pisces, which can only mean the orbit of Earth. Using a variation of Kepler's third law, if we take the cubic root of 7, which is ≈ 1.9129, and square it, the result is ≈ 3.6593. If this number is multiplied by 10, the result is, 365.93, which is very nearly the orbital period of Earth, 365.24 days. A coincidence? Likely no.

One final comment regarding the planetary intervals as defined by Timaeus. The intervals uncorrected and the planetary distances from the Sun associated with them are within or very close to the current calculations for the maximum and minimum distances from the Sun for the planets. Thus, the planetary intervals and their associated distances from the Sun, as set forth in Timaeus uncorrected, are still reliable as rough estimates for these measures.

e. The Astronomical Meaning of Timaeus' Text Summarized

The full meaning of the prior quoted text from Timaeus can now be addressed: "Now the motion of the outer circle he called the motion of the same, and the motion of the inner circle, the motion of the other or diverse. The motion of the same he carried around by the side to the right, and the motion of the diverse diagonally to the left. And he gave dominion to the motion of the same and like, for that he left single and undivided…"

The "motion of the same" is the right-to-left, or counter-clockwise motion of the planets as they orbit the Sun, from a top view perspective of the Solar System.

The "motion of the diverse" is the seeming contrary motion, or clockwise motion, of the seven planets as they cross in front of the Sun during their orbits, from a perspective outside of the Solar System. The diagonal motion is the apparent motion of the planets as they reach maximum elongation during their orbit and then appear to diagonally move toward the viewers perspective as they continue their orbit in front of the Sun. (Typically, this "motion of the diverse" is only demonstrated by Mercury and Venus, as they enter into inferior conjunction with the Earth, but all of the outer planets periodically demonstrate this apparent motion when the Earth "overtakes" them during its orbit.)

And continuing Timaeus' foregoing text, "…but the inner motion (i.e. "motion of the diverse") he divided in six places and made seven unequal circles having their intervals in ratios of two and three, three of each…", this, too, can now be addressed:

- The seven unequal circles are the orbits of: 1) Mercury 2) Venus, 3) Earth, 4) Mars, 5) Ceres, 6) Jupiter, and 7) Saturn.
- The six intervals or divisions are the Astronomical Units (AU) of: 1) Mars, 2) Venus, 3) Mercury, 4) Ceres, 5) Jupiter, and 6) Saturn. The first interval, that of 2, is as follows: 1) Mars is ½ the AU of Ceres, 2) Venus is ½ the AU of Mars, and 3) Mercury is ½ the AU of Venus. The second interval, that of 3, is as follows: 1) Ceres' AU is √3 X

√3 = 3, 2) Jupiter's AU is √3 X 3 =5.2, and 3) Saturn's AU is √3 X 5.2 = 9. The first interval, that of 1/2, is a derivative form of 2, and the second interval, that of √3, is a derivative form of 3.

f. Timaeus and the Pattern

The second Timaeus quote begins with language that clearly seems to raise the existence of the Pattern. In part, this text read, "After this he filled up the double intervals [i.e. between 1, 2, 4, 8] and the triple [i.e. between 1, 3, 9, 27] …" While these sequences are more abbreviated than those of the Pattern that was introduced in the first Part of this book, they clearly follow the same pattern of 2 and 3. But what are these patterns? Timaeus answers this in an earlier passage in Timaeus 35:

> "Out of the indivisible and unchangeable, and also out of that which is divisible and has to do with material bodies, he compounded a third and intermediate kind of essence, partaking of the nature of the same and of the other, and this compound he placed accordingly in a mean between the indivisible, and the divisible and material."

This "intermediate kind of essence" that he [god] placed between the "indivisible and unchangeable" [soul] and "…that which is divisible and has to do with material bodies…" is number and measure. Number and measure! This is how God organized the Solar System and the surrounding universe, and it is tempting to delve further into this line of thought. But I already have, in Part I of this book.

There is one more part of the first quoted section of Timaeus that must be addressed, and that is its concluding sentence.:

> "…and three [Sun, Mercury, and Venus] he made to move with **equal swiftness**, and the remaining four [Moon, Saturn, Mars, Jupiter] to move with unequal swiftness to the three and to one another, but in due proportion." (Author's emphasis added.)

On the face of it, this section appears to be discussing the orbital velocities of the various bodies of the solar system, an interpretation which the annotations clearly support. However, this cannot be the case, because the orbital velocities of the various bodies of the solar system move at different rates that are not even close to one another. The annotations are clearly incorrect in this regard, as the Sun has no orbital velocity and the orbital velocities of Mercury and Venus are radically different from one another. (Mercury, in particular, was noteworthy even in classical times for the speed of its orbital velocity.) There is another interpretation that can be made, and it is one that fits nicely with the foregoing discussion of the equilateral triangle formed from the Vesica Pisces, and the related Great Conjunction of Jupiter and Saturn and the equilateral triangle it creates.

The "three" discussed in the foregoing quote, in fact, is a reference to the Greatest Conjunction of Jupiter and Saturn, including the Earth, with Jupiter and Saturn in direct opposition. The three bodies are then, Earth, Jupiter and Saturn, which indeed move at "equal swiftness" in creating the triple conjunction of the Greatest Conjunction, where three conjunctions of Jupiter and Saturn occur within several months of one-another. The Greatest Conjunction is a singular occurrence, within the time of the cycle of the more frequent and regularly occurring, Great Conjunction, which occurs every 18-20 years, or 54-60 years in total to complete the equilateral triangle pattern of the Great Conjunction, though it is much less frequent as it occurs in a much longer time cycle.

That this is the correct interpretation is supported language found in a later section of Timaeus (39), which states:

> "And yet there is no difficulty in seeing that the **perfect number of time** fulfills the **perfect year** when all the **eight revolutions having their relative degrees of swiftness**, are accomplished together and attain their completion at **the same time**, measured by the rotation of the same and equally moving." (Author's emphasis added.)

This section clearly associates 'time" with "swiftness." Furthermore, it mentions "eight revolutions", which are also clearly associated with the measure of time. But what are they?

They are the eight great cycles of time: 1) solar day, 2) lunar synodic period (29.5 days), 3) tropical solar year, (365.24 days), 4) Saros cycle (lunar and solar) and its related Inex cycle (223 lunar synodic periods/242 lunar draconic periods), 5) the Venus inferior conjunction cycle (19 months or 584 days; to complete the five-point star pattern is 8 earth years, which is equal to 13 Venus synodical years), and its associated Venus transit cycle (243 earth years), 6) the great conjunction of Jupiter and Saturn (18-20 years), and 7) the precession of the equinoxes (the Platonic or great year of 25,920 years), and 8) the Phoenix cycle (the period between the occurrence of long-duration Saros series, which is 2 or 3 Venus transit cycles or 486 or 729 years).

The ensuing sections of Timaeus to those sections—34 to 37—that were quoted from throughout the preceding paragraphs seemingly contradict the entire argument that they describe a heliocentric solar system with its AU measures. However, there is no contradiction. The ensuing sections, including 38 and 39, clearly describe a geocentric plot, with the Earth at its center, but the wording in these sections also make it quite clear that they are speaking of the measures of time. For a variety of practical purposes, not the least of which is to facilitate comprehension of the measures of time, a geocentric plot of the solar system is far superior to a heliocentric one for directly relating the perceived motions of the Sun, Moon and planets to the measures of time. Imagine the complexities that would arise in trying to describe eclipses if a heliocentric system were used to explain them!

g. <u>The presence of Timaeus's fractions in the Step Pyramid complex</u>

The Step Pyramid complex of Saqqara contains several structural dimensions that are directly related to the first two of Timaeus second series of fractions, i.e. 3/2, and 4/3 (see Appendices 2 and 3):

i. The dimensions of each of the 40 columns of the main colonnade are: width at the base-2 royal cubits (3/2 X 4/3 = 2), which tapers to a width at the top of capital-1.3333 r.c. (4/3); depth at the base-1.5 royal cubits (3/2), measured tangentially from the front to the last reed of each column.

ii. The length of the main colonnade is 122.4 royal cubits, which compares with $\sqrt{3/2}$ or $\sqrt{1.5}$ = 1.2247 X 100 = 122.47. The columns of the main colonnade are divided by a low wall, such that there are 24 columns in the eastern portion pf the colonnade and 16 in the western; 24/16 = 3/2 or 1.5.

iii. The width of the Heb Sed area measured from the entryway of the Heb Sed courtyard, westward, to the wall at the far end of the "T temple", is the same measure: 122.4 royal cubits. The length of the Heb Sed courtyard is 177 royal cubits, which is 1.333^2 or 1.77 X 100 = 177. Both dimensions added together, 122.47 + 177 = 299.47, which is approximately 3.

So what is the significance of this? That measures of time, specifically those measures tracked by the Step Pyramid complex, have their origin in the Vesica Pisces, too, even though measures of time tend to be tracked with a geo-centric system, instead of the heliocentric system directly derived from the Vesica Pisces.

h. <u>The Timaeus fractions, 3/2 and 4/3, and Ancient Egyptian hieroglyphs</u>

There are two glyphs that are considered unique to the usual manner in which Ancient Egyptians normally depicted fractions. The usual manner was to use the symbol , Gardiner's sign D-21, as the numerator in <u>all</u> fractions. This sign, when used in fractions, has the value of "1" or unity. Thus, all usual fractions depicted in Ancient Egyptian renderings have the numeral 1. There are two profoundly important exceptions, 2/3 and 3/4, which are, respectively, represented by the glyphs: , Gardiner's sign D-22, and , Gardiner's sign D-23. As stated earlier these two fractions are unique to the normal manner of representing fractions in Ancient Egypt.[296] The inverse of these two fractions are: 3/2 and 4/3, respectively, and this may well be the further meanings, respectively, of these two glyphs. If these two glyphs are depicted in their normal manner, then D-22 would be 1/ (2/3) and D-23 would be 1/ (3/4), which in mathematics yields the inverse of the denominators in both instances, or 3/2 and 4/3, respectively, which are, of course, the first two Timaeus fractions.

There is a further aspect to the D-21, D-22, and D-23 glyphs, when used in Ancient Egyptian fractions, that is very noteworthy. All three have a shape that is striking similar to that of the almond-shaped center of the Vesica Pisces, even though these three signs depict the almond-shape horizontally, instead of vertically as is the case with the Vesica Pisces. It bears remarking that the usual meaning assigned to the D-21 sign is 'mouth.' Is it possible, then, that the fractions 3/2 and 4/3 arise from or are uttered by God in devising the Vesica Pisces? Is this the reason that they are unique to Ancient Egypt, because, as utterances of the Creator Himself, they are sacred? Remember from the foregoing discussion in section c that these two fractions manifest themselves in the Vesica Pisces, which, as Timaeus states, is the origin of the creation of the solar system. A final question arises: Does all of creation arise from the Vesica Pisces? It would seem logical that it does. Consider, for example, the Vesica Pisces formed by the North Celestial Pole and the North Ecliptic Pole, which is discussed in Chapter 60.

[296] Gardiner, Sir Alan; *Egyptian Grammar* (Griffith Institute, Oxford University, United Kingdom, 2005); § 265 Fractions, pp 196-197

PART V

THE WORD OF GOD AND CYCLES OF TIME; THE UNDERGROUND STELAE OF THE STEP PYRAMID

CHAPTER 63

THE STELAE; DESCRIPTION, LOCATION AND ORIENTATION

The Step Pyramid complex has a number of inscriptions in various locations. The largest group of inscriptions is found on six stone stelae found beneath the Step Pyramid and the South Tomb.

There are three stelae beneath the Step Pyramid and three beneath the South Tomb—six in total. The stelae are located some 30 meters below ground.

The primary narrative of the stelae pertains to God and creation, particularly: number and pattern, measure, and cycles of time—the very things that are the focus of this book. The gods of the Egyptian religion will be referenced frequently, which is in direct conflict with my assessment of the unreliability of such references in a previous Part of this book. And while this assessment is still valid, there is no alternative, when the stelae themselves speak of the gods. The principal problem with referencing the Egyptian pantheon is that the nature and attributes of the gods change with time, often markedly from earlier understandings of them, which are often vague and sometimes in direct conflict with subsequent understandings. However, there is no alternative, but to use the pantheon for reference purposes.

Another issue that arises in discussing the gods is their sexual attributes. This issue clouds modern minds as much as it did the minds of ancient Egyptians. The sexual nature of the gods depicted on the stelae is frequently ambiguous, and often androgynous. The gods depicted on the first three stelae, for example, are 'self-created' gods and they possess both male and female sexual attributes. But the gods are not cast in the image of man or woman; humans assign them these identities to achieve a better understanding of them.

The stelae designated 1, 2, & 3 in the illustration shown below are found beneath the Step Pyramid, and those designated 4, 5, & 6 are found near the South Tomb.[297] The principal image on each stela depicts a figure in left profile, facing toward the southwest. All the stelae have similar dimensions: approximately 1.04 meters X 0.6 meters, or 1.8 Egyptian royal cubits X 1 royal cubits. The stelae are roughly aligned along an axis that runs from northeast to southwest, starting from the Step Pyramid stelae and continuing with the South Tomb stelae, some 8-9° east of true north. The implied movement of the figure depicted on the six stelae is toward the southwest, where it would be logical to conclude he exits the underworld and appears on the southwest corner of the *temenos* wall. All the stelae are in low relief. The stelae have several common features, but each is unique, in its own way.

NOTE: There are a series of chapters included with this part, specifically chapters 68-74, that are not directly related to the stelae, but are topically related and bear on the overall interpretation of the stelae.

[297] Florence Dunn Friedman, "The Underground Relief Panels of King Djoser at the Step Pyramid Complex", (Journal of the American Research Center in Egypt; Vol. 32 (1995) pp 1-42). This reference provides a thorough discussion of the stela and of the various difficulties in interpreting them. The drawings of the stelae included with this article were done by Yvonne Markowitz.

Southern Stela

#3

Central Stela

#2

Northern Stela

#1

Stelae Under the Step Pyramid

Sketches are by Betsy Miller from: Cecil M. Firth and J. E. Quibell, *Excavations at Saqqara, The Step Pyramid*, (Originally published by Le Caire, Imprimerie de L'Institut Français D'Archéologie Orientale; 1935); Plates 15, 16 and 17, respectively, the southern, central and northern stelae; with permission of Martino Publishing © 2007

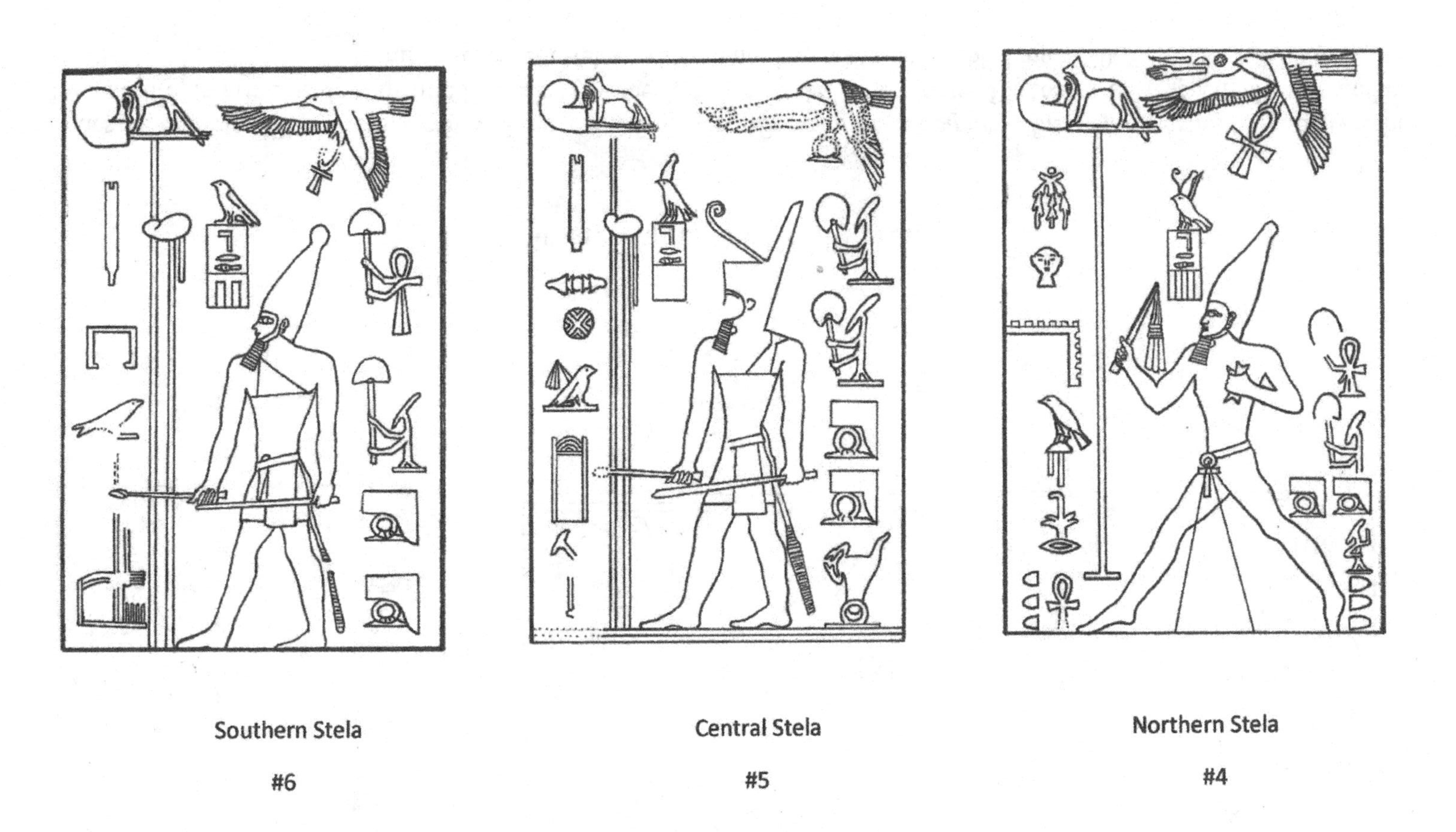

Stelae Under the South Tomb

Sketches are by Betsy Miller from: Cecil M. Firth and J. E. Quibell, *Excavations at Saqqara, The Step Pyramid*, (Originally published by Le Caire, Imprimerie de L'Institut Français D'Archéologie Orientale; 1935); Plates 40, 41 and 42; respectively, the southern, central and northern stelae; with permission of Martino Publishing © 2007

There is an order to the stelae that differs from that which is suggested by their physical locations and the direction in which they all face (roughly 9° west of south). The order starts with stela '4', the first 'runner', under the South Tomb and then progresses to the three stelae under the Pyramid, beginning with stela '3', then stela '2', then stela '1', and then returning to the South Tomb, with stela '6' and concluding with stela '5'. In every instance, including the return to the South Tomb stelae, they are always on the left side as the progression is followed. This progression imparts a counterclockwise or easterly, circular motion to the stelae, which, logically, form the six sides of a hexagon inscribed within a circle. This counterclockwise or easterly motion is the motion of the planets of the solar system. All of this gets ahead of the story, but it is important to keep this image in mind as the stelae are examined in detail, because they tell separate, but related stories, each being part of a whole, and this explanation is critical to an understanding of the whole, which is why it is presented now.

CHAPTER 64

FEATURES COMMON TO ALL SIX STELAE

Each of the stelae have the following features in common:

i. A *serekh* surmounted by a hawk, immediately in front of and slightly above the figure's face. The *serekh* symbol was used to identify the king by his Horus name, throughout Egyptian history. There are three symbols for the given name in the *serekh*, which is understood as Netjerikhet. (This name is discussed in depth in the following section.)
ii. An image of a figure, either running (on three stelae) or walking (on three stelae).
iii. A "standard" with a canine, similar to a grey-hound atop it, with a bag-like image and rearing cobra directly in front of the canine. This standard is always depicted in front of the figure, on the left side of the stelae. Egyptologists identify the canine as *Wepwawet*, the "opener of ways."
iv. A hawk in flight above the figure's head.
v. A pair of identical symbols behind the figure, with each depicting a circle surmounted by a curious glyph, similar to one half of the Egyptian glyph for sky.
vi. Also behind the figure, fans are depicted that are held by either an *ankh* symbol or a *was scepter.* There are five stelae with two fans and one with one fan.

a. <u>The Netjerikhet *Serekh*</u>

This is the name of the king associated with the Step Pyramid complex. On the face of it, the presence of his name in the *serekh* in front of the figure in each of the stelae identifies the depicted king as Netjerikhet. However, if the symbols comprising his name are analyzed more closely, a far different possible interpretation arises. The three glyphs in the *serekh* are shown below:

The first glyph is understood as Gardiner's sign R-8[298], and comprises the "Netjer" portion of the name. Gardiner gives its meaning "god" or "divine." The second glyph is understood as Gardiner's sign D-21, and is interpreted as "mouth"; however, it is understood by Egyptologists to be a "sound complement", which does not contribute to the pronunciation of the name. (Sound compliments are common in hieroglyphs, and are often used to help with the identification of the precise meaning of a word or to better arrange the actual writing of a word to make its depiction more visually appealing.) The third glyph is understood as Gardiner's sign F-32, and composes the "khet" portion of the name. Gardiner describes the khet glyph as an "animal's belly showing teats and tail" and gives its meaning as belly or body. (It is noteworthy for my argument that the body of the animal depicted has **six** teats, bisected along the animal's length by a line into **two** rows of **three** each, one set to each side of the animal. It is also noteworthy that the glyph depicts an animal with a tail.) Accordingly, Egyptologists understand Netjerikhet to mean, "Most Divine of the Corporation [of gods]" (bracketed phrase added by author).

However, if the "mouth" glyph has as its actual value the word "mouth" or "speak", then a far different meaning arises, and Netjerikhet would then mean, "God speaks and calls forth the corporation." Further, if the khet symbol is taken to be a created animal composed of **two** rows of **three** teats each, then Netjerikhet would mean "God speaks and calls forth two and three." This has profound ramifications for the overarching theme of this book, but it also means that the figure depicted in each of the six stelae, may not be a king at all, but most likely a god or several gods. (Hereafter, the image depicted will be referred to as a figure(s), more specifically, as 'runner(s)' or 'walker(s)', instead of a king. The reason for this is to avoid confusion, as it will become apparent later in this Part that the image is a god or gods.) As to the hawk atop each *serekh*, if the hawk is the expression of the heart of god, as discussed in an earlier chapter, then the *serekh* in total means, "God conceives in his heart, which took flight as a hawk (Horus?), and then God spoke and called forth two and three." The *khet* glyph has a tail to signify the mortality of His creation. (Recall the 2nd Timaeus quote in, "God and the Created Image of Eternity.")

b. The *Wepwawet* Standard

The "opener of ways" epithet of *Wepwawet*[299], the name of the canine sign (E-18 from Gardiner's list), is very interesting from an astronomical standpoint. This is an epithet that would readily describe the planet Mercury, whose motions define the upper and lower limits of the band of the ecliptic. Mercury also has the most rapid orbit of the Sun, and is often depicted

---

[298] All hieroglyphs shown and discussed in this Part reference: Sir Alan Gardiner's, *An Egyptian Grammar*, (Griffith Institute, Oxford England 3rd Edition); an online source for Gardiner's signs is also available, but it does not contain definitions of the glyphs: Wikipedia contributors, "Gardiner's sign list", Wikipedia the Free Encyclopedia, https://en.wikipedia.org/wiki/Gardiner%27s_sign_list ; all of the signs depicted are copied from the online source

[299] Toby A. H. Wilkinson, *Early Dynastic Egypt*, (Rutledge, New York, NY, USA, 1999), pp297-298. See also, Manfred Lurker, *An Illustrated Dictionary of The Gods and Symbols of Ancient Egypt*, (Thames and Hudson, London, England, English edition published 1980) pp 128-129.

as a lean canine in mythology. If this understanding is accurate, then the oval-shaped, bag-like glyph in front of the canine could represent the orbit of Mercury, which has the greatest eccentricity of any planet in the solar system, approximately twenty times greater than the Earth's, and the rectangular portion of the bag-like glyph would represent the band of the ecliptic. The meaning of the rearing cobra on the standard, then, can readily be understood to represent eclipses on the ecliptic (path of the Sun or centerline of the band of the ecliptic).

c. The Hawk in Flight

The hawk in flight above the god and slightly behind him is the planet Venus. As argued earlier, Venus was personified by the goddess Neith, in the Egyptian pantheon, whose usual symbol is crossed arrows, tips facing down, with flathead tips comprised of a broad point and two similarly-shaped barbs (rearward facing projections)[300]; alternatively, she is symbolized by two overlapping bows on a shield.

Two of the hawks clasp the ankh symbol in their talons (images '4' and '6'). This glyph is Gardiner's sign S-34, and is understood to mean 'life'

The other four hawks clasp the symbol for a circle, that is given the meaning of "cartouche" or enclosure for the king's name (Gardiner's sign V-9).

However, Gardiner explains that the circle can also be understood to mean, 'that which is encircled by the Sun.' This can be taken literally, or it can be taken as the apparent circular motion of the Sun and its connection with time, which latter interpretation is the meaning that I would give this glyph. More to the point, though, I would broadly define it to mean a cycle of time. In this case, the four hawks, with this symbol in their talons, are carrying a cycle of time.

There are also several glyphs above the hawk in stela 4. These are depicted below.

The first glyph represents a tusk or tooth and is Gardiner's sign F-18. The second glyph represents a hand and is Gardiner's sign D-46. The third sign represents the letter 't' in Egyptian and is Gardiner's sign X-1. The fourth glyph represents a crossroads, and has the meaning of town, and is Gardiner's sign O-49. The meaning of the glyphs is understood by

---

300 Wilkinson pp 291-292.

Egyptologists to be Behdet, or Horus in his form of Behdet, a winged-disc of the Sun, who was worshipped at the temple of Edfu in Upper Egypt. This is a sound interpretation; however, I would give it a different meaning altogether.

I would interpret the first glyph to be a determinative (a glyph that establishes meaning for the following letters) instead of a phonetic sound, which is a tooth or cusp, as Gardiner defines it. And I would interpret it as to be an ideogram (symbol) for the cusp or point of the five-pointed star pattern created by the Venus conjunction cycle. The meaning that I would give the next two glyphs is far more argumentative, as I would give the D-46 symbol the meaning of the letter 'd' in Egyptian, followed by the letter 't', or d-t, or *duat*, the Egyptian underworld. As the reader will recall, I identified the motions of the planet Venus with the *duat*, where the planet disappears after appearing as the evening sky and then re-emerging as the morning star. In this context, then, the *duat*, would be the region between the Earth and the Sun, where Venus regularly arrives at and an inferior conjunction between the Earth and Sun (5 times in 8 Earth tropical years and 13 Venusian years) and, at much less frequent times, transits the face of the Sun—the 243-year transit cycle that comprises 152 conjunction cycles. If this interpretation is accurate, then the final glyph can be understood to be symbolic of the Venus conjunction or transit cycle. Both these time cycles are richly attested in the surface structures of the Step Pyramid complex, which have been analyzed in-depth in earlier chapters. The full meaning of these glyphs would then be, the time cycles of the Planet Venus. This same series of glyphs also appears in the left margin of stela '1', but their meaning is limited to the conjunction cycle only, for reasons that will be given later. A more concise symbol for the Venus conjunction cycle is likely the enigmatic glyph (Gardiner's sign N-13) shown below, which has been discussed earlier. It is a composite glyph comprised of a tooth or cusp (Gardiner's sign F-18) and the five-pointed star (Gardiner's sign N-14), which I believe symbolizes one Venus inferior conjunction cycle, as was discussed in an earlier chapter.

That this is the case, seems to be supported by the angle of the hawk's wing spread in stela '4', which is approximately 94-95°—this is the angle of the combined spread between Venus's maximum eastern (47°) and maximum western (47°) elongations. It is at the mid-point of the elongations that inferior conjunction occurs, which is where the center of the ankh symbol held by the hawk lies in the image. The hawks depicted on the other stelae do not have this same angular spread between their wings.

d. <u>The Sky and Circle Glyphs</u>

There are twelve curious glyphs in the stelae, two on each stela. Each is a composite glyph, comprised of one half of the Egyptian sky glyph (Gardiner's sign N-1), and the circle (cartouche) glyph (Gardiner's sign V-9), which was described earlier and defined as a cycle of time.

The symbol seems to represent the 12 months of the year or the 12 signs of the zodiac, or both. The portion of the sky glyph shown below appears to be a separate glyph, Gardiner's sign Aa-16, especially when compared with the actual image shown in the stelae. (The second glyph is the same one, rotated 90° to the left, for easier comparison purposes.) The meaning accorded this glyph is 'side' or 'half.' If the Aa-16 glyph is the intended symbol on the stelae, then the meaning of the composite glyph likely is, one sign of the zodiac, and the cycle of time is 2,160 tropical years. Thus, the two composite glyphs on each stela represent two signs of the zodiac and a time cycle of 4,320 years.

A possible origin for the Aa-16 glyph is the leaf portion of one side of the equilateral triangle of the Vesica Pisces, which contains two such glyphs, laid end-to-end. Each of the six stelae, then, would comprise one side of the hexagon of the Vesica Pisces.

e. The Fan Glyphs and their Associated Symbols

Fan glyphs appear on every stela, but their accompanying signs and even the number of fan glyphs varies. In every case, though, the fan is held by either one or the other of the two accompanying glyphs. The glyphs used with the fans are depicted below:

The first glyph is the fan and is Gardiner's sign S-36, the second is sign S-40 and is named the *was* scepter, and the third is sign S-34 and has already been introduced as the *ankh* sign. The *was* scepter is a curious emblem or symbol and is often used when depicting deities or kings. Exactly what the *was* scepter represents is an unsettled issue in Egyptology, but it is used in other glyphs where gold is an attribute (ex: Gardiner's sign S-14*). It may be noteworthy that Gardiner's sign for necklace, S-12, is a component in signs S-14 and S-14*, which are, respectively, the signs for silver and gold). It would not be unreasonable, then, to assume that the *was* scepter represents either gold or the reflected light of gold, or, by logical extension, that it represents light, particularly the ambient light of Earth's atmosphere, as opposed to direct sunlight. This leads to an explanation as to the nature of the *was* scepter.

If the *was* scepter represents gold, then the *was* scepter could very well be the shank rod of the goldsmith, i.e. the rod that the smith uses to insert or retrieve a crucible from the fire used to refine or melt gold. This explanation accords well with the creative powers of God, who brings light to the world, or Who decrees this task to a lesser god. It also accords well when the *was* scepter appears with the *ankh* symbol, as they represent, respectively, light and life.

Returning, then to the fans, what is their meaning? The ready explanation is that they symbolize the wafting of "life and light" to the gods, which is supported by their definition in Gardiner. Less obvious, but also within the meaning of the glyph, they signify shadow, which raises the possibility that they speak to eclipses. This is a possibility.

The presence of the Saros and the associated Exeligmos cycle, and the Inex eclipse cycle are well attested elsewhere in the Step Pyramid complex, and are found on the stelae as well, which will be demonstrated later. However, it is the presence of the fan glyphs that will be dealt with here. If we accept that the glyphs are associated with eclipses, and that their supporting glyphs symbolize life and light, then the nature of the eclipses does not seem to be readily obvious. A defined eclipse cycle is represented here, but before it can be introduced, there will need to be further analysis of the glyphs.

Considering the fan glyphs and their supporting glyphs on all the stelae, there are similarities and differences between them. Stelae '2', '3' and '4'—those that depict 'runners'— are almost identical to one another in their composition (two fans—one each carried by an ankh and a *was* scepter), proximity of the two fan glyphs to one another (touching or almost touching one another) and the order of their supporting glyphs presentation (ankh atop the *was* scepter). The use of the ankh and *was* scepter glyphs together may not be significant, in that the probable intent in mixing the two is to signify that both life and light are necessary to creation—in this case the creation of the motions of the Earth and Moon, which cause eclipse cycles.

The fan glyphs depicted with the 'walkers', stelae '1', '4', and '6', differ markedly, when compared to the three previous stelae described, and when compared to one another. Stela '1' depicts only one fan glyph, which is held by an ankh glyph. Stela '5' depicts two fan glyphs, both of which are supported by *was* scepters. And stela '6' depicts two fan glyphs, one of which is supported by an ankh glyph while the fan glyph that appears below it is supported by a *was* scepter (similar to the arrangement and order of the fan glyphs found on stelae '1', '3', and '4'), but markedly different in their proximity to each other as there is a notable spacing between them. Summarizing these observations, there are three identical fan glyphs (stelae '2'. '3', and '4') followed by three dissimilar ones (stelae '1'. '5', and '6').

This is the pattern of the six-hundred-year, tetradia eclipse cycle, in which four total lunar eclipses—tetrads—occur with a spacing of six lunations between each of them. <u>Two full eclipses take place at each of the two lunar nodes</u>, and are separated by no more than 2 X 4.8° of celestial longitude at each node. (Recall that the entire Step Pyramid complex is oriented 4.5° east of true north, which is roughly the same angular distance of the umbral shadow from the center of the node that is required for the tetrads to occur.) These tetrads occur with some regularity during a roughly three-hundred-year period, followed by a roughly three-hundred-year period in which no tetrads occur. This is an eclipse cycle that occurs with regularity, although its exact duration varies somewhat from a nominal duration of six hundred years. This is a relatively minor cycle, when compared to the Saros, Exeligmos, and Inex eclipse cycles, but it is an important one.[301]

There is another possible explanation for the fan glyphs and their accompanying ankh and *was* scepter glyphs: they represent solar and lunar eclipses, with no pattern or eclipse cycle associated with their appearances on the stelae. The ankh glyphs represent lunar eclipses while the *was* scepters represent solar eclipses.

The ankh and *was* scepter also appear elsewhere in several of the stelae, but these appearances will be dealt with separately.

[301] Jean Meeus, *Mathematical Astronomy Morsels III*, (Willmann-Bell, Inc. Richmond, Va. 2004) pp123-140, Chapter 21 "Lunar Tetrads".

CHAPTER 65

FEATURES COMMON TO THE THREE STELAE DEPICTING RUNNING FIGURES

The three stelae depicting running figures share certain features. Each of the figures wears the white, *hedjet* crown that is associated with Upper Egypt. Each carries a flail with three tails in the right hand that is raised up in front, and a curious baton-like object in the left hand. Each of the figures is minimally clothed, with figures '4' and '3' wearing nothing more than a string like device around the waist, while the third has a loin cloth and an affixed tail. The flail is Gardiner's sign S-45 and the baton-like object is sign N-36, which are depicted below.

The baton-like object will be discussed first. This object is defined by Gardiner as *mr,* with a meaning of 'canal', 'channel', and, variably, 'love.' In the several stelae that depict this object, though, it has noticeable, swallow-tail ends, but this may not be a material difference. In light of the meanings that Gardiner attaches to the object, then, it seems logical that this is a vagina. Such a crude and vulgar interpretation may seem totally out of line, until one recalls that Atum was a 'self-created god', and that his left hand (the female hand) was literally the means of his self-creation. The several meanings attached by Gardiner to the object are entirely consistent with this interpretation. Each of the running figures, then, is a self-created god, but further interpretation is necessary, which will be done momentarily.

The flail with three tails is such a common-place piece of royal iconography that its meaning may seem obvious: it is a symbol of authority. However, if the object is carried by a god, then it may have an entirely different meaning. This is borne out by the flail's association with Min, a fertility god from Egypt's remotest past, and this is the meaning that it also carries in the glyph. However, if the god is carrying the female aspect of life in his left hand, then the flail in his right hand must represent the male aspect of fertility or potency. This is an entirely consistent interpretation and furthers the argument that the running figures depicted on these three stelae are each a 'self-created god.'

So, who are the gods depicted? In order, they are: Atum or Atum-Re (stela '4'), Horus (stela '3') and Thoth (stela '2'). Repeating an earlier passage quoted from Egyptian religion:

> "In the form of **Atum** there came into being heart and there came into being tongue. But the supreme god is **Ptah**, who has endowed all the gods and their Ka's through that heart [of his] which appeared in the form of **Horus** and through that tongue [of his] which appeared in the form of **Thoth**, both of which are forms of **Ptah**."

It is this religious tenet of the ancient Egyptian religion that is expressed in the three figures depicted on these three stelae, but more on this in a moment.

On each of these three stelae, there are two sets of three 'hill-like' glyphs, drawn lying on their sides, which are stacked one on top of the other—one set to each side of the running figure, depicted on these three stelae. There are no known glyphs that explain them. However, they appear to set limits to the runner, which offers a clue as to their meaning. Egyptologists have identified them as the boundary markers for the king's *Heb Sed* run, and I am in full agreement with them. They are, indeed, markers of a sorts, but they are more than this. Each set of three such glyphs composes the three Exeligmos cycles of a Saros eclipse series. (The shape of the hill-shaped glyph is similar to the shadow zone cast by the eclipse across the face of the Earth. See the figure shown below, but note that the eclipse shadow area is white while the area not affected by the eclipse is grey.[302])

[302] Fred Espenak, NASA, GSFC; "Periodicity of Lunar Eclipses"; U.S. National Aeronautics and Space Administration (NASA) web site, https://eclipse.gsfc.nasa.gov/LEsaros/LEperiodicity.html (See section 1.4, Figure 1)

Figure 1. Lunar Eclipses from Saros 136: 1932 to 2022

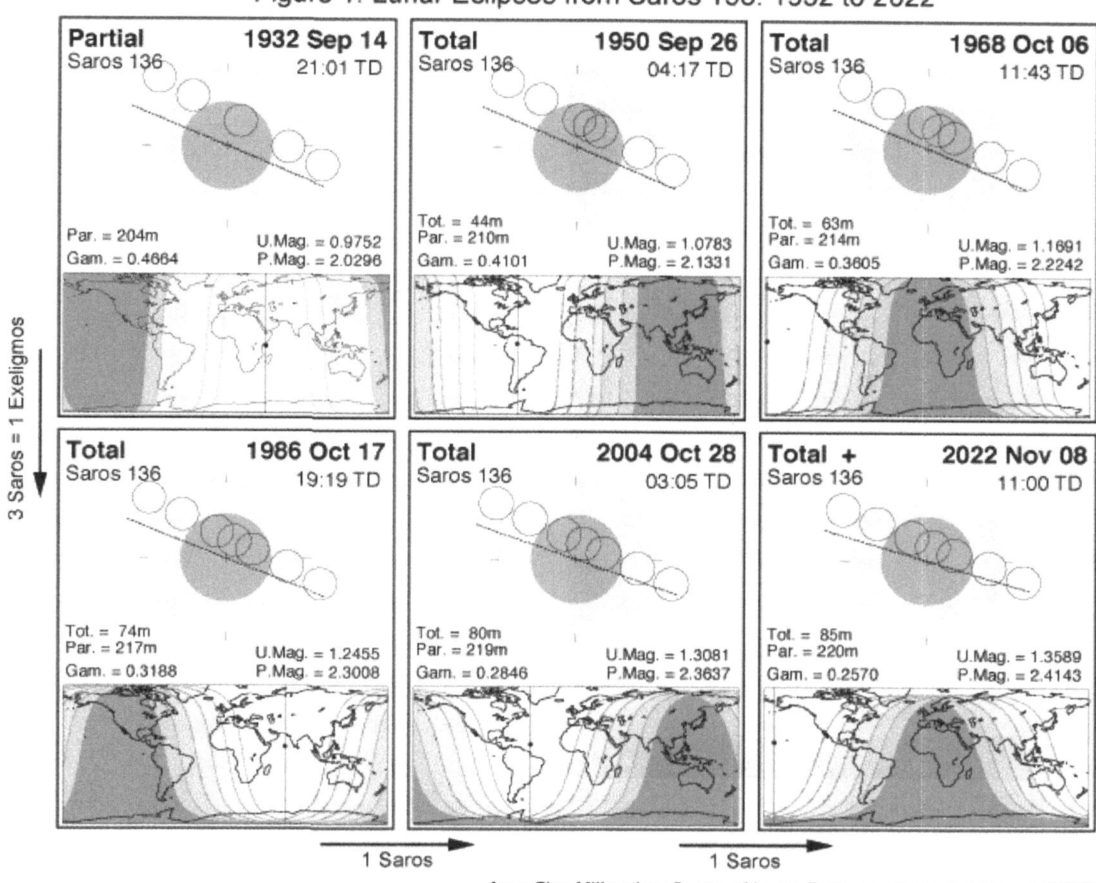

from *Five Millennium Canon of Lunar Eclipses*, Espenak & Meeus (2009)

And, if each of these sets of three markers represents a Saros series, then the space between them that the runner crosses represents an Inex cycle, of 28.945 tropical years or approximately 30 eclipse years (346.62 days each). This is consistent with both the known facts of the *Heb Sed* festival, with its 30-year cycle, and the time-period between the end of one Saros series and the beginning of another one on the same longitude, approximately 30 eclipse-years later, at the opposite pole, which is the Inex cycle. Thus, all three of the major lunar eclipse cycles—Saros, Exeligmos, and Inex—are depicted on each of these three stelae.

There are also three triangles depicted on these three stelae, three on stela '4', and two each on stelae '3' and '2'. The spread of the runner's legs in each of the stela depicts a 45° right triangle. The angle between the handle of the flail and the center of the flail's three tails on each of the stelae forms the 30° angle of a 30°—60°—90° triangle. And the ray-like figure emanating from the navel area of the runner in stela '4' depicts the 26.5° angle of a 26.5°—63.5°—90° triangle, which is the triangle formed by the diagonal of a 1 X 2 rectangle. This last triangle is noteworthy in that it is commonly used to generate the golden mean (1.618). (All angles are taken from the actual photos of the stelae, and not from the drawings made from them. However, in the case of stela '4', depicted below, the drawing made from the photo is also accurate.)

Each of these three triangles has received extensive analysis earlier in this book. Also, the irregular numbers derived from them, √2 or 1.414, √3 or 1.732, and 1.618 have also received extensive analysis elsewhere in this book. These three triangles and their associated irregular numbers are critical to measure, and to the order (Pattern) that lies gives rise to it, as was amply demonstrated earlier. All are at the very heart of creation.

CHAPTER 66

FEATURES COMMON TO THE THREE STELAE DEPICTING WALKING FIGURES

Similar to the three stelae depicting 'runners', the three stelae depicting 'walkers' have certain features in common. These include: a 'standard' on a pole preceding the figure, clothing (including an attached tail), and devices held in their hands (rods and a club). Each also has a crown, with two wearing the same white, *hedjet* crown as the runners, and one wearing the red, *desret* crown of Lower Egypt.

Many Egyptologists believe the 'standard' mounted on a pole immediately in front of each walker depicts a placenta, and I am in broad agreement with this. However, I would further refine the description: it is the lining of a uterus (including the placenta) that is partially distended by an implied fetus, with an attached umbilical cord. Broadly speaking, it is afterbirth material and it signifies human birth, implying that the figure is born of a human Mother. This, of course, begs the question, is the figure depicted human or god? Likely, both, for reasons that will be discussed momentarily.

Each of the figures wears clothing, including: a close-fitting halter with a shoulder strap across one shoulder, a short skirt with a belt, and a narrow apron that reaches from the belt to just below the hem of the skirt. In two stelae, '6' and '1', the figure carries a knife, which is absent in the third. All three figures have tails. The shoulder strap of the halter on the figures on stelae '6' and '1' (the same ones wearing the white crown and carrying a knife) is worn across the right shoulder (male side), while the shoulder strap of the halter on the figure on stela '5' is worn across the left (female) side.

The rods carried by all three figures are likely measuring rods, which may or may not have actual values for length. Unfortunately, there are no known definitive measures of either the stelae or the rods, although several Egyptologists have made measures of the stelae that were accurate to within 0.5 inches, which is satisfactory for general purposes, but not for precision measurement to compare with dimension measures, such as the Egyptian foot (299.43 mm), nautical foot (308 mm at the 30th Latitude), etc. The same problem arises when trying to determine whether the length of a figure's stride has any metric significance. The club is likely the glyph for 'white', Gardiner sign T-3, and, curiously, is carried in the figure's left hand, but behind.

Determining the nature of the figures—i.e., Do they depict a god, a human, or a being with a combined nature? —is a matter of weighing the significance of the dress and accoutrements of the three figures. All three are likely gods, by virtue of the *serekh* (discussed in the foregoing section on features common to all six stelae) that is immediately in front of each of them, but they are likely lesser gods with animal-like attributes, by virtue of the tails they wear. The placenta standard implies that all three figures are born of woman. The figures on stelae '6' and '1' are males or have predominant male-like qualities; the figure on stela '5' is female or has predominant female-like qualities.

Who are the gods depicted? Stela '1' appears to be Osiris, or the god who would later carry that name. The determining factor in this regard is the glyph in the lower right-hand corner of the stela that shows a scorpion being restrained in the arms of or dispatched by an anthropomorphized *was* scepter glyph. The scorpion is Gardiner's sign L-7 and is identified with the goddess Selket, and has the meaning, 'she-who-relieves-the-windpipe.' An interesting passage from the Pyramid Texts has it that, "My mother is Isis, my nurse is Nephthys, she who suckled me is the *Sh3t-Hr* cow, Neith is behind me and Selket is before me."[303] An interesting formulation, but it is the final part of the statement that is the most compelling for our purposes, "…Neith is behind me and Selket is before me." Another way of stating this is: Birth is behind me and death is before me."[304] The scorpion, then, may also be symbolic of death.

It would be easy to conclude that the figure depicted on this stela is Osiris, which would be a satisfactory answer, but there is a distinct possibility that the figure is the goddess Neith, or that the figure is a composite of both gods. (The connection of Neith to Venus will be discussed in detail in a subsequent section of this Part.)

The figure depicted on stela '6' is hard to identify for the moment. It seems to be male, or to have male-like attributes, but beyond this nothing more can be said. An analysis of the glyphs in front of this figure is required before proceeding further. The same is true of the figure on stela '5', which appears to be female or to have female-like attributes. However, it can be observed that both of these figures have animal-like attributes (the tail) and both were born of woman (the uterus/placenta standard).

A noteworthy curiosity on stela '5', though, is the re-appearance of the scorpion glyph in the same position as it appears on stela '1', but in this instance, it appears with a time circle and a device that seems to be the double of the device on the two glyphs depicted above it, which were earlier determined to likely be signs for a great age. It is also noteworthy that, unlike the

[303] Manfred Lurker, *An Illustrated Dictionary of the Gods and Symbols of Ancient Egypt,* (English language edition 1980, Thames and Hudson, London) page 106, entry for Selket.
[304] Ibid, page 106, reference to Pyramid Text No. 1375, under entry for Selket.

other five stelae, no ankh (life) glyph appears anywhere on this stela. And it may be further significant that the white, *hedjet,* crown is worn by the falcon on the *serekh* glyph on this stela, whereas no falcon on the *serekh* glyphs in any of the other stelae wears a crown, except for the *serekh* falcon depicted on the first 'runner stela, '4'; in this latter case, the falcon wears the combined crowns (Gardiner's sign S-5). Does this mean that a cycle of time has ended and death has ensued? An interesting question to ponder, but there is nothing further that can be said about it for the moment, until the rest of the glyphs on the stelae are analyzed.

All three of the walking-figure stelae appear to be concerned with measure, by virtue of the rod(s) carried by each of the figures. Unfortunately, an analysis of the measures that may be incorporated in these stelae is not possible because there are no definitive measures of the stelae's dimensions available.

CHAPTER 67

THE COLUMN OF GLYPHS IN FRONT OF THE FIGURES

Each of the stelae presents a column of glyphs along the left margin, in front of the figure; no two are alike, but some glyphs appear on several stelae. The stelae are analyzed in the order presented at the beginning of this Part: 4, 3, 2, 1, 6, 5. The first three are the stelae with 'runners' and the final three are those with 'walkers.'

a. Stela '4'

The following glyphs appear on this stela, from top to bottom:

They are, respectively, Gardiner's signs: F-31, D-2, O-14, G-5, S-36, and M-24. However, on the stela, sign G-5 is on top of sign S-36, and they are depicted as one. Gardiner defines: sign F-31 as, 'gives birth'; sign D-2 as 'upon' (See Gardiner §165, page 127); sign O-14 as, broadly speaking, 'wall' and in this case, because of the orientation of the stela to the complex as a whole, specifically, the 'southwest wall.' Gardiner defines sign G-5 as 'horus', and sign S-36 as 'shade or shadow'; however, as indicated earlier, these signs are not separate, but combined.

The combined sign, G-5 and S-36, is most likely sign R-14 (depicted below), which Gardiner defines as, 'west'. This sign—with several variations, including one where a falcon is depicted in place of the feather, and another that has both a feather and a falcon—is also associated with the goddess Imentet (Varr. Amentet).[305] Imentet is associated with the west and with necropolises, most of which were located on the west bank of the Nile. The association of sign R-14 is an important one, because it is not the 'west, per se, that is

[305] Wikipedia contributors, "Imentet", Wikipedia the Free Encyclopedia, https://en.wikipedia.org/wiki/Imentet, Accessed July 29, 2017

intended, but the westward, or clockwise movement on the Celestial or Ecliptic Sphere of several, extremely important astronomical phenomena, all of which pertain to the measure of time.

All of the following astronomical phenomena move toward the west, as measured in longitude on the Celestial of Ecliptic Sphere, in a clockwise or retrograde motion, which is counter to the counterclockwise motion of the solar system planets: 1) precession of the equinoxes (the great year of 25,920 years), 2) precession of the lunar nodes, 3) the Exeligmos cycle members of a lunar eclipse Saros series (eclipse shadow), and 4) the orbits of Mercury and Venus, when their orbit reaches elongation and is entering or departing inferior conjunction with the Earth and Sun, and **appears to be in westward or retrograde motion** (actual motion is easterly or counterclockwise). This apparent westward motion of Mercury and Venus is why they and their respective Egyptian personifications, Wepwawet (see earlier discussion in this Part) and Khentiamenti (Osiris) were referred to as 'westerners'; they literally returned from the west, the land of the dead, whereas the 'dead' traveled to the west, never to return.

The retrograde motion of the astronomical phenomena listed in the foregoing paragraph are difficult to understand, but this is the reason that the *temenos* wall of the Step Pyramid complex measures the time cycles associated with them in a clockwise direction. And it is also the reason that the actual direction of the 'runners' and 'walkers' on the six stelae is in the southwest or clockwise direction, the same direction as the time cycles memorialized on the stelae.

Sign M-24 is defined by Gardiner as 'South' or 'dream.' South can be Upper Egypt, but it can also be another way of describing the world below as opposed to heaven above (North). However, earlier I identified the plant portion of this sign with the number 5, more specifically with patterns that are based on the number 5, and most importantly with the Venus inferior conjunction and transit time cycles, and the phoenix cycle. (I am assuming that the Phoenix cycle is associated with the periodic reoccurrence of long-term lunar-eclipse Saros series'. The Venus transit and phoenix cycles are based on time cycles that are numbers from the Pattern (i.e. 243 and 486), but each is composed of inferior conjunction cycles that are based on the pattern of 5, or 2 + 3.) The mouth portion of the sign M-24 likely means 'calls forth.'

The meaning and intent of the glyphs on stela '4', as a whole, are likely:

God (Atum) speaks and creates upon His wall the patterns of those 'westerners' who are cycles of time in the pattern of 2 X 3; He speaks again and creates the patterns of those 'westerners' who are cycles of time in the pattern of 2 + 3."[306]

This is a clear and concise statement, but it is most profound one and its ramifications are almost beyond reckoning. However, to fully understand it requires a great deal of further analysis, which has already been performed in the earlier Parts of this book, including sections of this Part. (In every sense of the term, then, we have come full circle!) The 'wall' is likely the circle in the heavens created by Earth's daily rotation and its orbit, and its precession about the North Ecliptic Pole. The North Ecliptic Pole would be Atum.

b. Stela '3'

The following glyphs appear on this stela, from top to bottom:

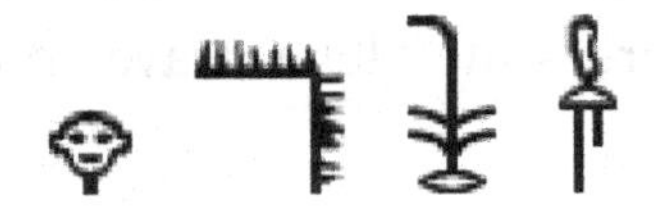

They are respectively, Gardiner's signs: D-2, O-14, M-24, and R-14. These signs were all depicted on stela '4' (see foregoing), but their order is slightly different. The order for signs M-24 and R-14 is reversed on stela '3', perhaps to suggest that neither takes precedence over the other. It is also noteworthy that the falcon on the R-14 sign on this stela faces the figure (as do the falcons and birds depicted in the column of glyphs on all the other stelae), whereas the falcon on this sign on stela '4' faces away from the figure. This may indicate that it is the primordial creator god, Atum (the figure on stela '4'), who creates the patterns of time, and is speaking to the other creator gods (the figures on stelae '3' and '2') and directing them to comply with these patterns in their acts of creation. However, the meanings of the glyphs on stela '3' are the same as they were on stela '4.'

On stela '3', the author of the glyphs is Atum-Re, not Atum, who spoke on stela '4'. The meaning and intent of the glyphs on this stela, as a whole, are likely:

God (Atum-Re) speaks and creates upon His wall the motions of those 'westerners' who have cycles

[306] 2 X 3 is the Pattern, or Ma'at, as described in the first Part of this book; 2 + 3 is the pattern created by the Fibonacci series, (recall that the Venus inferior conjunction cycle is comprised of these numbers from the series: 5—8—13. The 2-3 component of sign R-14 may incorporate both patterns, and it is further noteworthy in this respect that a feather is often included in the glyph, which is a symbol for the goddess Ma'at, who symbolizes order.

of time in the pattern of 2 X 3; He speaks again and creates the motions of those 'westerners' who have cycles of time in the pattern of 2 + 3."

In this instance, the 'wall' is likely the wall implied by the orbit of the Earth about the Sun as personified by Atum-Re.

c. Stela '2'

The following glyphs appear on this stela, from top to bottom:

They are, respectively, Gardiner's signs: T-3, O-11, G-36 (repeated three times), an 'ape' (no sign with the ape in this squatting position was available) O-9, and O-17. However, on the stela, sign T-3 is superimposed on sign O-11, and they are depicted as one. Sign G-36, the sign for a swallow, is repeated three times, but it is depicted on the steal facing in the opposite direction and toward the figure. The 'squatting ape', a symbolic representation of the god, Thoth, is depicted sitting atop sign O-17, with one hand resting on sign O-9, which is also depicted atop sign O-17. Thus, signs T-3 and O-11 are one composite glyph and the final three signs are also one composite glyph.

Gardiner defines: sign T-3 as, 'bright' or 'white'; sign O-11 as a 'palace with battlements', which when combined with T-3 suggests a meaning of 'white palace' or 'white walls.' Sign G-36 is defined as 'great', which is repeated three times. Gardiner defines sign O-17 as a 'gate with a cobra frieze' and O-9 as the symbol for the goddess '*Nephthys*.'

The first combined sign, the 'white palace with battlements', can be understood as, the manifestation on Earth of that which is in the heavens above, meaning the *temenos* wall of the Step Pyramid complex, which tracks the cycles of time. It could also be understood as the 'palace of light of Ptah", which may be related to *hikuptah*, the ancient Egyptian word for Egypt[307].

Following the first glyph, are the three swallows, which literally translates as three times great or thrice great. This is an epithet of the god Thoth, though it is one that is thought to have originated much later in history.[308] The association of this epithet with Thoth on this stela is

[307] Wikipedia contributors, "Ptah", Wikipedia the Free Encyclopedia https://en.wikipedia.org/wiki/Ptah (Accessed August 1, 2017) See entry for 'Legacy' at the bottom of article

[308] Wikipedia contributors, "Hermes Trismegistus", Wikipedia the Free Encyclopedia https://en.wikipedia.org/wiki/Hermes_Trismegistus (Accessed August 1, 2017)

strongly by the presence of Thoth's ape in the glyph immediately following. However, the 'thrice great' glyph could also be symbolic for the three heavenly bodies whose motions are the sources for measuring time, i.e. the Sun (relative motion), the Moon, and Venus.

The glyph with Thoth's ape that appears to be talking and resting one hand on the sign for the goddess *Nephthys*, while sitting atop a gateway or wall with a cobra frieze, is more difficult to interpret, mainly because of the presence of the symbol for *Nephthys*. Thoth, is the god of astronomy, geometry, measures of time, writing, etc.[309], and the depiction of him sitting atop a gateway surmounted by cobras on this stela, is contextually to be expected, but what of *Nephthys*? Perhaps, in this context, she is associated with form and dimension, both of which are suggested by the appearance of a rectangle in her symbol. Was Nephthys originally associated with Thoth, a creator god as depicted on this glyph, or were her attributes originally his attributes? It is impossible to say with certainty, but the presence of her symbol, if indeed it does signify form and dimension, does add to the meaning of this combined glyph. The meaning of this glyph, then, is most likely, Thoth decrees form and measure from the cycles of time, and then records them on, or in the enclosure of, the wall of cobras[310], i.e. "white walls" or the Step Pyramid complex.

On stela '2', the author of the glyphs is Thoth. The meaning and intent of the glyphs on this stela, as a whole, are likely:

> "On the walls of His Palace of Light, God (Thoth) decrees the motions for the three great ones,[311] initiating the cycles of time, and then calls forth form and measure from their motions."[312]

In this instance, 'the walls' refers to the circles of light created by the heavenly bodies in apparent orbit about Earth's North Celestial Pole.

---

[309] Wikipedia contributors, "Thoth" Wikipedia the Free Encyclopedia , https://en.wikipedia.org/wiki/Thoth (Accessed August 1, 2017)

[310] Interestingly, the image of the cobra can also signify 'green' or 'vigorous.' See Gardiner, *Egyptian Grammar*, page 560, entry for *w3ḏ*. The association of cobras (Saros lunar eclipse series) with green and vigorous may also explain the presence of the blue-green tiles that are affixed to the walls surrounding the stelae, and in several other nearby rooms.

[311] It seems likely that the "Three Great Ones" refers to the three celestial bodies whose motions are the principal sources for measuring time: Sun (relative motion about the Earth), Moon, and Venus.

[312] The *Temenos* Wall of the Step Pyramid complex.

d. Stela '1'

The following glyphs appear on this stela, from top to bottom:

The first sign is a variation of Gardiner's sign O-28, who defines it as '*iwn*', which means column or pillar. This word is closely associated with Heliopolis, which was called '*iwnw*' in ancient Egyptian, meaning Column or Pillar Town. Sign O-28 is also very similar to sign V-36, the sign for the *henti* period of 120 years, which was discussed earlier (figure is depicted below).

The two signs are very similar in appearance to one another, and I believe that they are related. Sign O-28 has a vertical line between the two cusps of the crown, while V-36 lacks this characteristic. Both signs have a band, but the band on O-28 is lower than the one on V-36, which also has a side curl extending off of the band. These are the only distinguishing characteristics, between the two, which are, otherwise, identical. Perhaps the key to the difference is to be found in another meaning that Gardiner assigns to V-36, which is, 'end.' Could this imply that sign O-28 is the sign to be used during the *henti* period, and sign V-36 signifies its conclusion? I believe it does. If so, then the signs are more than related; they are variations of the same sign, both of which are 'pillars.'

The significance of finding the pillar sign on this stela is that it refers to *iwnw* or Pillar Town or Heliopolis. More than this, though, it refers to the time measurement practices of Heliopolis, in general, and specifically to the practice of measuring the *henti* period of 120 years (i.e. the Venus transit half-cycle of 121.5 years and the double *henti* period of 240 years, which is the Venus transit cycle of 243 years).

The second through the sixth signs are enclosed in a box, because they have been discussed earlier as they also appear on stela '4', except that the falcon is in flight on this stela, whereas it is standing on stela '3.' The meaning, however, is largely the same, "The Venus inferior conjunction cycle." But with the falcon standing, the implication is that its flight is ended, and the appearance of the final glyph below it, which is Gardiner's sign O-19, meaning 'Great House' (see figure depicted below), confirms this.

There are certain critical differences between sign O-19 and the sign that appears on the stela. There are no vertical lines on the stela's glyph and there are five 'cusps' mounted on the front of the 'house' and not four as there are on sign O-19. The glyph on the stela signifies that five cycles have been completed, which when reviewed in context with the aforementioned glyphs associated with the Venus inferior conjunction cycle, means that an inferior conjunction series has come to an end.

There is one other detail on this glyph that must be addressed, and that is the 'step' that appears in front of it at the base. This glyph is Gardiner's sign Aa-16 (depicted below), and is often associated with expressions of 'truth' and 'order.' Sign Aa-16 is also symbolic of the goddess Ma'at, who symbolizes truth and order. Also carved on this glyph, and as shown on the drawing of stela '1', are three, small vertical bars, arranged in order by their descending height. It is not known what they represent.

There is one final detail on the stela that must be discussed before interpreting these glyphs as a whole, and it is one that suggests an identity for the god depicted on the stela. The *was* scepter, the symbol for light, that is depicted in the lower right-hand corner is wrestling with and subduing a scorpion, a symbol for death (see earlier discussions of these two signs). This, coupled with the fact that the god depicted on this stela is the first in the order of the stelae series to be shown walking and with the uterus/placenta-umbilical-cord standard in front. This is also the only god to be depicted carrying the T-3 sign—a mace—which means, 'bright' or 'white.' The simple explanation is that this is a son of a great god (Atum?)—making him a lesser god—who is born of a human mother, and who knew but conquered death. And, in this context, the T-3 sign likely means that he can, in his own right, create lesser gods himself.

The Egyptian god that most readily matches all of these characteristics is Osiris, except for the human mother aspect. However, this is a problem that is easily disposed of, as having a human mother is essential in order to expose the god to death. Therefore, the god depicted may be Osiris, and this identification also comports well with the Venus inferior conjunction cycle, where a god (Venus) appears and then disappears in the heavens, only to appear again, or arise from the dead. In Egyptian mythology, Osiris does not re-appear, but he arises in the form of his son, Horus.[313] The only problem with equating the god on this stela with Osiris is that Osiris was not attested by name in history until much later than the generally accepted chronology for the Step Pyramid complex.[314] This is not a significant problem, though, because the attributes of the god who is attested on the stela are what matter, and these fit the attributes generally attributed to Osiris. The nature of the god on this stela also suggests Jesus Christ, the Son of God, but exploring this further would be way too distracting for my purposes.

[313] Wikipedia contributors, "Osiris", Wikipedia the Free Encyclopedia, https://en.wikipedia.org/wiki/Osiris

[314] Wilkinson, page 292, entry for Osiris.

However, that a God who knows and conquers death, has been here before under a different name, should come as no surprise.

On stela '1', the author of the glyphs is Osiris. The meaning and intent of the glyphs on this stela, as a whole, are likely:

"At Heliopolis, God (Osiris) initiates the great time cycles of life and death[315]."

The meaning of the clothing and other accoutrements of this god on the stela is more problematic. This is the first figure in the order of the Stelae series to appear clothed, and is also one of two equipped with a knife. The figure is also uniquely equipped with a walking stick or measuring rod. The clothing may be best explained as a further indication of the god's dual nature (god and human), which would also explain the presence of the knife, because, possessing a human nature, the god is capable of violence. Defining the walking stick as a measuring rod seems to be the best interpretation of this piece of regalia.

e. Stela '6'

The following glyphs appear on this stela, from top to bottom:

The first glyph is Gardner's sign O-28, and has been discussed in detail under stela '1.' Its meaning here is the same: Heliopolis and time measurement. The second glyph is Gardiner's sign O-1, which means 'house.' The third glyph is Gardiner's sign G-36 and has been discussed in detail under stela '2'; its meaning is 'great.' Signs two and three are read together on the stela, and mean 'shrine' or 'great house.' The fourth glyph is Gardiner's sign O-19, except that its appearance on this stela differs significantly from Gardiner's depiction of it. The differences between sign O-19 and its actual depiction on this stela are critical. Gardiner's sign shows four cusps mounted on the face of the structure, while the glyph on the stela may depict these same four cusps or a 'loop device', instead. The glyph on the stela also differs in that it includes four cusps standing upright in front of the shrine.

The key to interpreting the glyphs on this stela is the unique detail found on glyph four, specifically, the vertical lines and the four standing cusps in front of the shrine. The four

[315] The time cycles memorialized in these glyphs likely include: the Venus inferior conjunction cycle and the related, but longer-term *henti* cycles. The latter include: 1) the *henti* period of 120 years, which is an abridged expression for the 121.5-year, Venus half-transit cycle, and 2) the double-*henti* period of 240 years, which is an abridged expression for the Venus transit cycle of 243 years.

standing cusps are four *henti* periods, of 121.5 years each, for a total of 486 years. This is the Phoenix cycle, that was discussed earlier in this book, and detailed in Appendix 1. And, as was also discussed earlier, this time-period is associated with the appearance of the longer-term, lunar Saros periods, which are represented on the shrine by the four vertical lines depicted on the front of the shrine. What the 'loop' or 'four cusps' shown mounted on the face of the shrine represent is not known for certain.

The identity of the god depicted on this stela is unknown, but the glyph of the hawk in flight above the god's image may hold a critical clue. The hawk holds the *ankh,* symbol of life, and it is one of only two hawks depicted on the six stelae to hold this symbol (the other being the hawk n stela '4'). The hawks depicted on the other four stelae all hold the circle device for time. Does this mean that the hawk is renewing life? It appears so, and if this is correct, then the god may be Horus.

On stela '6', the author of the glyphs is uncertain, but may be Horus. The meaning and intent of the glyphs on this stela, as a whole, are likely:

"At Heliopolis, God (Horus) calls forth the great ones[316] and initiates the phoenix time cycle of rebirth."

The meaning of the clothing and other accoutrements of this god is problematic. This is the second figure in the order of the stelae series to appear clothed, and is also the second equipped with a knife (the other being the one on stela '1'). The figure is also equipped with two measuring rods. The clothing may be best explained as a further indication of the god's dual nature (god and human), which would also explain the presence of the knife, because, possessing a human nature, the god is capable of violence.

f. Stela '5'

The following glyphs appear on this stela, from top to bottom:

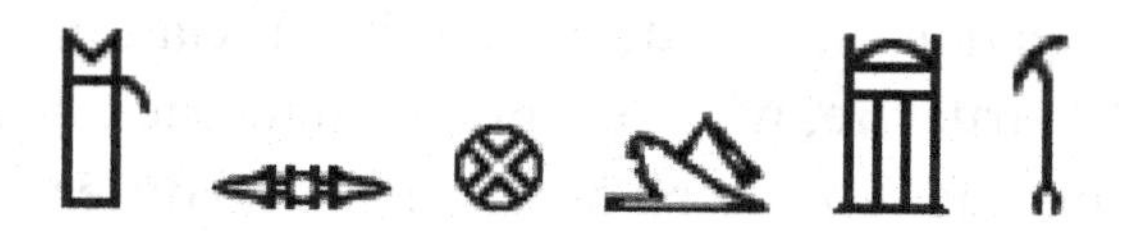

These first glyph is Gardiner's sign V-36, the symbol for Heliopolis and the *henti* period of 120 years, as previously discussed. The second glyph is Gardiner's sign R-23, which Gardiner defines as the symbol for the goddess Min and also of the city of Letopolis (see earlier discussion of Letopolis and its association with the measure of the minute). The third glyph is Gardiner's sign O-49, which, as discussed above in the section on stela '1', has an alternative meaning of, 'an astronomical conjunction or transit.' The fourth glyph is Gardiner's sign G-12,

[316] The long-term, Saros lunar eclipse series.

which Gardiner defines as, 'divine image.' However, the glyph's depiction of a falcon at rest, holding a flail, suggests a different meaning that holds astronomical significance, which will be further elaborated on later.

The fifth glyph is similar to Gardiner's sign O-20, which means' 'sanctuary' or 'shrine', but differs in that the glyph on the stela does not seem to have the internal 'paneling' depicted in Gardiner's rendering of the glyph, and the glyph on the stela has a series of concentric arcs in the 'roofing' of the shrine, whereas Gardiner's rendering shows only one arc. The concentric arcs of the roofing suggest that this detail defines a separate glyph, similar to Gardiner's sign N-28, which Gardiner defines broadly as, 'the sun rising over a hill.' This sign is depicted below.

The rendering of this glyph on the stela clearly has multiple concentric arc, while Gardiner's rendering of it has only one or possibly two concentric arcs. If this glyph, as depicted on the stela, is a variant of Gardiner's sign N-28, then a possible meaning of the variation is that the sun is shown gradually rising above the hill by degrees. This alternate meaning has astronomical significance, which will be further elaborated on momentarily.

The final glyph depicted is the *was* scepter, Gardiner's signS-40, which as was discussed earlier, means 'light.'

The first two glyphs define the measure of a minute of time, as measured by successive solar observations taken at Heliopolis to Letopolis. (See earlier discussion of this measure.) The timing of these solar observations is defined by the fourth and fifth glyphs, which define that the observations are taken when the Sun has come to 'rest', i.e. at local apparent noon (LAN). This is the meaning of the 'resting falcon' in the fourth glyph, when the Sun actually appears to 'rest' briefly at its maximum height at LAN, which is the meaning of the gradually rising concentric arcs depicted in the roofing of the fifth glyph. Two observations would be taken at LAN from each location, Heliopolis and Letopolis. The first, when the Sun stops its rise and rests, and the second. when it resumes its motion and declines. The precise timing of the beginning of the Sun's 'rest' at Heliopolis would be recorded there and simultaneously signaled to Letopolis, where the timing of the beginning of the subsequent 'rest' at that location would be recorded and signaled back to Heliopolis. The differences in time between the two observations define the length of a minute of time, when corrected as described in the earlier discussion of this measure. Similar observations would be done of the Sun's resumption of motion and decline. Thus, two measures of the Sun's relative motion, with its relationship to the measure of a minute of time, would be taken at LAN at Heliopolis and Letopolis. The importance of the measure of a minute of time cannot be overstated; it is a most fundamental measure, as was detailed in Part I.

The third glyph in the series defines the 'conjunctions' that arise along the line of measured distance from Heliopolis to Letopolis, where the 'conjunctions' are the consecutive LAN measures taken at Heliopolis and Letopolis, which intersect this line at right angles. The purpose served by depicting the *was* scepter, the sixth in this series, is not clear, unless it symbolizes that the signaling between Heliopolis and Letopolis was accomplished with reflected light. i.e. mirrors.

Preliminarily, then, the meaning of these glyphs, as a whole, is: God spoke and called forth the measure of a minute of time at Heliopolis and Letopolis. The identity of this god, though, is difficult to determine. The figure depicted on this stela is markedly different than those on any of the other five stelae. Significantly, the figure stands on a plinth (platform), which is a symbol of the goddess Ma'at, who symbolizes world order and harmony.[317] Ma'at is usually associated with the god, Thoth, who is depicted in stela '2', as discussed above. It is likely very significant that Thoth is the only other god depicted in the six stelae, who stands on a plinth.

The accoutrements of the figure are also markedly different. The shoulder strap of the halter passes over the left shoulder, which is the side typically associated with femininity. The crown worn by the figure is the *nt* crown, the 'red crown of lower Egypt' (see Gardiner's sign S-3, depicted below). (The red crown was also called *dsrt*.) The name of this crown, *Nt*, is also the name of the goddess Neith, although her symbol is either overlapping bows (Gardiner's sign R-24, depicted below) or crossed arrows. The *nt* crown was later associated with the goddess Neith. However, the third glyph shown below, Gardiner's sign D-15, has previously been discussed, and may be linked to the goddess Ma'at, through the number six and the Pattern detailed in Part I.

So, who is the goddess depicted on this stela, Ma'at or Neith? The best answer is that she likely is a composite of both. Neith is attested much earlier in history[318] and is closer to the accepted chronology of the Step Pyramid complex than Ma'at, but the attributes of Ma'at are more closely associated with the themes of order and measure, which are the entire foci of the Step Pyramid complex and of the other five stelae. While the goddess depicted is likely a composite of both Neith and Ma'at, the attributes of Ma'at are more in accord with these themes of order and time. Therefore, the goddess is most likely Ma'at.

The crossed bows and crossed arrows that are the symbols of Neith are further dealt with in a following section. While they do <u>not</u> appear on any of the stelae, the crossed arrows do appear elsewhere in the Step Pyramid complex.

---

[317] Wikipedia contributors, "Maat", Wikipedia the Free Encyclopedia, https://en.wikipedia.org/wiki/Maat

[318] Toby A. Wilkinson, *Early Dynastic Egypt*, pp 291-292.

On stela '5', the author of the glyphs is Ma'at. The meaning and intent of the glyphs on this stela, as a whole, are likely:

At Heliopolis and Letopolis, the Goddess (Ma'at) observes the motions of the Sun (Atum-Re), then decrees the measure of his progress and calls forth time and order on Earth.

There is no ankh (life) sign on this stela, while there is a scorpion, which, as was argued earlier, is a sign of death. This may indicate that this stela ends the series, or it may indicate that all of the creatures on Earth are subject to time and will find death. It is also noteworthy that the falcon on the serekh on this stela wears the white crow—while on at least three other stelae, possibly four, the falcon does not wear any crown—but the significance of this is unknown.

CHAPTER 68

THE GRANITE VAULTS OF THE STEP PYRAMID AND SOUTH TOMB, REVISITED

The vaults of the Step Pyramid and South Tomb have already been discussed at length, but they have more than a passing connection to the six stelae discussed in the foregoing sections. The vault of the Step Pyramid starts or initiates the 'westward' or clock-wise motion of the stelae, while this motion culminates at the vault of the South Tomb, and then rises to the doorway of the mastaba structure of the *temenos* wall, where all time measures begin. Thus, time or its concept originates in the vault of the Step Pyramid and then manifests itself in its various particulars as it progresses along the stelae, until it comes into form at the vault of the South Tomb, where it rises up and appears on the wall.

The dimensions of the two vaults were discussed earlier and were determined to be:

Step Pyramid vault:	57 cubic royal cubits
South Tomb vault:	24 cubic royal cubits
Total:	81 cubic royal cubits

Also, as was discussed earlier, if the number 57 is understood to be the number of ecliptic years (EY = 346.62 days) in the Exeligmos lunar eclipse cycle, then this number equates to 54 tropical years (TY = 365.25 days), and, 54 X 24 = 1,296 years ≈ the number of tropical years in the average Saros series. Also, 81 X 3 = 243, which is the number of years in the Venus transit cycle. Further, 81 X 6 = 486, which is the number of years in the Phoenix cycle. The number 57 is also the root number of the 5,700,000 years in the Easter repetition cycle, discussed earlier.

As discussed in an earlier section, there are 'star' blocks placed around or in the near vicinity of these two vaults, which are found nowhere else in the Step Pyramid complex. Each of these blocks is inscribed with two or four, five-pointed stars. The blocks with two stars have the stars carved side-by-side on the same face. The blocks with four stars have two stars carved side-by-side on one face and two on the opposite face, also carved side-by-side. As argued earlier, these blocks symbolize the two 112.5-year intervals of the Venus transit cycle of 243 years. But in the following section, they will also be shown to symbolize Neith and Osiris. The star that appears on each block is Gardiner's sign N-14, depicted below.

CHAPTER 69

THE GODDESS NEITH AND HER SYMBOLS

a. The Crossed Arrows

The symbol of two, crossed-arrows is one of the more common symbols for Neith. The tips of the two arrows, as discussed earlier, are almost always depicted pointing downwards, and have a point and two rearward facing barbs, three points in all. If two such arrow heads are laid atop one another, they create an interesting pattern that immediately calls to mind the five- pointed star. This is not a matter of chance or coincidence. The five-pointed star is also an optical illusion. (See five-pointed star below below.) If one focuses on the top cusp and the second and fourth cusps simultaneously, an arrow head appears. Similarly, if one focuses on the top cusp and the third and fifth cusps simultaneously a second arrow head appears. These two arrowheads are positioned such that they cross one another. The five-pointed star, then, is composed of crossed arrow heads, which invites a ready analogy to crossed arrows with the same points. It is no great stretch to envision that the ancient Egyptians conveyed the thought of the five-pointed star by depicting crossed arrows, instead.

The five-pointed star is the symbol for Venus, and by extension, and in the form of the crossed arrows, the symbol of the goddess Neith, as well. Neith, then, is the anthropomorphic manifestation of Venus. Re-visiting the issue of the identity of the god depicted on stela '3', the best answer—in light of this discussion of Neith—is that the god is a composite of the attributes of both Neith and Osiris.[319]

There is further support for this argument in that the principal deity worshipped at Abydos in pre-dynastic and early dynastic times was Khentiamentiu, the 'Foremost of the Westerners.' This name, or title, later became one of the principal epithets of Osiris when Abydos became the main cult center for the worship of Osiris, who was not specifically mentioned by name until the 5th Dynasty.[320] As discussed in an earlier section, the 'westerners' are an astronomical feature of: 1) the western movement of the equinoxes in response to precession, 2) the western movement of the Moon's orbital nodal points on the ecliptic, 3) the planet Mercury, when its western orbit takes it into inferior conjunction with the Earth, and 4) the planet Venus when its western orbit takes it into inferior conjunction with the Earth. Mercury has already been identified as Wepwawet, the 'Opener of Ways.' The epithet 'Foremost of the Westerners', then, can only refer to the planet Venus. And Osiris, as the 'Foremost of the Westerners', would clearly be associated with the planet Venus. Neith and Osiris, then, are best understood as the male and female aspects of Venus, with Neith likely symbolizing the motions and time keeping functions of the planet, and Osiris likely symbolizing the apparent death and resurrection (re-birth) of the planet as it reaches inferior conjunction with the Earth.

b. The Overlapping Bows

Another common symbol of Neith's consists of two, overlapping bows, either shown by themselves or positioned across a shield. As was the case for the crossed arrows, this symbol, Gardiner's sign R-24, is invariably interpreted to signify the war-like qualities of Neith. There is another interpretation, though.

The sign can also signify two overlapping crescents that could very well symbolize the octaeteris cycle, where a crescent Moon is in conjunction with and in close proximity to the planet Venus, when the planet is also in its crescent phase.[321] This is one of the most spectacular astronomical displays and occurs every 584 days, which is the same time cycle

[319] Interestingly, Herodotus mentions that there was a substantial adjunct to the temple of Minerva (sic), (Greek Athena, Egyptian Neith) at Sais, where certain secret rights or mysteries were conducted in the name of a god that he was reluctant to mention. Could this god have been Osiris, whose rights were considered mysteries by the Egyptians? If so, then it is very significant that Osiris was also worshiped in a temple dedicated to Neith. Herodotus, *The History*, translated by George Rawlinson (University of Chicago, The Great Books, Chicago, Illinois, 1952), paragraphs 170 and 171.

[320] Wilkinson, page 288, entry for Khentiamentiu.

[321] Wikipedia contributors, "Phases of Venus", Wikipedia the Free Encyclopedia, https://en.wikipedia.org/wiki/Phases_of_Venus (Accessed August 19, 2017)

for the Venus inferior conjunction cycle. The octaeteris cycle, then, features a display with two heavenly bodies—the second and third brightest in the heavens—that are each in their crescent phase, a display that could easily and accurately be symbolized by two overlapping bows. (Significantly, the planet Mercury, like Venus, also has crescent phases, as it progresses on its western orbit, and the planet's Egyptian god, Wepwawet, is also symbolized by a bow.)[322]

[322] Lurker, page 128, entry for Wepwawet.

CHAPTER 70

THE NUMBERS 5 AND 6, AND THE INSCRIPTIONS ON THE LINTELS OF THE DOORWAYS TO THE STELAE

The significance to measure of these two numbers has already been discussed extensively. However, they appear again in the stelae, in the form of all six figures depicted on the stelae wearing crowns, with five figures wearing the white crown of Upper Egypt and one wearing the red crown of Lower Egypt. These same two numbers are inherent in the crossed arrows and five-pointed star symbols, where two arrow heads, each with three points (tip and two barbs), can be overlapped to create the five-pointed star. These two numbers also are at the root of the *nsw* and *bty* glyphs (depicted below), which are Gardiner's signs M-23 and L-2, respectively, that together constitute the nsw-bty name of the king of Egypt. As explained earlier, *nsw* is 5 and *bty* is 6.

It is probably in the sense of 5 and 6 that the *nsw-bty* glyphs appear on the lintel of the doorways to each of the stelae, to the immediate left and right of center. On each lintel, these glyphs are followed, to either side of the lintel, by the nb-ty glyph (depicted below), Gardiner's sign G-16, that constitutes the second name of the king of Egypt. As explained earlier, the cobra represents the Saros cycle and the vulture represents the Inex cycle (Heb-Sed).

This glyph is followed, to either side of the lintel, by the glyphs for, Netjerikhet, which as discussed in an earlier Section of this Part, means, "God spoke and called forth 2 and 3", i.e. 2 X 3 = 6 (the Ma'at Pattern of measure) and 2 + 3 = 5 (the Fibonacci series of life). This is immediately followed, to either side, by the gold symbol Gardiner's sign S-14 (depicted below), and then finally, by the circle symbol for time. I am unable to explain why the gold symbol appears in this context and in this order.

There are also three combination glyphs that combine Gardiner's sign V-39 (variation of S-34) and R-11, depicted below, but with V-39 superimposed on R-11 in the actual rendering on the lintel. This glyph appears three times on each lintel: 1st, in the center of the lintel, between the nsw-bty signs, and then 2nd and 3rd, between the Netjerikhet signs and the circle sign, on each side of the lintel.

The R-11 sign is the *djed* column, that in an earlier Part was identified with the Earth. The V-39 sign is the *tet* symbol that is usually identified as the 'blood of Isis.' The combined sign is usually translated as 'life.'[323] But what is the 'blood of Isis?' Horus? The revitalized and resurrected Osiris, who in later Egyptian history was identified with the *djed* column? There doesn't seem to be a clear answer, but possibly, the 'living Earth' would work best. But, if the combined sign was also identified with Osiris, then, was the Earth itself raised from the dead, too? If so, what exactly does that mean? The questions must remain unanswered.

323 Lurker, page 72, paragraph titled, 'Blood of Isis.'

CHAPTER 71

THE GLYPHS ON THE STATUE BASE AND THEIR MEANING

The base of a statue found just south of the temenos wall has glyphs on it that are relevant to the discussion of the stelae. [324] The glyphs are found on the top of the base, and consist of nine bows, depicted in a vertical pattern directly under the feet of the statue, which is standing atop them, (Gardiner's sign T-9) and three *rḫyt* birds (Gardiner's sign G-24) that are beneath the bows and immediately in front of the statue's feet. The Glyphs and their arrangement are depicted below.

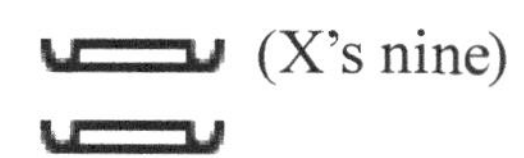

The nine bows are typically interpreted as the nine traditional enemies of Egypt, but in light of the foregoing discussions, they can also signify Venus in its crescent phase, which is assumed to be the case here, because of the statue's surroundings and context. If the bow or crescent signifies the Venus inferior conjunction cycle, 584 days, then the nine bows likely imply nine such cycles, or 9 X 584 = 5,256 days. In Earth tropical years of 365.25 days, this is equal to 5,256/365.25 = 14.4, which has as its root number 144. This number is the 144° counter-clockwise shift in ecliptic longitude that the Venus inferior conjunction experiences on the sphere of the ecliptic as it rotates through the five conjunctions in the inferior conjunction patter: 5 conjunctions —8 Earth years—13 Venus years (orbits about the Sun).

Also, 9 X 54 (the implied number from the cubic dimensions of the vault under the Step Pyramid, where the number of the actual cubic measure of the vault, 57, is taken as the number

[324] Cecil M. Firth and J.E. Quibell, with plans by J.P. Lauer, Excavations at Saqqara, The Step Pyramid, Vols, I and II, (Service des Antiquities de Egypte, Le Caire, Imprimerie deL' Institute Francais D'Archaeologie Orientale, 1935; and republished by Martino Publishing, Mansfield Center, Conn. USA, 2007), photograph plate 58

of ecliptic years in an Exeligmos cycle, which is also equal to 54 tropical years) = 486 tropical years, which is the Phoenix cycle. The three *rḫyt* birds, then, are interpreted as three Phoenix cycles or 3 X 486 years = 1,458 years, which is the number of panels in the eastern wall of the *temenos* wall surrounding the Step Pyramid complex. And, if 1,458 is divided by 9, the result is 162, or the approximate root number of phi, the golden mean (1.618) to within .001% accuracy.

So, why is the statue standing on the nine over-lapping bow glyphs? The best answer would seem to be that it is present in the midst of time, as measured by the Phoenix Cycle.

CHAPTER 72

THE MEASUREMENTS OF THE GRANITE VAULTS RE-VISITED

If the combined cubic dimensions of the two vaults, 81 royal cubits^3, is converted to cubic Egyptian-feet, the result is: 81 royal cubits^3 X (524 mm^3) X (524mm/299.42mm)^3 or $(7/4)^3$ or $(1.75)^3$ = 81 X 5.35 = 433.3 cubic Egyptian-feet. This number is within .003 of a number from the Pattern, 432. The number 432 can be rendered into two, 6 cubic Egyptian-foot cubes. If the two cubes are placed side-by-side, the resultant solid has longer sides that are 1 X 2 rectangles, each of which, as discussed earlier, generates the golden mean, 1.618. The combined cubic dimensions of the two vaults, 81, is also ½ of 162, which is within .002 of the root number for *phi*, 1618. Both 81 and 162 are numbers form the Pattern.

Another set of numbers can be derived from the measures of the two vaults if their measures are considered linear and factored in Egyptian feet instead of royal cubits.

Step Pyramid vault: 57 linear royal cubits X 1.75[325] = 99.75 ≈ 100 linear Egyptian feet
South Tomb vault: 24 linear royal cubits X 1.75 = 42 linear Egyptian feet

If the cubic measure of the Step Pyramid vault is 100 cubic Egyptian feet, then the root number is one, which is unity. (There are 10 granite blocks in the roof of the vault, which also expresses unity.) Then, if the cubic measure of the South Tomb is 42 cubic Egyptian feet, this number is approximately the same as the number of active Saros series in existence at any one time. The implication of the two numbers is that the 42 Saros series derive from unity or one, meaning...god? If so, which god? In the Hall of Judgement, where Osiris presides and the heart of a decedent is being weighed against Truth, the decedent must turn to each of the 42 assessors and deny he has committed 42 separate sins. If he cannot, then his heart is devoured by a beast and the decedent dies a second and final death. If he survives the ordeal, he passes to the realm of the blessed dead.[326]

[325] 524mm (royal cubit)/299.42mm (Egyptian foot) = 7/4 = 1.75

[326] Lurker, page 61, entry for "Heart."

The volumetric measure of the Step Pyramid vault is 57 royal $cubits^3$. If this measure is itself cubed, 57^3, then the resulting number, 185,193, which approximates the speed of light, 186,234 miles/second, to within .0056%. Is this significant, does it associate the speed of light with God? I don't know, but the possibilities are intriguing. The number 57, as discussed earlier, is also a number associated with repeating dates in the Gregorian calendar's 57,000,000-year cycle. Again, I don't know the significance of this.

CHAPTER 73

THE SIGNIFICANCE OF THE STAR BLOCKS AROUND THE VAULTS

Given their restricted locations to the area surrounding the two vaults, the star blocks signify Venus, in its aspects as an Evening Star (vault under Step Pyramid) and a Morning Star (vault under South Tomb). Furthermore, as already argued earlier, the Venus inferior conjunction is a personification of the god Osiris, who is Venus in its aspect as an Evening Star. Horus, then, is Venus in its aspect as a Morning Star. The star blocks symbolize death and resurrection or rebirth.

The star blocks around the burial vault also signify the end or death of a Venus transit cycle or series, or both, while the blocks around the vault at the south tomb signify the start or birth of a new cycle or series. It is logical, then, to assume that the intervening space between the two vaults represents the *duat* or region of transformation.

CHAPTER 74

ROSTAU

Rostau or *R-sṯȝw* is an archaic name for the broader region that encompasses Giza, Busiris (modern name: Abusir), and Saqqara—an area that roughly comprises the entire necropolis of the West Bank near ancient Memphis. It is a name often encountered in ancient Egyptian religious texts, such as, the Book of What is in the Duat, the Book of Gates, etc.[327] *Rostau* is an interesting subject, in and off itself, but the only reason that it is brought up here is because the name may possibly be comprised of two words, *res* and *tȝw*. *Res* is Gardiner's sign T-13, which he defines as, 'wakeful' or 'vigilant.' And *tȝw* is defined by Gardiner as 'earth' or 'land.'[328] (Note: While these are Gardiner's definitions for the two individual words, as presented, Gardiner does not support such an interpretation of *R-sṯȝw*, which he defines as the 'Necropolis of Sokar'; furthermore, it is not supported by any other source that the author is aware of either.) However, there is a logic to the interpretation that is relevant to the broader subject matter of this book. Using the two-word interpretation, then, *Res-tȝw* would mean, 'land of the wakeful' or 'land of the vigilant', a meaning which could readily be extended further to, 'land of the watchers', a likely reference to astronomers. In light of the entire thrust of this book, this meaning, 'Land of the Watchers', makes perfect sense. Interestingly, *rs* is also related to the Egyptian word *rswt*, which means to 'dream'—a meaning that may open vast cosmological and metaphysical vistas, yet unconsidered.

Did the vault under the Step Pyramid at one time contain the body of a god? Or is this god number—perfect, immutable, eternal? And in his form of number does he lie there, yet—sleeping and dreaming, dreaming in the perfection of pattern and number? And in that dream, does he call forth all the measures of time and distance that give form to our world, and to our perception of its order and reality? And is our perception, but a manifestation of his dream; and we, but a part of it?

Deo Gratia

[327] Robert Temple, *The Sphinx Mystery: the Forgotten Origins of the Sanctuary of Anubis*, (Inner Traditions, Rochester, Vermont, 2009) numerous references under 'Rostau'; a comprehensive treatment of the term.
[328] Gardiner, *Egyptian Grammar*. See definition of *tȝ* found on page 598, upper left column, top entry.

APPENDIX I

THE PHOENIX CYCLE

While there is no apparent astrophysical connection between Venus transits and their cycle, and the Saros-Inex series and their respective cycles, arguably there is a rough harmonic connection, whereby the occurrence of periodic transits of certain Venus transit series does seem to roughly coincide with and herald the onset of the long-lasting Saros series—those with 80 or more eclipse members. All such series cluster together in groups of 3-5, although there are clusters with as few as 1 and as many as 6. Since the time period between these transits ranges from 486 years (2 · 243 years) to 729 years (3 · 243 years), this cycle is similar to the Phoenix cycle of mythology, which purportedly involved a time period of 500-1,000 years. Also, the time period between the onset of the final long-lasting series in a cluster and the first one in the ensuing cluster (see shaded cells in the table below) average 474 years, which, too, is a number close to the myth's time period of 500 years.

Depicted in the following table are the Venus transit dates that roughly correspond to the first series in each cluster of long lasting Saros series, along with the identity of the transit series they belong to.[329] Some of these dates precede the onset of the long-lasting Saros series, while others occur years after their onset and in the midst of the cluster. Also shown are the dates that the Saros series with the highest total number of eclipses first started producing eclipses.[330] Only Saros series with 80 or more eclipse members are considered to be long lasting series, unless the series is bracketed by long lasting series.

Venus Transit Date	Venus Transit Series #	Time Period between listed Transits	Saros Series	Number of Eclipses in Series	Year Series Began
-3,213	#2		-15	87	-3,322
			-14	88	-3,233
			-13	87	-3,192

[329] NASA Web Site, "NASA – Catalogue of Transits of Venus"; http://eclipse.gsfc.nasa.gov/transit/catalog/VenusCatalog.html

[330] NASA Publication/TP-2009-214172, "Five Millennium Canon of Lunar Eclipses, -1,999 to +3,000 (2,000 BCE to 3,000 CE) Fred Espenak and Jean Meeus; Tables 5-3 to 5-8. Available on-line at NASA Web Site, http://eclipse.gsfc.nasa.gov/SEpubs/5MCLE.html

			-12	84	-3,199
			-11	83	-3,134
			-10	83	-3,069
					(445 yrs.)
-2,727	#2	729 yrs. (3213-2727)	6	86	-2,624
			7	89	-2,595
			8	86	-2,494
					(463 yrs.)
-1,998	#2	486 yrs. (2727-1998)	24	85	-2,031
			25	87	-2,038
			26	85	-1,919
			27	85	-1,926
					(463 yrs.)
-1,512	#2	486 yrs. (1998-1512)	43	85	-1,463
			44	76	-1,199
			45	85	-1,351
			46	76	-1,358
			47	86	-1,275
					(553 yrs.)
-783	#2	729 yrs. (1512-783)	63	82	-722
			64	84	-783
			65	86	-736
			66	84	-671
					(463 yrs.)
-669	#3	#2 series ends; #3 transit series begins			
-183	#3	486 yrs. (669-183)	82	84	-208
			83	84	-197
			84	84	-96
					(456 yrs.)
303	#3	486 yrs. (183+303)	101	83	360
			102	84	461
			103	82	472
					(463 yrs.)
789	#3	486 yrs. (789-303)	119	82	935
			120	83	1,000
			121	82	1,047
					(517 yrs.)
1,518	#3	729 yrs. (1518-789)	138	82	1,512
					(448 yrs.)
2,004	#3	486 yrs. (2004-1518)	156	81	2,060

			157	73	2,306
			158	81	2,154

Observations on the Phoenix Cycle:

1) The Phoenix cycle is not a particularly reliable indicator of when long lasting Saros series will begin, but there does seem to be a harmonic between the Venus transit cycle of a particular transit series and the onset of these Saros series. In the past the transit series involved was #2, but this series ended in -540, and was last active in the Phoenix cycle in -783, while series #3 has been active in this cycle ever since.
2) A problem with the Phoenix cycle is that 2 transit cycles, 486 years, don't always herald the onset of the long lasting Saros series; instead, there are sometimes 3 transit cycles (729 years) between the onset of a cluster of long lasting Saros series and the onset of a succeeding cluster. There is no discernible pattern as to when 2 transit cycles will intervene or 3 transit cycles. However, there is no apparent impact on the use of the Phoenix cycle, since if an expected onset of a cluster of long lasting Saros series does not occur after 2 transit cycles, then a third cycle is added to the count to readjust it. In any case, the Phoenix cycle always involves either 2 or 3 Venus transit cycles of 243 years each, and all of these transits are always members of a single Venus transit series.
3) A '500-year festival' to celebrate this event could be held on the exact occasion of the transit, which would be approximately 486 years since the previous transit marking the onset of the long lasting Saros series, or the festival could be held 14 years after the transit event in order to mark an exact 500-year anniversary. Either approach would work and neither would impact the records of the underlying Venus transit and Saros series.
4) The Phoenix cycle, as defined herein, is an extreme anachronism. The Phoenix myth and its cycle appear to be active in certain early cultures far earlier than any recorded knowledge of and use of the Saros series as an eclipse predictor. And, to be certain, there is no equivalent Saros-Inex Panorama from these early dates, nor are there any records documenting the length of Saros series. Furthermore, there is no known records of early cultures having been aware of or to have observed Venus transits, or to have systematically recorded their occurrence for analysis and for forecasting future events. So how, then, could the Phoenix cycle have been observed and described for use as a forecasting tool, if indeed there is an actual Phoenix cycle underlying the Phoenix myth? The answer to this seemingly irresolvable conundrum will become apparent in Part II of this book. In the interim, I ask the reader's indulgence and forbearance in accepting that the Phoenix cycle is a legitimate forecasting tool and that certain ancient cultures were aware of it and used it.
5) The Phoenix cycle is a compelling myth that describes a very real astronomical phenomenon. After living for 500 years (486-729 years using the transit cycle) the

bright plumed bird (planet Venus) descends into a conflagration (transits the Sun) and is consumed, only to arise from its ashes (remerges from the transit) and begins a new life and reinvigorates life everywhere (lifting up and reinvigorating the Saros eclipse cycles), much like a heartbeat begins anew.

APPENDIX II

MEASUREMENTS OF THE STEP PYRAMID COMPLEX OF SAQQARA

"Research on the Measures Employed"[331]

We have sought to determine as exactly as possible the length of the measuring unit employed for the construction of the monuments of King Zoser. Significantly, we have found on the foundation of the base of the southernmost wall of the court of the Monument of the South (chapel of the Princess Hetep-her-nebti) at m .90 of the southwestern angle of the court., some indications of a measure with a red feature, among which two small features have vertical spaces of 1 m 04 and 7 to 8 millimeters. That dimension would thus correspond exactly to 2 royal cubits of about m .524. We then measured the dimensions of the courts, the solid masses and the principal monuments together, while converting them into cubits m .524 and we will elaborate and comment on the results obtained.

1. Action plan of different projects of the Pyramid.

Let us point out there were, for the Pyramid has the same degree, five successive projects. An initial mastaba was increased twice successively, which gives us the projects M1, M2 and M3. Then, over these mastaba projects it was decided to build a step pyramid of contour P1 that increases then according to contour P2.

For M1 and M2 we know in a precise way only their measurements on their east and west sides, that is to say their width. M1 would have been 123 cubits across (of width) and M2, which consisted of an addition around the circumference of M1, with a surface section of 7 cubits thickness, would thus have had a combined measurement of 123 + (2 x 7) = 137 cubits across. The M3 project comprised enlarging the M2 project, but only on the north and east

[331] Jean-Phillippe Lauer, "Étude sur Quelques Monuments de la IIIe Dynastie (Pyramid à Degrés de Saqqarah)", *Annales du Service des Antiquities Volume XXXI, 1931.* Section titled, "Recherche des Mesures Employees", pages 59-64. Reprinted with kind permission of Dr. Zahi Hawass, Secretary General, Supreme Council of Antiquities, Zamalek-Cairo, Egypt, November 23, 2010 (e-mail). Translated from the original French by Joanne Fischer.

faces. On the north we have found the line of its surface behind the serdab of the king. On the east we have also found it and we see that the width of the dimension of the mastaba is augmented by 16 cubits. The width of M3 was thus 137 + 16 = 153 cubits. Its [M3's] length, which we can also measure, knowing the line of its surface in the north and the south, was 190 cubits.

It is also plausible that the dimension of the extension added by project M3 on the earlier project M2 was equal on both the north face and on the east face. Thus, in supposing that the dimension of the extension to the length of M2 was of the same dimension as that of the extension to its width, the length of M2 would have been 190 – 16 = 174 cubits, while the length of M1 would be: 174 – (2 x 7) = 160 cubits.

We will now consider the two projects P 1 and P2 of the step pyramid. The only dimension that we have measured with precision is the width in the North-South direction, because we observe that toward the west the base of the pyramid was covered by a solid mass, and did not comprise a coating.

For project P1, we have found the bases of its coating on the north, the east, and on the south. Its North-South width was thus approximately 104 m 75, or 200 cubits (measured as m.524 corresponds to 104 m. 80). The East-West length cannot be measured in a very exact way, but approximately it seems to have been of 230 cubits.

As for the final project P2, it comprised an increase in P1 on the three faces north, east, and south: on the north, an increase of 5 cubits; on the east and on the south an increase of 3 cubits only. The dimensions of P2 would thus have been 200 + 5 + 3 = 208 cubits and 230 + 3 = 233 cubits.

In summary:

The project M1 = 123 X 160 cubits
M2 = 137 X 174 —
M3 = 153 X 190 —
P1 = 230 X 200 —
P2 = 233 X 208 —

Other Principals of remarkable measurements:

The length of the enclosure is about 544 m. 80, which is 1040 cubits (which would give exactly 544 m. 96).

Its width is approximately 276 m 85, which is 528 cubits (which would give 276 m. 67).

Let us notice that we find exactly the length of 1000 cubits, that is to say of 524 meters, from the face of the coating of the southernmost wall of the large court of the south, in front of the [south] tomb of Zoser, to the end of the northern exterior wall of the enclosure. The thickness of the wall of the south enclosure, coatings and bastions included, is thus: 1040 - 1000 = 40 cubits; the width of the superstructure of the [south] tomb on this wall is of 25 cubits.

The dimensions of the South Grand Court: width 107 m. 40, which is 205 cubits; length to project M1: 183 m. 290, which is about 350 cubits (=183 m. 40); length to project P2: 175 m. 25, which is about 335 cubits (= 175 m. 54).

Distance from the north face of the altar to the first solid mass of the 'B' form [structure]: 100 cubits approximately. Width of B form : 10 cubits. Length: 21 cubits approximately.

The width of the large court of the south from its south-eastern angle until the angle formed by the projection of the solid mass containing the vault of the [south] tomb of Zoser: 123 cubits.

The projection of the solid mass itself: 33 cubits and a few centimeters.

Thickness of the solid mass east of court of the south, from the south-eastern angle of this court up to the wall of the enclosure, bastions included next to the east dimension: 73 m. 20, is 140 cubits (=73 m. 26).

The length of the colonnade, taken until the center hinge of the door of its western end until that of the door with only one hinged section at its end is: 64 m. 40, therefore 123 cubits) = 64 m. 45).

The width of the solid mass of construction situated at the lower court of Heb-Sed (small low walls included), measured from the western face of the hardcore solid mass of the enclosure wall in front of the access to the south-east of this court: 33 cubits and a few centimeters.

The width of the court of Heb-Sed : 33 cubits and some centimeters.

The width of the solid mass of construction located to the west of the court of Heb-Sed (small low walls included) until the court of the temple T: 33 cubits and some centimeters.

These three dimensions, similar in total, thus give us100 cubits, and if we add there the width of the court of temple T, which is of 23 cubits, we find the figure of 123 cubits which we already have discovered has three different reprises.

The length of the court of *Heb-Sed*: 177 cubits. We note that this figure, 177, is the compliment of 123, which compares with 300. We also discover this precise measure of 123 cubits again along the southern wall of the court of the *Heb-Sed*,, [as measured from the entryway to the court to] the [eastern] wall of the little [structure that faces] the hypostyle room which is situated at the western end of the colonnade.

The width of the little temple T (including the torus roll): 8 m. 83; or 17 cubits = 8 m. 91.

The length of the little temple T (including the torus roll): 19 m. 83; or 38 cubits = 19 m. 91.

We note here that on each of the two dimensions, width and length, 8 centimeters are missing for the exact correspondence in cubits. This is undoubtedly due to the fact that the ancients had taken their measurements before the rough-casting of the torus rolls on the edges were removed, which would then have removed a peripheral layer of approximately 4 centimeters of thickness.

The distance of the simulacra of the door, situated at the south-east angle of the Pyramid until the first wall meets towards the east: 44 cubits.

On the east face of this same wall until the Pyramid: 40 cubits.

The width of the court situated in front of the monument of the south (chapel of the Princess Hetep-her-nebti): 70 cubits.

The thickness of the solid mass of the wall of the enclosure (coatings and bastions included), to the east of this court: about 20 cubits

The width of the face of the Monument of the North (chapel of the Princess Int-ka-s): 17 m. 50; or 33 cubits 1/3 = 17 m. 47.

The width of the solid mass bordering this court at the west (coatings included) also approximately 33 cubits.

The width of the court called "du Serdab": 50 cubits

The length: 61 m. 40; or 117 cubits = 61 m. 31.

The length of this court at the time of the first project of the P1 pyramid : 122 cubits.

The length of this court at the time of the first project of the mastaba M1 : 150 cubits.

The distance of the front face towards the south of the northern wall of this court to the facing of the large wall which limits fine limestone construction towards the north: 25 cubits.

Still, let us note the following observations.

The width of the bastions of the enclosure wall is 6 cubits and their intervals are of 8 cubits. The thickness of the coating of the enclosure wall, bastions included, are 9 cubits. The entry passage giving access to the colonnade is 12 cubits in length.

The thickness of the majority of the coatings of the solid masses varies between 1 m 50 and 1 m. 60, which is approximately 3 cubits. The steps of the walls of the large court of the south or the court of the monument of the south have just a broad cubit. It is the same way for the pavement around the court of *Heb-Sed* and the temple T. All the small niches shown are a square cubit. The diameter was at the beginning of the fibrous columns is of 2 cubits. That of the fluted columns of the two monuments of north and the south, known as vaults of the princesses, is approximately a cubit.

In short, we note that one rather frequently finds complete measurements by m. We have noted thus the number of cubits: 1040, 1000, 350, 230, 200, 160, 150, 140, 100, 70, 50, 40, 20, 10.

We have come across, in another area, seven different reprisals of the figure of 33 cubits + a few centimeters. However, 33 cubits and 17 cm. .5 is exactly 1/3 of 100 cubits. Perhaps the architects of the 3rd dynasty employed some extension of cubits 1/3 length for their measurements, as we ourselves employ bands of 10, 20 or 30 meters? In this case wouldn't they have preferred to defer exactly this length for certain divisions important to their plan layout?

We find, in another area, five times for the important measurements, the figure of 123 cubits, in particular for the width of the first project of mastaba M1 and for the length of the colonnade. Let us notice that this number is rather characteristic, since it is the form of the continuation of the first three digits. Wouldn't this characteristic have conferred a special virtue to him [Zoser]?

Let us note finally some figures end in 5 like: 335, 205, 25 (twice).

We see thus with the 3rd dynasty the architects preferred to employ in the majority of the cases, for direct measurements, either of the figures ending by zeros or sometimes by five, or characteristic figures such as those which we have just described, and which do not seem to have been the result of chance. Sometimes however we meet unspecified figures; we then note generally that they are not dimensions first, but the result of additions or of subtractions made of these characteristic dimensions, such is the case, in particular, for the figures obtained for dimensions of the projects M2, M3 and P2, or of the P1 pyramid. Finally for the average measurements ranging between 1 and 40 cubits approximately, the Egyptians have sought generally to have integers of cubits, and it is only, it seems, for the moldings or the very small elements that they employed fractions of cubits.

APPENDIX III

PLANS OF THE STEP PYRAMID COMPLEX OF SAQQARA[332]

Plans (copied with kind permission of Martino Publishing), in following order:

Plate 1. General plan of pyramid temenos. (J.P. Lauer)
Plate 34. Enclosure wall. Elevation. (J.P. Lauer)
Plate 47. Great South Tomb. Sections and plan. (J.P. Lauer)
Plate 48. Great South Tomb. Plan of upper structures. (J.P. Lauer)
Plate 52. Colonnade. Top: Transept from west; Bottom: Main colonnade from east
Plate 54. Plan of S.E. quarter, including colonnade (J.P. Lauer)
Plate 67. Plan of Heb-sed court and "T" temple (J.P. Lauer)
Plate 77. Façade, Tomb of 'Northern Princess.' Restored elevation. (J.P. Lauer)
Plate 79. Façade, Tomb of 'Southern Princess.' Restored elevation. (J.P. Lauer)

[332] Cecil M. Firth and J.E. Quibell, with plans by J.P. Lauer, Excavations at Saqqara, The Step Pyramid, Vols, I and II, (Service des Antiquities de Egypte, Le Caire, Imprimerie deL' Institute Francais D'Archaeologie Orientale, 1935; and republished by Martino Publishing, Mansfield Center, Conn. USA, 2007)

Step Pyramid. Pl. 1

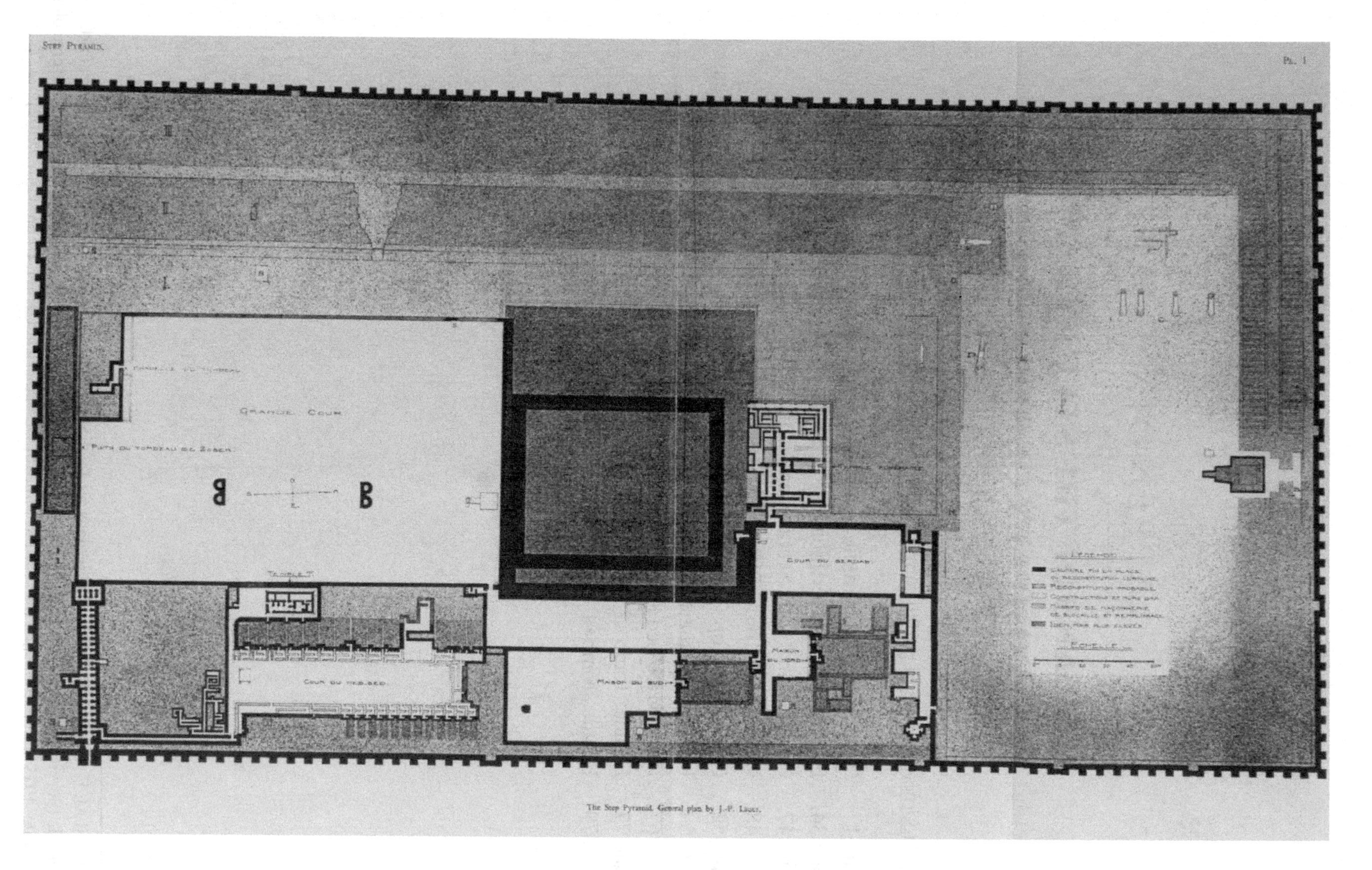

The Step Pyramid. General plan by J.-P. Lauer.

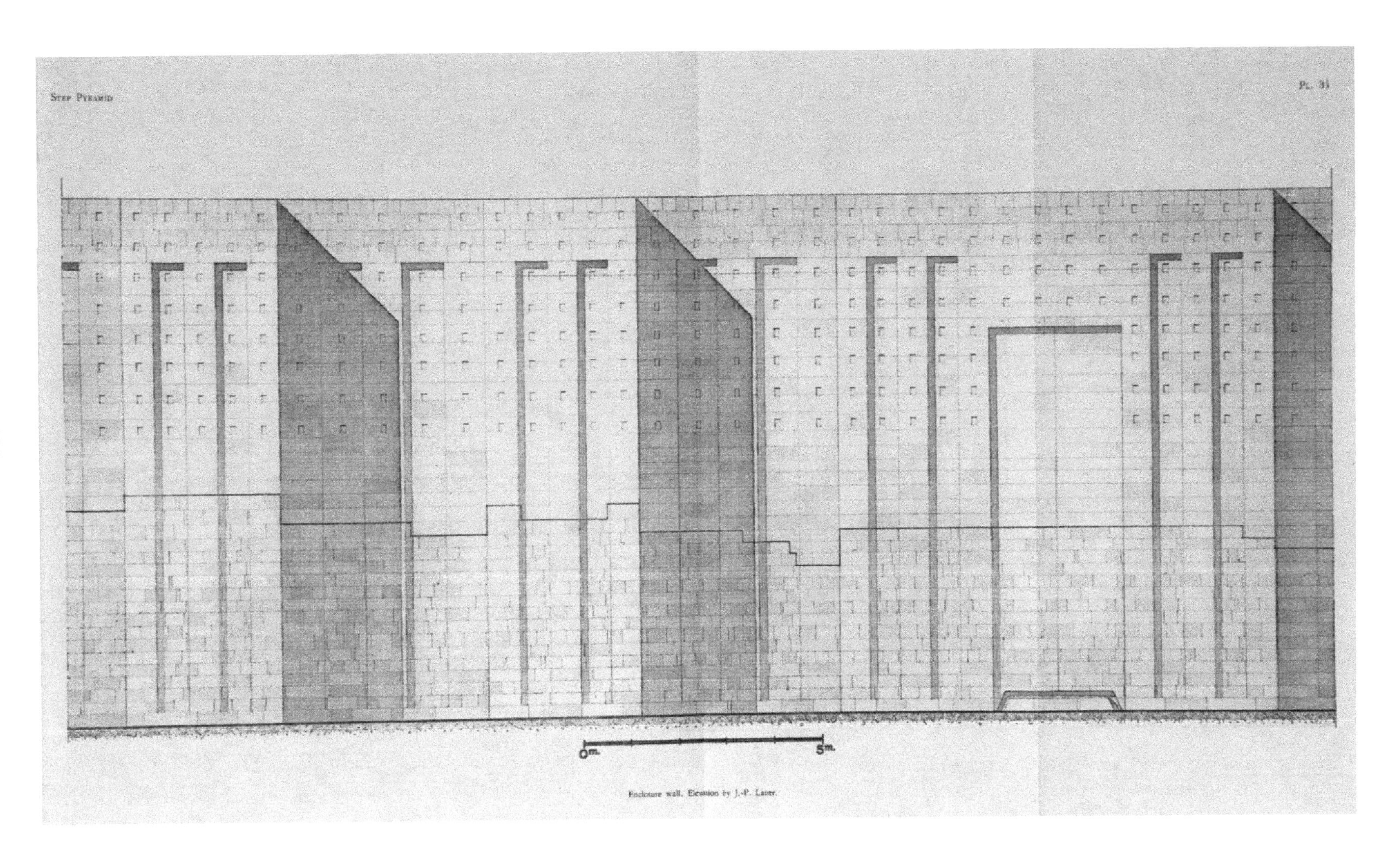

Enclosure wall. Elevation by J.-P. Lauer.

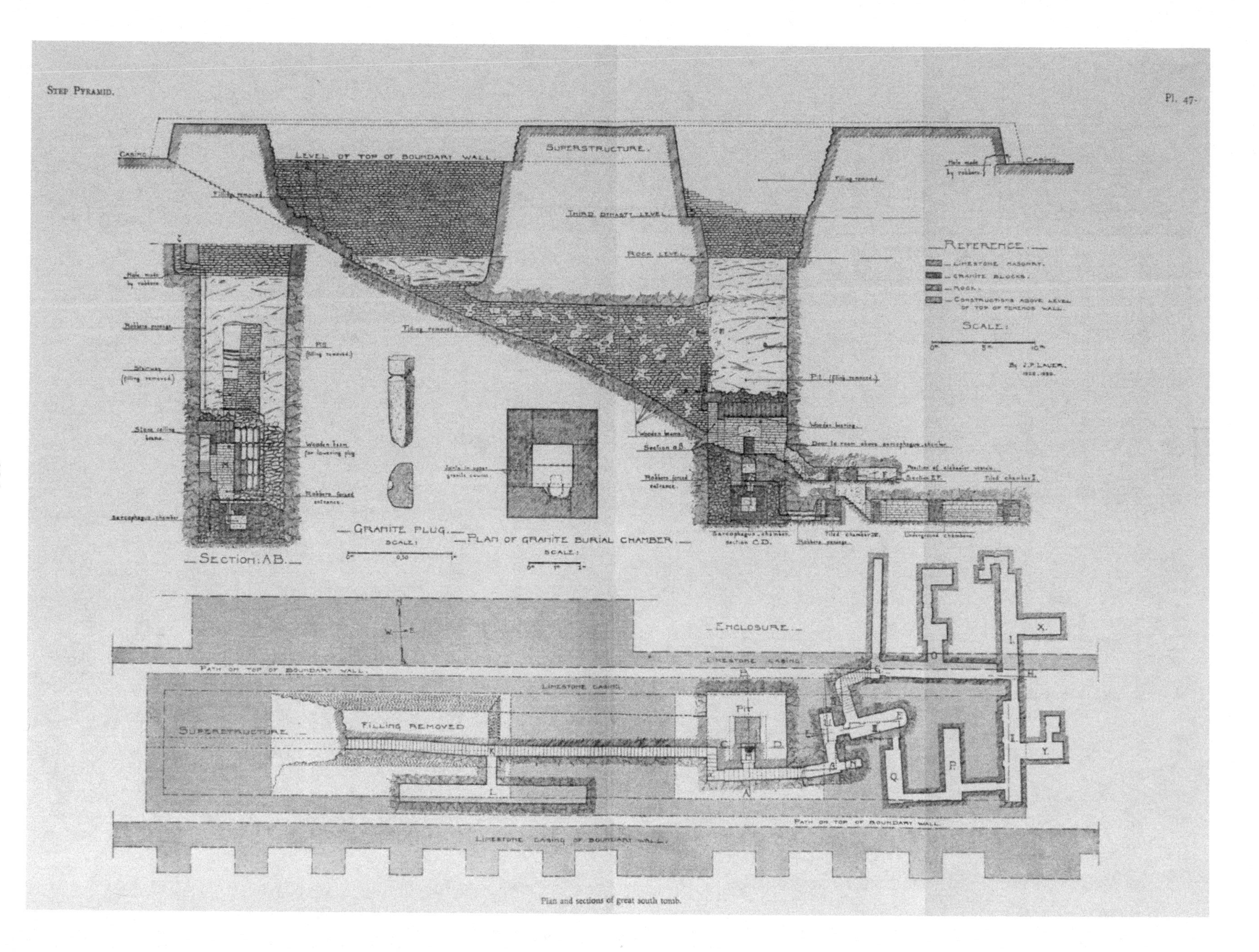

Plan and sections of great south tomb.

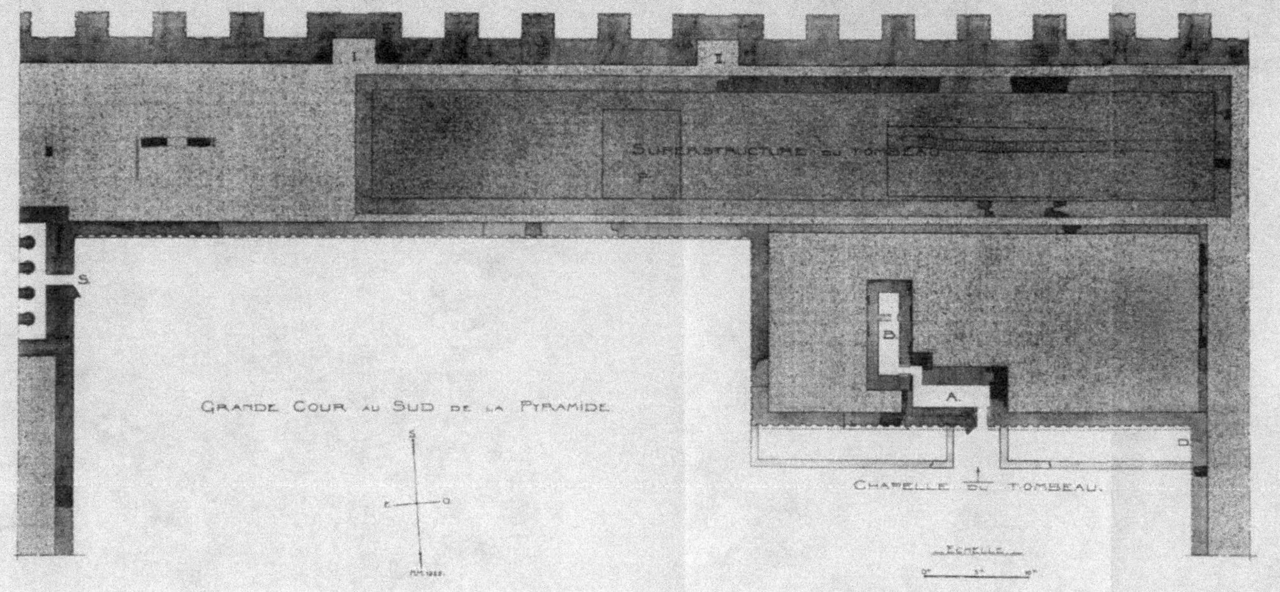

South Tomb. Plan of superficial structure and chapel, by J.-P. Lauer.

Colonnade. West transept.

Colonnade. From east.

COUR DU HEB-SED

Plan of colonnade and south-east quarter, by J.-P. Lauer.

Step Pyramid.

Pl. 67

Hebsed court: plan by J.-P. Lauer.

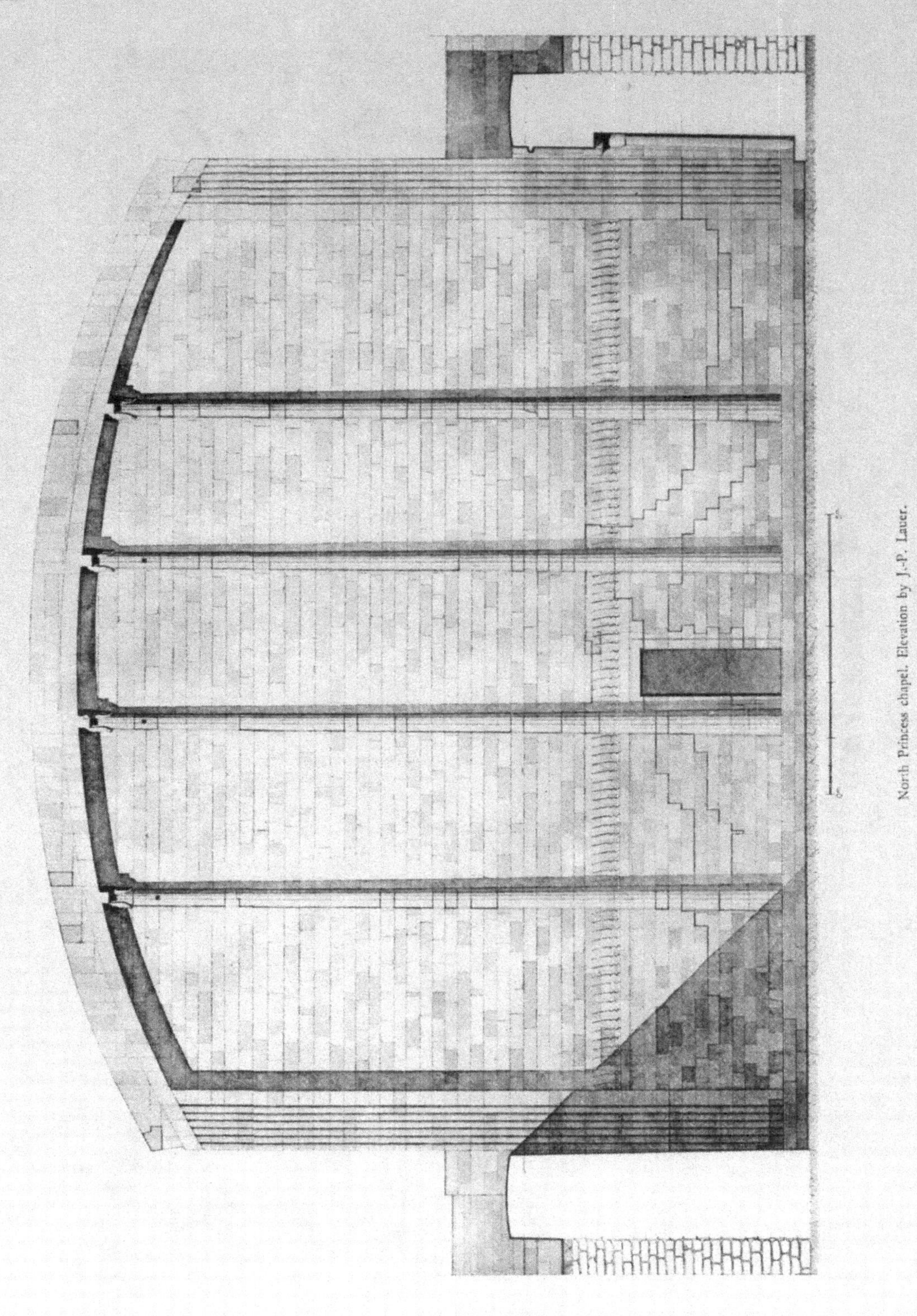

North Princess chapel. Elevation by J.-P. Lauer.

South Princess chapel. Elevation by J.-P. Lauer.

Imp. G. Bouan - Paris

APPENDIX IV

THE IMAGE OF THE EARTH IN THE GREAT PYRAMID

The measures used to generate this image are based on the internal structures of the Great Pyramid. These structures are shown in the figure below.

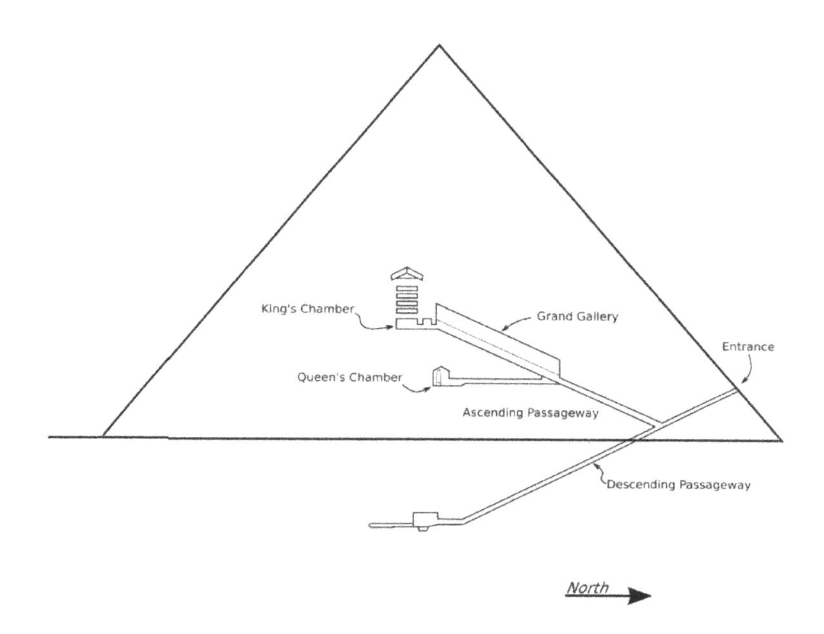

The following measures, and their sources, were used in the generation of the Image:

Horizontal distance from Pyramid's centerline to the centerline of the King's Chamber: **21 royal cubits**. (Petrie survey[333]; also, Maragioglio and Rinaldi survey[334].)

---

[333] Flinders Petrie, Tour of Egypt (published 1883), on line version is found at: Petrie Tour Egypt: http://www.touregypt.net/petrie/

[334] Maragioglio, V. and Rinaldi, C., *L'Architettura Delle Piramidi Menfile* Part IV Piramide di Cheops, (Rappolo, 1965) Tavola (Table) 3. Copy for downloading is available online at: http://www.gizapyramids.org/static/pdf%20library/maragioglio_piramidi_4tav.pdf

Vertical distance from the base of the Pyramid to the center of the ceiling of the fourth relieving chamber is: **114.52 royal cubits** (Maragioglio and Rinaldi survey[335]). This is based on:

a. Vertical distance from base of Pyramid to floor of the king's chamber: 81.81 r. c.
b. Vertical distance from floor of king's chamber to ceiling of chamber: 11.15 r. c.
c. Vertical distance from k. c. ceiling to ceiling of 4th relieving chamber: 21.56 r. c.

114.52 r. c

335 Ibid Tavola (Table) 7

Printed in the United States
by Baker & Taylor Publisher Services